DRAGONWARS

DRAGONWARS
Armed Struggle &
the Conventions of
Modern War

J. Bowyer Bell

Routledge
Taylor & Francis Group
LONDON AND NEW YORK

First published 1999 by Transaction Publishers

Published 2017 by Routledge
2 Park Square, Milton Park, Abingdon, Oxon OX14 4RN
711 Third Avenue, New York, NY 10017, USA

First issued in paperback 2018

Routledge is an imprint of the Taylor & Francis Group, an informa business

Copyright © 1999 by Taylor & Francis

All rights reserved. No part of this book may be reprinted or reproduced or utilised in any form or by any electronic, mechanical, or other means, now known or hereafter invented, including photocopying and recording, or in any information storage or retrieval system, without permission in writing from the publishers.

Notice:
Product or corporate names may be trademarks or registered trademarks, and are used only for identification and explanation without intent to infringe.

Library of Congress Catalog Number: 99-20843

Library of Congress Cataloging-in-Publication Data

Bell, J. Bowyer, 1931–
   Dragonwars : armed struggle and the conventions of modern war / J. Bowyer Bell.
     p. cm.
   Includes bibliographical references (p.   ) and index.
   ISBN 1-56000-357-X (acid-free paper)
   1. Low-intensity conflicts (Military science) 2. World politics—1989– 3. United States—Military policy. I. Title.
U240.B44 1999
355.02—dc21   99-20843
                                                                                       CIP

ISBN 13: 978-1-138-50931-3 (pbk)
ISBN 13: 978-1-56000-357-1 (hbk)

For
Sigfriedo Maovaz
and
Ely Tavin

Contents

Foreword   ix

Part I: Introduction to Dragonwars

1. The Revolutionary Ecosystem and the Armed Struggle   3
2. Dragonworld: Lebanon   17

Part II: The Dynamics of the Underground

3. The Dragonworlds: Revolutionary Ecosystems   61
4. The Design and Determinants of the Dragonworld   83
5. The Dynamics of the Armed Struggle   127
6. Responses to the Armed Struggle   177

Part III: America and the Dragonwars

7. America and the Conventions of War   231
8. Unconventional War, the Cold War, and the American Way of War: Greece to Vietnam, 1945–1966   265
9. Vietnam and the American Way of War, 1967–1972   311
10. Conflict on the Margins of the Cold War   327
11. The Postmodern World and Dragonwars: Into the Next Century   373

Epilogue   417
Sources   431
Index   447

Foreword

October 23, 1983, the day that the massive car bomb went off in the midst of the United States Marines on the edge of the Beirut airport is slipping away into history. Yet, amid the seemingly endless terrorist spectaculars and horrors, that particular moment still remains symbolic, the epitome of America's confusion and despair in the face of revolutionary fanaticism. Vietnam was too complex, open to too many varied definitions and responses. Vietnam means many things. Not Beirut. There Americans seeking only peace were murdered by a zealot. No one has forgotten. Few understand. Nearly every American who remembers Beirut is still outraged. This indignant anguish has been long with the nation in large part because the nature of the contemporary revolutionary and their special world are alien. To Americans such killers seem mad or bestial, fanatics without reason and beyond easy retaliation. The IRA volunteer, the teenage killer prowling the the streets of Africa, the bombers and depraved defenders, the hijackers all blend into the awful, the inexplicable. Such America analysis, arising from the virtues and vices of the American character, has for decades tended to be uncertain, often intense but also often counterproductive.

American intervention in the revolutionary world has been ineffective, neither eliminating irregular challenges to legitimate authority nor encouraging subversion and rebellion. Too often whatever Americans do seems wrong: an overreaction or if not, then a toleration of provocation. What works best has seemed to be great force swiftly deployed and a swift withdrawal: Antigua and Panama and the Dominican Republic. And there is always the danger that there can be no withdrawal as in Vietnam or with honor as in Somalia. Once committed in Haiti or Bosnia, Washington is riven with doubt. Can the troops disengage? Should they have been sent? Is it futile to become involved in such complex, intractable missions?

Americans deploying cunning and stealth and the advantages of technology have indicated only the difficulty in special operations. American efforts to manipulate the alien, even when militarily effective, have often not produced the desired result. The zealots often make poor pawns and those purchased require regular payments and often produce no results. So amid all the alarms and challenges of the postmodern world, the revolutionary, the irregular, the terrorist and the gunman remain as they have for generations, a provocation that has not engendered an effective American response.

Americans are not about to change, gain sudden insight into the violent world of the gunman no matter how appropriate the cause. In this the establishment and the responsible are one with the country. The unconventional are uncongenial. Washington may have supported the Contras in Nicaragua and Afghan mujahedeen, funded potential Iraqi rebels and armed Angolan guerrillas in central Africa; but such support for irregulars often seems to produce more problems than solutions: a tarnished image, ruined policy, ruined reputation, clients transformed into enemies. Each and all of these unconventional liabilities were unforeseen by the pragmatic advocates of special operations. Time and again covert initiatives engendered unpleasant returns. American technique is often elegant but the result somehow not as intended no matter the skill and craft displayed.

The result has been a long-lived dilemma. Whatever Washington does is wrong and doing nothing is often wrong as well. All the nation's virtues are changed to handicaps. All experience leads to error and confusion. All those involved, angry with the unconventional, persist and so, having learned nothing, repeat previous error or evade the prospect by recourse to worse solutions. And at the end those responsible rationalize that they meant well. Americans always mean well and somehow their friends in Vietnam pay or their allies are revealed as untrustworthy or simply unsavory, arms given to freedom fighters are sold to bigots or used against friends. As in chaos theory any small intervention, a general bought or a check forwarded, may have enormous, unexpected, and dangerous consequence. In open battles there are winners or losers, the big battalions do well, training, experience, and elegant weapons are apt to win. Yet experience, training and elegant weapons guarantee nothing in unconventional wars, in asymmetrical conflicts between those with power and those dependent on revealed truth or old intractable gods, on the habits of the past or a vision of the future. Then America finds orthodoxy confounded by an alien world.

This particular examination of such underground worlds, Dragonworlds, and the American response assumes as given that neither the irregulars nor the Americans will be greatly transformed in the foreseeable future. The underground has run to rule for over a century no matter the agenda, the arena, or the response. Tribal revolts, ethnic defenders, warlords, and bandits are this century much like those of the previous century. These irregulars may lack faith but deploy irregular skills, are driven by an agenda and assumptions not at all American. Even less so are those zealots who seek to impose a vision on the real world. And Americans are Americans. Rebels will fashion their world as always and Americans will remain, virtues and vices intact, true to their national inclination. The future will not be all that different from the past, filled with surprises but arriving more or less on schedule in matters of protracted, low-intensity conflict. In fact, all indications appear to suggest that insurrection, terror, and insurgency will not only remain intact but also supply most of the wars of the future as they have in the past. The end of the Cold War merely gave the irregular more visibility and less easily acquired patrons. The unconventional conflicts simply continued within a changed global arena.

Dragonwars are actually a growth area. The break up of the old, bipolar, Cold War confrontation released small ambitions, tribal greed, and flexibility for the small player. Nuclear terror may be less fearful but terror seemingly more likely and certainly more visible. The post-modern world is composed of the same intense national interests, new faiths and new technologies, more of the same grievances and quarrels. Zones of security are small; even democracy and prosperity in open society only attract zealots. So if past is prologue, America will still find distasteful challenges by the irregular and still seek conventional responses to unconventional problems. The American national character, the American way of war, the habits of the past will continue to make any response to unconventional challenge halting. America can, however, do better and not betray national ideals or change the character of the people. The problem is there is no solution, only incremental adjustments to advantage for Americans as long as the national ethos persists. And none would want the nation to turn monster to counter the monsters of the Dragonworld.

This tour through the dynamics of the Dragonworld that shapes the world's armed struggles-the Dragonwars American finds so difficult will end, as such tours often do, with a prescriptive tail. The prescription is not a cure, not a solution, but a means to reduce the persistent policy predicament. The recommendations on matters of unconventional conflict do not require vast institutional reform. The Pentagon, that most American of institutions, is apt to continue to be as immutable to radical change as is the American character. In general, the most effective response to unconventional provocation is the least response: do as little as possible and that cunningly adjusted to the arena. Do no harm. Deny one's self, do less rather than more. These are uncongenial to Americans who instead desire to deploy great force, quickly, to achieve a swift and final result. What is needed is a shift in the perception of those responsible in Washington: less can be more, others are not like us, and a neat and tidy world is not worth the cost.

In general and particular the struggles against persistent irregular challenge are apt to be protracted. There are to be the certain conflicts of the future. Each arises from immutable conviction not easily waged to quick conclusion, each is special, and yet all have similar dynamics. And those dynamics are still operative as they have been for a century. Those who dispatched the car bomb in Beirut still pursue their Jihad. Those like them have since bombed in America, taken American weapons from Afghanistan on missions against the West, turn Algeria into a charnel house and the Palestinian peace process into cause for atrocity. They roam the world as transnational terrorists. There are always new zealots, new faiths to feed on grievance and so new undergrounds. There have always been long irregular campaigns in the outback at the margins. This was how the West was won. The center often does not hold and new despots arise or anarchy exist. Chaos is not easy to counter, not everywhere, not every time. The orthodox, the Americans preferred a decisive victory, the triumph of the orthodox, Gettysburg or San Juan Hill. Yet the violent frontier generated Indian wars that engaged the country for three centuries. Many unconventional struggles are apt to be protracted but few so extended as the American experience. Yet America, the powerful and bold, determined, technically elegant and driven by enterprise and reason, often neglects history, learns what is sought, and so seeks a future amenable to the great national virtues not the habits of the frontier or the patience and cunning that small and protracted wars require.

And so, appropriately, the gestation of this work has also been protracted, a long march to publication with somewhat unconventional results and means. I first became interested in the "new" American response to the next generation of insurgency when, soon after the end of the Vietnam war, there was almost none in Washington who wanted to contemplate the irregular. One Vietnam was ample. Even thinking about the unthinkable engendered professional distaste in the Pentagon and much of orthodox Washington. If the military had no irregular capacity, no doctrine, no systems, then none could ask for another adventure into a jungle, into a Dragonworld where somehow tangible assets had not assured even a draw. Many felt more tangible assets would have made the difference, their opponents argued that nothing could have made an unjust war successful. Very few in Washington, however, wanted to reconsider the unconventional. Mao again went unread, special operations went unfunded, and future unorthodox challenges were ignored. Then ever so haltingly a few in Washington suggested that America should at least consider the appropriate response to the unconventional: denying the irregular was a recipe for disaster.

The single great shadow across theory and doctrine was that of Vietnam—a trauma that would last the life of the involved. Most soldiers then and later, most everyone, never wanted to contemplate any revolutionary war after Vietnam. Each new unconventional crisis for a generation was shaped to Vietnam. After 1975 no one wanted to hear of jungles or special operations or anti-insurgency again, ever. And for years no one in authority wavered. Those who were concerned even changed the name of their subject to low-intensity warfare to erase the evil image. They and their advocates noted the rise of new modes, like terrorism, and new villains, like Muammar el-Gaddafi of Libya. The orthodox establishment was not interested. The focus was on shaping a conventional military, a NATO military, carrier task forces, new delivery systems, new weapons systems, better morale, and so a force prepared to repel the conventional forces of the communist empire.

The friends of special operations arguments only hardened Pentagon hearts. And seemingly for good reason given the chaos and confusion that emanated from the White House basement during Colonel Oliver North's watch. In any case, if el-Gaddafi were a problem, then an orthodox airstrike would be effective. Even the collapse of the Soviet Union could not overcome orthodox suspicion for special operations and special missions, when conventional force could be deployed. No one in the Pentagon wanted to interdict smuggling, pursue terrorists, or convoy refugees. Such a distaste was not novel nor especially American; for the professional military was in the business of war, preparing for war—a war that had not yet come in recognizable form but one the Pentagon knew was the real threat. The refusal to contemplate any other threat, any other mission, was not viable as a basis for policy. Every American administration deployed limited force at a distance, often in response to unconventional challenge and at times for unconventional purpose. The Pentagon was repeatedly asked to respond to unwanted missions and persisted in responding conventionally—nothing learned, nothing forgotten.

What fascinated me was that the advocates of an unconventional response deploying special units and low-intensity tactics still seemed as innocent of the nature of modern revolutionaries and irregulars as their superiors who wanted to concentrate on main-battle tanks and replaying D-Day. Tradecraft deployed by special forces did not turn Americans into revolutionaries or zealots. Many of those who waged small wars, the terrorists and gunmen, where not shaped by craft and skill but faith. Americans in and out of the Department of Defense believed the world peopled by those not unlike Americans, believed that the unconventional could be given rules, taught in class, made congenial. If all would not reason together, any could still be defeated by superior craft or great force. In the national security establishment, especially the Pentagon, each new essay into the irregular taught few new lessons. And all the lessons were filtered through American monitors, shaped to American assumptions and agendas.

For a generation I had wandered the dangerous edges of the contemporary revolutionary world, attending wars and coups and discussing old campaigns with gunmen and terrorists, some retired, some not. In the course of the years, I piled up what later became adventures but at the time seemed irritants and interruptions, the natural cost of dealing with subversives. I was asked to leave some countries for doing unauthorized research, prohibited entry in others, followed, arrested, harassed, even kidnapped in Jordan because I refused to listen to Ba'athist fedayeen's explanations of reality—no difficulty in meeting terrorists for they come with a gun and collect you. I had trouble with bandits in Yemen and the British army in Ulster. And from time to time, not often, I was shot at by a variety of people, usually arbitrarily for being in the wrong place with the right people but on rare occasions personally and close up: field research in free-fire zones or what a friend at Transaction calls political ethnography. It was a generation spent inside or on the margins of Dragonworlds with a frayed visa and a list of questions few wanted to ask, else I could have stayed back at one university or another and read what I needed to know, gone to Washington and talked to generals. Instead I had to venture into the real world of the unconventional.

Out there no one much seemed to believe that I could be standing at the end of some mean lane in Belfast or Aden solely for academic purpose, but practically everyone seemed in the end willing to talk about the cause. Their world cast a cold shadow over the more conventional analytical world when I returned from the outback to seminars and journals. At times it seemed that I existed in both worlds only on sufferance. While I was clearly engaged in analysis, even at times academically valid analysis, the real world always seemed to be not Columbian nor Harvard but that of the lethal dreamers conspiring in seedy offices at the far ends of the world. So I have spent much of my professional life associated with these dreamers who possess the absolute truth and smell slightly of cordite. And there were those even more irregular without dreams but possessed of appetites and automatic weapons. Few fit easily the models of orthodox analysis, although all shared certain dynamics.

Mainly, I have been concerned about how revolutions, whatever the flavor, work in practice—their dynamics. And, of course, each case is special. This is the one universal conviction among those who are underground: their cause is unique. Certainly wandering in strange places, this often seemed to be true. What did a Catholic, working class, IRA volunteer off the Falls Road in Belfast share with a Palestinian fundamentalist in a small room in Gaza? Only after years and years did the similarities become more obvious, so compelling was the proffered truth and peculiar historical arena in each case. Each dream was different, even when universal. Each campaign was special. Arabs were not Irish, Zionists not Maoists, South Arabia not like Rome or Berlin. All Dragonworlds are different but all are also the same. The dynamics that drive the underground, that shape each armed struggle are much the same. To be covert, committed to the gun, opposed to a state, illicit and illegitimate, to be driven by a dream determines the nature of the unnecessary ecosystem and the dynamics of the struggle. And if so, all responses can rest on basic assumptions about the nature of the underground. There are other irregulars, warlords and tribal levies and mutineers, who also share some of the dynamics of the faithful, and they too are often irregular in regular ways if not those imagined at the military academy.

Since these underground ecosystems are a constant, the obvious conclusion is that they could be approached for whatever purpose with guide book to hand. Washington, however, has always seemed to be using irrelevant texts. The orthodox wanted to crush the rebel with brute power, bring up battleships off the Lebanese coast and shell the fanatics. The American military advocates of the unconventional seemed to feel that superior fieldcraft is the key to counter-terror, wear black pajamas in the forests of the night and as a better soldier crush a rebel. And the more daring wanted to encourage the irregular, set American irregulars or foreign clients to unorthodox mission. America really had no doctrine for small wars but a text of techniques that engendered much Pentagon suspicion. No one in Washington seems to accept that all those underground have been telling the truth when they insist that the tangible power will in time lose to will and zeal. The faithful accepted as axiom that you can kill the rebel but not the dream. Killing rebels with better fieldcraft can no more kill the revolution than battleship salvos. Buying a underground is a contradiction in terms. The underground deploys will, not tangible assets. Not only are Dragonwars asymmetrical but so too the American response: orthodox assumptions and strategies to deal with the unorthodox, to impose orthodoxy. Things and systems are deployed against perceptions. And the tangibles usually win but not always.

Very few understood the dynamics of the underground ecosystem, the nature of a world fashioned by rebel perceptions quite beyond the everyday. For the rebel what is real is what they perceive to be real, not what Washington can see or model or count. And Americans took the Vietnam experience, all their experiences, and modeled and counted and saw not. Americans, never having been in love, could not grasp the nature of romance. And so as time passed, Vietnam and previous unconventional campaigns continued to be mined for lessons to bolster the conventional desire for big battalions in small wars or to bolster the desire for special units to win small wars by craft.

And with no great tangible enemy after 1989, the military drifted, expressing interest in missions they did not want and could not fulfill. Somalia was a disaster, Haiti a disappointment, Bosnia a great risk, and the terrorists and narco-criminals an imposition. The Pentagon did have an opportunity to deploy as doctrine and wont indicated in the Gulf War. For the Pentagon it was a perfect war run at half-speed with world support against a tyrant and a war with an assured victory. It was a Pentagon war: hi-tech systems, logistics, an isolated battle arena, splendid full-color coverage, ease of preparation, and dispatch in pursuit. It was also unique.

Mostly the challenges remained irregular, unexpected, often almost trivial but always outside the basic doctrine of dispatching conventional force or underwriting unconventional forces. Again no matter, for dispatching the orthodox was always first choice—in Antigua or Panama—and imposing American perceptions and agenda on the irregular. When fanatics detonated a car bomb in front of an American military barracks in Saudi Arabia, the Pentagon invested huge sums in seminars and programs, adjusted security requirements, funded intelligence, and deployed assets, but barely understood either the zealots or the American vulnerability-the war was asymmetrical, things opposed to perceptions.

In a sense, as long as the rebel ecosystem remained unknown, romantic, mysterious, or appalling to Americans and so to the responsible in Washington, any policy reaction would continued to be frustrated. The virtues of the American character assure frustration in responding to the unconventional. And understanding of the nature of the armed struggle will not shift vital American perceptions but can indicate means to adjust to effect.

So I amid other projects decided to adjust my ideas about the nature of underground systems to American responses. As I have before, I went to see several old friends. The late Frank Barnett at the National Strategy Information Center along with Bill Bodie, who reemerged as my program officer at Smith Richardson, offered aid and comfort. And at the Earhart Foundation, Dick Ware had retired to be replaced with David Kennedy with Tony Sullivan always in place, once again with considerable grace had the courage of my convictions. I thus received two Earhart grants on unconventional war. Progress, however, tended to be interrupted by the contingent and unforeseen and not the least by my attendance at on-going wars.

After decades I have grown to have a proprietary interest in certain armed struggles. Thus in Ireland some IRA volunteers I first noted a generation ago in prams now contemplate with dismay as their own children arrive at the vulnerable age when the Irish Republican truth makes converts no matter the horrors of the last years. In Italy, on the other hand, the children seem largely immune to the rebel vocation that led their fathers to seize the heart of the state and find the nation did not care—no small gift from a generation of gunmen now in prison, penitent, retired but all discredited. In the Middle East with the *Intifada*, a new generation crafted another unconventional strategy that led to the Gaza-Jericho agreement that in turn provoke the new Islamic generation to pursue their Jihad. So my revolutionaries distracted me from finishing my first draft with any dispatch. Those who supported or funded all these investigations gave comfort and on occasion aid, in particular Devon Gaffney, Bill Bodie and the directors at Smith Richardson, and Karen Colvard and her board at the Harry Frank Guggenheim.

In time my indefatigable agent Tony Outhwaite found a charming and concerned editor in Robie Macauley at Houghton Mifflin. Like the subject, the script proved sufficiently protracted in completion that Robie retired before his suggested revisions were incorporated—delighted, I suspect, to escape a subject and characters he found as distasteful as most Americans. The project had by then grown as protracted as most armed struggles.

Adrift among wars and other analytical distractions, I let my dragons rest. When it was time to begin again the text had to be rewritten and finally ended with Irving Horowitz and Mary Curtis at Transaction, safe home at last. Irving sensibly pointed out that while the dragons and their world had stayed the same, the real world had changed and so must the existing text. The result was a new book, rewritten yet again. This time the text arising from the underground effort of previous years deployed my more recent conclusions: still the same dragons and still the same response, a different arena and a maturity of insight. And this time at the insistence of Transaction more care was given to structure, not that even Irving Horowitz can transform a grasshopper into an ant, anymore than an American general empathize with a gunman, but he did insist that less was more.

Over the years it has been the dragons that have determined the analysis not the usual list of fellow scholars, research centers, and congenial libraries. Irving Horowitz can perhaps be cataloged as a guerrilla guide, a zealot for logic and argument. Much of the time, I have had the benefit of forthcoming colleagues, a congenial research center, and, if need be, efficient libraries. For the first half of the Dragonwars project I was associated with the Institute of War and Peace Studies at Columbia University; but, as has often been the case, largely in absentia and finally not at all as I moved into another project on the Irish Troubles. Beyond the universities a variety of the respectably involved, generals and statesmen and observers, as well as local witnesses have contributed to my work: over a generation, thousands and thousands of people focused on their wars, their dragons, their experience.

Thus this work is distilled from a generation of investigation and endless discussions with both the activists and the analysts, so there is no hope of a tidy list of those who have given their time. There have been tens of thousands of hours of interviews, formal and informal, thousands of hours of analytical discussion, mostly informal, various academic conferences and analytical seminar in America and abroad. Always, even at present when I have given up wars and time spent on the dangerous edge, those involved matter most. I may have given up gunmen but they have not always given me up. In fact, one old but hardly retired gunmen recently pointed out that he could no longer spare time from the armed struggle to do marginal comments on my text—a strange sense of priorities but typical for those underground. There no one has time for analysis but only for the dream and the operations of the day, and rarely a few hours with me.

What follows on these pages is my accumulated assumptions and conclusions at the moment. It is not conventional social science and the format is not shaped to most academic expectation; the text follows my own devices and desires. There are thousands of texts on unconventional conflict, some of which are referred to in the final section on sources. My analysis addresses what interests me, not an academic agenda or the practicalities of waging revolutionary war. My own analysis is simply a special way to examine a general phenomena. The conclusions may appear to violate the common wisdom, but I do not find them especially novel. The analysis may be expressed differently than Mao or Giap, be less quantitative and rigorous than some social science approaches, but it is more rather than less a variation on the common wisdom on revolutionary matters. It is, however, drawn from a unique experience and detailed in novel ways. There is this, however, as my friend the late Bill Fox—Professor William T. R. Fox, Director of Columbia's Institute of War and Peace Studies—noted: the Japanese never lost anything by being second. So much that follows has been noted before, as well it should, for armed revolution during much of this century has flourished.

Along the way several people, old friends not easily dissuaded by my subversive acquaintances, have regularly offered aid and comfort. In Dublin Oliver and Cliodna Snoddy and Charles and Carmel Murphy, Sigi and Francesca Maovaz in Rome, Dr. Ely Tavin and Miriam in Tel Aviv. Neither Sigi nor Ely are available for the last chapter, so this book is for them—one who lived his life in Rome as a painter and the other who came to Rome but once to bomb the British Embassy—rather the two poles of my own concerns. And in Washington there have been far too many people over the years, but special thanks to Dr. T. W. Adams, Dr. Neil Livingstone, Dr. George Tanham, and always Professor Amos Perlmutter, my old Harvard colleague who has a guest room crammed with books and a mind filled with questions both academic and pragmatic.

And for much of the march to Dragonwars, my Kerry wife Nora Browne made all things possible.

Beir bua

J. B. B.
New York
Dublin
New York

Part I

Introduction to Dragonwars

1

The Revolutionary Ecosystem and the Armed Struggle

For centuries international order has been troubled by small wars, insurrections, revolts—low-intensity conflict. During the decades of the Cold War, these struggles were shaped to great power interest, seemingly a by-product of a strategic nuclear stalemate that usually reduced conflict to the edges of events. Idealists assumed that after the implosion of the Soviet empire such violence could be eradicated, justice institutionalized, and wrong answered. And Americans were apt to be the most optimistic of all. A few even felt that history had ended in 1989, that all just grievances could be accommodated. The open society, the democratic system, increased education and productivity would become the norm: if history was not quite over then the future order boded well.

Instead the new world order has been filled with turmoil, pogroms and terror, famine, civil war, and rebellion. Some had always accepted that such was a normal aspect of any global system and largely beyond remedy. In any case there had been conventional wars, the Israelis and Arabs, India and Pakistan, even the United Kingdom and Argentina. And there had been unconventional conflicts arising from turmoil on the margins, unresolved quarrels, the impact of development. The pessimists anticipated more of the same: the old world turmoil but without the strategic nuclear confrontation.

Americans were by nature optimistic and America dominated the counsels of the new world order. It could impose force at a distance. In theory such power could be effectively deployed against those determined on deploying violence for undesirable or impossible ends. In practice the cost was often greater than the returns. In post-Vietnam America no one wanted to right Asian wrongs or to pacify Africa. America, NATO, the West was rarely willing to sacrifice for a fair, just, and tidy world. At times intervention for special purpose or the general good was, indeed, initiated by Washington, by NATO, by regional interests, or by the United Nations. Such exercises seldom evolved quite as planned. So the world was seemingly still filled with uncompromising zealots, often armed and dangerous. Even famine and plague were often beyond reach of decency. At times stability could be imposed as in Haiti or the unsavory deposed as in Panama but just as often, in Lebanon or Somalia or Southeast Asia, results were not as intended.

Some violence was seemingly inevitable, had deep roots, resulted from implacable aspirations, high ideals as well as greed and malice. Some lands, like Palestine, had been twice promised. Some states, most states in Africa, were charades incapable of imposing order. And then, even within the heartlands of the West, zealots had appeared seeking radical change at once by means of terror. None of this was new, only more visible to the media and to Americans in the last decade of the century.

There had always been wars, the unexpected appearance of radicals and zealots, and the problem for the powerful in responding to unconventional armed challenges. For two centuries dissenters to the system had sought not an end to history but to drive it in a different direction. The faithful had repeatedly found the truth and employed force in its name. For some, Islam or Mao revived was the answer. Others wanted to create new nations or protect old ones, bring the people to power or punish hereditary enemies. Some had no truth but only the habits of the past or the needs and greed of the present. None of this was really new. Certainly, with the rise of popular nationalism during the French revolution, the global state system had been prone to violence. Since then there were those who pursued their dream of a new nation or a liberated class, those who sought to change history. These were rebels eager to seize the state, colonists opposed to empire, armed advocates of classes or causes. The era of nation-states was an age of violence, discrete, local, regional, and general, great wars and small.

A constant in modern times was recourse to limited force. There were small wars between states and small wars against states and even

The Revolutionary Ecosystem and the Armed Struggle 5

transnational rebels waging war of a sort against whole systems. There were wars for the tribe, for the faith, for the people: all small, often offering a muddle of causes, each different and each much the same. So for two centuries there have been low-intensity conflicts, Dragonwars. The collapse of the Soviet empire not only changed nothing but the context of such wars, but it also initiated a new set of conflicts within the successor dynasties.

Traditionally such little wars come and go. Only a few are protracted. Only a few trouble the great powers or even the historian. Most attacks on the system simply disappear, changes made or the country pacified, the empire withdraws from the margins, nothing much changes. In recent times times those involved on the margins have emerged in the capitals of the post-industrial world bringing terror to the world stage. The developed world of instant communication, mass politics, a post-industrial economy, advanced technology has in turn generated rebels with impossible causes and absolute convictions. No matter, most often the rebels end in shallow graves, footnote to local history. The weak rarely win. The center holds and things do not fall apart.

On the other hand, in some cases the center does not hold. The weak grow strong or the strong have no staying power. Revolution and insurrection may be dubious ventures but have imposed spectacular change, turned China communist, help destroy the great empires of Europe, replaced the weak and vulnerable in the Third World.

Some conflicts are even less ordered that the classical insurgencies. Beyond the core of the post-modern world, there are still bandit zones, provinces beyond the law, and new nations where the regime does not rule but exploits, does not engender loyalty and cannot impose order. The revolutionaries and rebels, the fanatics and the transient terrorists, the irregulars of the Third World generate the violent muzak of the new world order but rarely if ever great change.

Almost all of these low-intensity campaigns are pursued with similar unconventional tactics. The unconventional means—insurrection, the way of the guerrilla, urban or rural, the use of terror—become convention. And so the unconventional generated a broad spectrum of response: experts, libraries, academic careers, military academies, tactics, techniques, and special units. These are focused on the means the weak used to achieve power. To counter the deployed assets of the conventional the classic underground shapes a congenial ecosystem that guarded the revealed truth, allowed the faithful to persist but imposed severe penalties on capacity. This underground is a perceptual creation. Those underground may spend their lives seemingly no differently than before, attend school, hold a job, deliver the milk, but all the time are armed agents of change, driven by revelation. They, of course, may be guerrillas in the outback, soldiers of a bandit army, irregulars in a bush war, but are as often part of the everyday world—gunmen only on call, defenders only when attacked. Their world is not only the visible but also that beyond easy reach, an unannounced agenda, priorities that are not those of the conventional, habits and customs arising from the faith and necessity. They pose their own perceptions against real power.

While many go underground as idealists, the faithful, possessed of the truth that could be deployed as advanced ideology or a return to the religious basics, thus forming classical revolutionary ecosystems, many others pursue armed struggle as vocation: traditional bandits of the outback or vengeful tribesman, defenders at the crossroads or rural agitators burning barns. Some are driven underground by the power of the center. All, however, to some extent, were and are shaped by aspirations denied. All find a necessity for recourse to violence. And all must contend with the capacities of the system, usually the state system. All share the necessity to operate out of weakness, against the strong, operate often in secrecy, in opposition without recognized legitimacy or often hope of swift victory. Their war is irregular, their cause denied, their hidden world safe only by recourse to unconventional tactics. Thus the faithful enter the underground, create their own hidden world, both real and imagined, and so engage in a protracted armed struggle. This is an asymmetrical war that opposes conventional power and assets with the will of the dedicated. For those underground, victory is promised by history not by tangible power. To win, their faith must persist until the will erodes the capacity of the center.

Not all those within a underground belong to a new and vital faith. Many who deploy unconventional tactics are conventional. Those without ideals but appetites or habits deploy not perception but tactics and commitment that generate conflicts asymmetrical in assets as much as in perceptions. These are the bandits or warlords who rely on custom, lack of options and tradecraft to persist. Persistence often is the great convention of such an unconventional campaign, persist for gain, persist in loyalty to the old, persist for lack of alternatives, and by necessity persist out of sight in the outback or underground. There even the mad and the criminal may find haven.

Those in control, those with legitimacy, recognition, power and responsibility, tangible assets, and possession of the state tend to prevail: not always because then tomorrow would always be like yesterday. At times some at the center lack capacity and so can be defeated. When those in charge are defeated in a swift coup or by simple intimidation, conspiracy and audacity will do rather than an underground. Then no protracted struggle need be contemplated. Even when the underground persists, most armed struggles fail, simply later rather than sooner. Truly protracted campaigns like that of the Irish Republican Army or the Palestinians are the exception and offer classical examples of the armed struggle—an evolving challenge to existing order by those too persistent to ignore but not powerful enough to triumph. Their tactics mimic those of classical irregulars, the martial tribes, warlords, truculent clans that seek to preserve history rather than accept change just as the bandit can be taken as a rebel.

The chaos of great wars and the failure of will to maintain empire meant that in this century the vocation of the irregular—despite the risks—was not without triumphs. At times the successes came against those who felt no need to defend the existing order, but at other times—Ireland, Algeria, Vietnam, or Cuba—against substantial opposition. Mao in China did indeed change history. Some struggles found accommodation, as in South Africa, or moved on to an end game as in Palestine and Northern Ireland. And some of the traditionalists, the irregulars of Yemen or Afghanistan or Somalia, managed to defend history against the center. No matter the particular outcome, the global order, old or new, with or without the Cold War, generated thousands of underground movements dedicated to an armed struggle in thousands of irregular campaigns.

For a century each decade has seen special examples, particularly rich arenas of disturbance. Africa or India gripped by ethnic and religious slaughter, once Latin America and before that China, province of warlords. The filter of the Cold War simply hid a continuing phenomena: the struggle for old causes or new, in the outback or in the capitals of Europe, unconventional violence by those without countervailing power but compelling interests opposed to the state. If some of these rebellions are no more than bandit forays, others are modern classics: armed struggles arising from deep ideological commitment and displaying all the techniques and tactics of modern insurrection. These may threaten the center and often trouble international order. A few escalate into conventional war and some are chronic. Visibility often depends not on scale—tens of thousands may die unnoted in the south Sudan or in the mountains of central Asia without notice—but on perception, the media, or the priorities of the center. This is obviously true when the campaign takes place at the center, in Rome or in New York. Yet, the perception of national strategic interests has from time to time provoked a massive conventional response to a marginal arena as in Vietnam or Algeria or Afghanistan.

The most classics examples are wars pursued by those who, possessing the absolute truth, would transform the future. They attack from the protection of a perceptual ecosystem beyond reach of all but the righteous. There underground, in a Dragonworld quite unlike the everyday, committed to a dream but confined by incapacity, the faithful persist in a long war that rarely produces the future imagined but most often assures in the fullness of time despair, defeat, and ruin. For at least two centuries, each classical example has been shaped as an armed struggle: revolutionary violence by the few seeking to impose a vision on the many, seeking to change objective reality by recourse to the gun and a dream. For analysis the grievance and the goal—a communist society, the return of the king, an Irish republic, or a fascist regime—are less significant than the energy supplied by commitment. Revolution is not the province of one special ideology, even if for most the language of the Left has predominated since the arrival of the industrial age, nor best understood by the physical nature of the arena, a special history, or the societal factors of the time. What matters is the faith that generates energy that in the service of an armed struggle against entrenched state power can only be deployed in certain universal ways.

The dynamics of all armed struggles are much the same. In Italy the new fascists and the post-Mao communists were very different in aspiration but quite similar in their dynamics, in their underground existence, in their craft and assumptions, different but the same. And the Italian gunmen and terrorists shared much with those whose armed struggle was isolated in the *altiplano* of Peru or the streets of Beirut; even the defenders of Bosnia or the clan warlords of Somalia shared much more than guns and craft. There have been all manner of revolts. Alone those against the British Empire after 1944 included Arab and

The Revolutionary Ecosystem and the Armed Struggle 9

African nationalists, Zionists, conservative and pious Greek irredentists on Cyprus, Chinese communists in Malaya, Irish republicans, and elsewhere idealists, clansmen, socialists, liberals, and adventurers. And all of these armed struggles against the Crown—from the Mau Mau in Kenya to the National Liberation Front in Southern Arabia—were if different also were the same: energy protected by a perceptual defense and released in certain ways. Their prospects, as was the case with all underground regardless of the enemy, were everywhere limited by power and by the very nature of the protected ecosystem of the faith and persistence assured by special circumstances and the nature of the underground.

The great underground asset is the will founded on revealed truth. Maoist thought is more important than arms: the former is vital and the later can always be found, stolen, bought, manufactured. The faith, on the other hand, is revealed—one faith for Mao and another for the Moslem. Some have faith in the past or the tribe, are less incandescent in conviction but reassured by habit and ethos. The truth may vary, is often contradictory, at times can be found in in competition with other revelations. The intensity of the committed also may vary, can be found pure in *Brigate rosse* or the leadership of the *Khmer rouge* and only as a trace element in the anarchy of Liberia. There are all sorts of low-intensity conflict, some classical, some arising from greed or parochial issues. Even the classical revolutionary campaign is not rare. Over much of the century at any given moment, there have been hundreds of revolutionary undergrounds and many classics. The phenomena is common.

To the faithful, sooner rather than later, the only viable means to change the future appears this resort to the gun, to revolt. Some traditional defenders or tribal levies come with such a vocation: the gun as heritage and tool. The modern rebel is apt to be a convert to the gun, a volunteer to the underground. There in a world shaped by perceptions the faithful trust that will can overpower the state, for it is largely the state that controls the future. At times there are universal enemies but even then a state is apt to be prime enemy—the Great Satan dominated the West and so is enemy of Islam. Mostly an armed struggle is arena-bound, state opposed, may speak for a universal but kills the particular in one place, seeks Islam in Egypt or communism in Italy. The dream may be universal but the enemy is in the palace, the police, the regime, the empire.

An armed struggle may vary in intensity, vary in the numbers involved or the degree of turmoil. What is apt to matter the most is what the involved assume matters. The perception of observers may be in error as to the degree of intensity and of threat. Terror spectaculars may be the province of the few but add enormously to the sense of action. Killing in the outback may have little impact in bandit societies but the murder of Europeans by the Mau Mau shattered complacency in colonial Kenya. The direction of the campaign, the intensity of the struggle, the shape of the future is not determined by numbers, by data deployed, but by what is assumed is real, what matters matters.

To those underground revelation promises all. They are willing to take on an empire with a dozen faithful and a few revolvers. What those who must operate underground seek is a means to act strategically on the future by operations on the morrow. The tangible means can be acquired, the strategic promise of the truth is a gift achieved by analysis. This analysis usually imposes the responsibility to pursue a military campaign.

If the military action can be protracted, if the underground can protect the faithful, a campaign evolves, again each special yet again each driven by general currents. A rural post-Maoist insurrection has much in common with a neo-fascist terror campaign even if the action seems quite diverse. All those engaged in pursuing an armed struggle are involved in a thrust to orthodoxy, seek to deploy a legitimate army in the service of a recognized government that successfully controls the arena. Mao moved from a few gunmen fleeing capture in time through various stages to command real armies engaged in a real war. The hidden galaxy of the faith does not want to be hidden. All want to be able to discard the secrecy of a hidden ecosystem, not to rely on will and the legacy of history but on real power. They want to be an army with banners, a party with power, a movement that controls the future and the past. An armed struggle thus always seeks to control, to control by right as well as by power, to control the arena, control events, control time and place, people and history. This revolutionary will, no matter how intense, can seldom defeat reality imposed by tangible power. The index of books on revolution are filled with the acronyms of guerrilla movements long forgotten by all but specialists. The weak seldom win. Secrecy, a safe haven, a congenial hidden ecosystem of the faith, great cunning, dedication, talent, and audacity can and often do produce another failure. The Palestinians had almost

The Revolutionary Ecosystem and the Armed Struggle 11

everything and ended with almost nothing, and even this was *something* for mostly the center holds or at the end of empire withdraws only overt control and development expense to maintain interests in other ways. Still, there have always been those driven to arms by the requirements of their vision.

In the classical case those transformed by the dream persist until the will triumphs over overt power. Such an lethal asymmetrical dialogue is easily found in the modern world, if unevenly distributed. If the power at the center is too great, too brutal and efficient, too legitimate, there can be no revolt. Almost no one rises to achieve a blood sacrifice. And if center is too feeble to resist the faithful, there is no need for an armed struggle, an underground. A coup will do or simply riotous display or civil disobedience.

Since the few can achieve spectacular violence and impose terror, the result is that stable systems, legitimacy, a complex of institutions, associations, civility and trust, can be perceived endangered. And those institutions, parties, and loyalties, those decent and responsible, can at times not satisfy everyone, not those in the grip of revelation or at times those committed to schismatic habit or old values. There are always some discontent because the present is not like the past or is not the future promised or is simply disagreeable. A brutal and efficient regime does not even try to achieve decency and consensus and is so less vulnerable to challenge by the idealists or the marginal. Coercion is the first response to provocation of any sort. Open, prosperous democratic states try to cope, try to address grievance, try to find accommodation, and since open are vulnerable to the few who deploy terror for a dream. Surprised and horrified at each new affront but essentially secure, democracies need only reassure their own not adjust to the frantic. Most of the world is not open, prosperous, and democratic, not so structured, not fair and not efficient, not even effectively brutal. Much of the world is thus an arena open to the gun if not to the victory of the gunmen.

Some gunmen denied real prospects by the powerful of the center simply seek out vulnerabilities that allow vengeance as redress of grievance if not prologue to power: the Great Satan can be punished, the system can be damaged, and the struggle so pursued until the will grows stronger. Some must go underground to defend the existing reality, the vulnerable system, the village or the old cause: vigilantes of the systems, the posse, the night riders, the paramilitaries for the past. Others pursue power without a dream, without need of avowed responsibility, pursue control for narrow purpose, wage a campaign because a campaign is possible and agreeable. The warlord makes war because it is his nature, the times do not matter, the arena is congenial, inherited, to be exploited. So there are all sorts of low-intensity conflicts driven only in some part by faith. All share similar dynamics and a paucity of resources.

The center, those in the palace, those in the capital, seemingly have all the assets. To achieve the dream, even to begin, the faithful often must seek refuge underground. And underground is not one place and has no address, but it is the secret world of the committed, a galaxy of the faithful visible only at certain rituals and during operations. There at the core of the system are the totally dedicated, surrounded by those less committed, and further out the sympathetic or concerned, a great whirling, invisible, perceptual galaxy. There within the center of their universe, they are shaped to conspiracy and soon dedicated to an armed struggle that may escalate but must persist. An underground is inefficient in all matters but always can offer sanctuary and so persistence.

The limits of the underground are so apparent to the involved that every effort is made to shape tangible control to secure real recognition, to become conventional: control the jungle, appear at international conferences, rely on irregulars not guerrillas or guerrillas not assassins. What the underground does offer is haven so that a long war, a long march can begin, can move toward the realization of the dream. And if there is no dream those underground can pursue other aims fueled by habits, old aspirations, new greed, perceptions less intense but often as effective as those of the ideologues.

Most times, however, logic prevails. Perceptions cannot outlast the holder. Tangible assets often can be deployed swiftly and to effect by the threatened. The dream is snuffed out quickly. The first gunmen are usually the last. Few undergrounds survive to pursue an armed struggle and those that do often win or lose within a few years. Truly protracted struggles operate under special rules that limit coercion—Northern Ireland—or in special places that encourage persistence—the Latin American *altiplano*. Mostly classical campaigns are rare and victories rarer still. That the weak can win over the strong, have done so, that will can defeat power, has done so, inspires those exhilarated by revelation but without prospects, without means. Remember Cuba and Nicaragua and Vietnam and the Ayatollahs of Iran.

The Revolutionary Ecosystem and the Armed Struggle 13

The commitment to the dream supplies the enormous energy necessary to purse the struggle against most odds and so is the single great asset. The revealed faith explains all and emboldens the faithful to act against the odds, against reason. The committed are protected not only by their own perceptions of reality but also by the secrecy of the underground and the assets of the galaxy of the faithful. Those with access only to a small revelation or dedicated traditional aspirations must rely on habit and stealth, experience, the old ways, cunning, a congenial arena, and the chaos of the day. A tribal agenda is no small matter nor the defense of the village. These will fuel a campaign. Often, the less compelling the vision the more familiar the struggle: bandit wars, tribal forays, village defense, and sectarian murder—low in intensity, low in aspiration. The greater the dream the more spectacular and persistent the results: transnational terror, the IRA bombs in London and Manchester, murder in the cities of Italy, or the long wars of Latin America.

What makes a classical armed struggle is the dream, the revealed truth that requires action. Those dedicated to the dream shape about them a world unlike the everyday, perceive a new reality, are endowed with revelation and responsibility. Their world is intense, demanding, isolated from the ordinary. The underground is not a place but a nexus of perceptions, a haven for the faithful, a base for action, a necessary but quite unromantic way station to the future. None underground seek a vocation, few even seek competence, but all want power as a means to move into orthodoxy, seize the center, and be recognized as legitimate. Their underground career is transitory, the gun a means not an end, their dearest desire to be orthodox, conventional, recognized.

At times the great cause, the resistance, or the emerging orthodox armies attract those with conventional motives. The few are faithful: who else would go with Che Guevara into the jungle or seek to kill the Tsar for the People's Will? The many who joined Mao's Long March in route or signed on with the Palestine Liberation Organization had more complex ambitions, not quite as pure if not as traditional as the village defenders or the pirates off the Gold Coast. Thus there is an arc of commitment as well as degrees of combat intensity: some compelling and classical dreams are held only by a handful of gunmen, while some very narrow aspirations engender the collapse of the center and a descent of the arena into the chaos found in Liberia or Mozambique or Afghanistan. Defenders may be found in Montana or Texas. In Switzerland or Los Angeles, there are lethal cults shaped about bizarre revelations and self-interest. Each engenders a special response but in a similar pattern.

Most of those who analyze such threats are apt to focus on the dream as desirable or not and to the techniques as effective or not. How the dream works to shape the challenge is often less clear. How the underground works and so how the armed struggle is waged become less important than motives and techniques. Few at the center are driven by revealed truth—not always, the mullahs had to contend with the mujahedeen. Few at the center can imagine power other than tangible or legitimacy other than their own. Those who seek to influence hearts and minds within the battle arena seldom understand either the power of the underground dream. An armed struggle, except in the rarest of cases, is a minority matter, the cause often flawed, legitimacy dubious outside the galaxy. The truth is not to be co-opted by decency or reason.

* * *

Perception determines reality for those involved in such small wars. The faith will inevitably triumphant or will not. The pragmatic and orthodox in the West are apt to find the underground intangible, outrageous, inexplicable. Americans in particular find underground perceptions alien. An armed struggle is not only an asymmetrical struggle between will and assets but also asymmetrical in analysis. The center often assumes that the rebellion fails not for want of capacity but because an armed struggle is illegitimate. And in any case faith cannot move real mountains. Americans have faith in mountains, in practicality, majority votes, in development, tangible structures, the virtues of open society, and democratic norms. In America the revealed truth has long been institutionalized except for the weird, the misguided, and fanatic. Car bombs in the World Trade Center or in front of a federal office building in Oklahoma City only reinforced the American commitment to reason. Americans have faith in their system as real and doubt that faith displayed by the gunmen is faith at all but rather delusion, not apt to move mountains.

The Americans in this century find revolutionary zealots alien. The American way of war has difficulty focusing on the gunman and the guerrilla, where perceptions counts as power and the big battalions cannot easy be brought to bear. Americans prefer wars that can be won by the commitment of conventional power—swift, full, technologically elegant and permanent solutions to provocation. If the underground challenge does not fit an appropriate pattern, the American response is apt to seek to impose a conventional pattern on the unconventional: oppose irregularity with regulars. More than most, the Americans are apt to find their involvement in an asymmetrical dialogue uncongenial, unpopular, and so to be avoided—certainly avoided after the grim Vietnam experience. In a real war, analysis is a matter of weighting countervailing assets and factoring in morale. In an armed struggle, the Americans cannot easily find opposing assets and do not understand the intensity of the faith—not a matter of majorities or tangible rewards or logic. Yet for two centuries Americans have engaged in a whole spectrum of small wars from the Indian campaigns to contemporary expeditionary intervention in the Caribbean, Africa, and the Middle East. The consensus in Washington still remains that such small wars, if unavoidable, must be shaped to be real wars.

And so in Lebanon a generation ago, for example, the Americans did not understand the nature of the various visible and invisible power centers, the harsh edges of the underground, the warlords, and village defenders, all the shapes and variants of lethal faiths and interests. The Americans were not alone; few in the West had previous exposure to anarchy. Few understood the rules of Lebanese chaos or the nature of the overlapping and competing Dragonworlds, the ambitions of the factions, or the decay of order and restraint. The same would be true a generation later in Africa with the Somalis or in the Middle East with the latest tide of Islamic fundamentalists. Americans in particular imagine their perceptions to be universal. So few understood the Lebanese arena and fewer the power and logic of the revealed truth of ancient agendas and revolutionary aspirations. Few could credit simple spite, malice, and rancor as policy imperatives, or the faithful eager to die in order to make the world pure.

More than any of the other arenas, Lebanon was replete with the entire spectrum of Dragonworlds that had engendered small wars, terror campaigns, random violence, chaos in one small place that had no real center. In the 1980s Lebanon had a flag but no legitimacy, a capital but one divided into factions, citizens who were united only in fear and loathing. No external power, no matter how configured, coped very well. Lebanon was as close to anarchy as the postmodern world could then offer, something horrid for everyone: terror and deceit as system. In Lebanon the Americans in particular failed to understand the nature of such a reality and so failed to pursue advantage effectively. Force was improperly deployed, ineffectually protected, and vulnerable to those with limited assets and enormous conviction. And at the end Americans were little wiser only bitter.

2

Dragonworld: Lebanon

> *"Lebanon is a political riddle that time tries to solve, but my Lebanon is hills rising in splendor toward the blue sky."*
> —Khahil Gibran

It was a glorious, Lebanese, July day in 1958, summer at its most pleasant. The Beirut beaches were thronged despite the political crisis, a crisis that was seemingly deepening into civil war. Suddenly, unexpectedly, the sunbathers, the beach boys, the tan women in bright bits of Parisian swim-wear began to stand up, one by one, group by group. The beach ball games stopped and the urchins straggling along the Tripoli road began to move down onto the strand. From out of the west, suddenly in real life, a full-color war movie seemed to be underway. Splashing into the beach came wave after wave of United States Marines, grim-faced, martial, arms at the ready.

Arriving twenty-four hours early, the Marines had been sent to prop up Western interests. They were to foil any coup like the one in Iraq on July 14 that had brought radical officers to power and death to the king and many in his government. The Marines were ready for war, for armed resistance by Nasserite communists. Instead, they were engulfed first by startled sunbathers and then by holiday crowds. Ice-cream vendors pushed their carts through the sands to reach the beachheads. The small boys, ever alert to commercial opportunity, tagged along, surrounding patrols, offering chewing gum or candy. The Marines had landed in vacationland and were soon deployed about Beirut to the delight of the merchants, the hordes of urchin salesmen, many conservative politicians, and all the friends of the West.

In three months, three rather pleasant months for them if not for the harried diplomats and warriors patching together a Lebanese settlement, they would be withdrawn, not a shot fired. It had been a moment of mid-summer delight in the midst of a cold war: Beirut at its best, an ice-cream war, 1958 a vintage Marine year.

* * *

A generation later Lebanon fell apart into chaos and anarchy, became an arena for unconventional conflict, regional disorder, external intervention, and great power concern—no ice cream wars. Lebanon was not unique. Revolts, attacks, civilian deaths, raids, all were the usual in the Middle East. All were the usual in the global arena where all wars are local. The Lebanese were simply one more clump of the angry and discontented engaged in irregular conflict for uncertain cause. In the 1980s as in the 1970s Lebanon was no novelty really, except that the Americans returned pursuing decency and advantage.

In Lebanon by 1983, time had long run out without an answer to the local political riddles. Lebanon had for years been wracked by civil war, foreign invasion and intervention, and seemingly mindless violence that ruined each new hope for a solution. After the Jordanians expelled the Palestinians in 1970 following the Black September fighting, Lebanon had to absorb the armed refugees. Tensions grew as the Palestinians moved into South Lebanon, provoking Israel. Over the same period the growth in the unrepresented Shi'ite Moslem population also put strains on the always weak Lebanese government. Then another Israeli invasion in 1982—Operation Peace for Galilee—internationalized the crisis.

Activists in the new Reagan administration in Washington began to consider whether the correlation of Mid-Eastern forces was congenial to American an intervention similar to that which had led to the Camp David accords between Israel and Egypt. Perhaps a presidential initiative would be timely, welcome? Perhaps 1958 could be repeated? Mostly, however, Reagan's key advisors had other interests. Few at the center knew anything of Lebanon. Secretary of State Alexander Haig was a military aide raised to power. His successor Charles Shultz at State and Caspar Weinberg at the Pentagon, despite their backgrounds in international business, knew little of the politics of the Mideast. Those who did know Lebanon and the Middle East—staff officers in the White House basement or the Old Executive Office

Building, desk men at State or the CIA—were a long way from the political players around the President where Lebanon was a distraction that was fitted into Cold War priorities and then related to Israel's interests.

Still, after the Israel invasion that led to the withdrawal of the PLO from Beirut, the President authorized an American initiative. The Marines were sent into Lebanon in August 1982 to buttress the forces seeking peace. Even those specialists in and out of the government who doubted the wisdom of involving the United States militarily—and many of those specialists were to be found in Pentagon offices—agreed that something should be done. At a moment of flux and confusion, there was an opportunity to support the uncertain center, to act for peace as a member of the Multinational Force of Western military units. And the Marine Amphibious Unit landed as ordered and as expected by September 10 was soon back on the ships sailing for Naples—withdrawn without complications. Complications soon began, however, and the Marines again returned to Beirut seeking to aid peace, stability, order—a task that kept them in place for over a year without tangible returns.

> *"Don't get small units caught in between the forces of history."*
> —General John W. Vessey
> Chairman, United States Joint Chiefs of Staff

Beirut 04:50 23 October 1983

The Beirut dawn promised to be gorgeous on October 23, 1983, as it had been the day before and as it would be the next day. Just before five in the morning, the hills rising in splendor toward the east cut off the first light. Beyond the mountains and the Bekaa Valley lay more mountains and then Damascus. The weather was always gorgeous in the Lebanon in October, clear, bright, warm with a light breeze. By 1983 not much else was as pleasant, not for the Lebanese after nearly a decade of war, not for the Syrians in the east or the Israelis in the south drawing small returns on their investments, not for the United Nations force embedded in Israeli positions, a symbol, vulnerable and futile, not for the dozens of private armies, imported crusaders, sleeping warlords, and tribal militias uncertain of the day's prospects, and most especially not for members of the Multinational Force, MNF, the Americans, British, French, and Italians who had returned to Lebanon in September 1982. The Marines were scattered on the edge of the international airport, but after a year more target than presence.

The dispatch of the MNF was a Western attempt to bolster the new, vulnerable government of the Christian Phalangist Amin Gemayel after his hasty election to replaced his assassinated brother. The immediate crisis in September 1982 had followed the revelation of the massacre by Christian militia of an estimated 460 Palestinians in the Sabra and Shatila refugee camps. The MNF was in theory international, neutral, a peacekeeping force, without military mission. In practice the MNF was viewed by the locals as one more Lebanese player with an uncertain role. The Marines had followed orders, landed in Lebanon on September 29, 1982, as part of the MNF. They had accepted an offer of the solid ferro-concrete building, evacuated by the Beirut International Airport executives, as headquarters. At first, anyway, they met an enthusiastic local welcome. Most felt why not. The Marines had landed to keep the peace that everyone professed to want.

Beirut 05:00 23 October 1983

At just on five, on Sunday morning, October 23, a single Mercedes truck with a yellow stake-bed and a gray cab was noted by several of the Marine guards moving alongside the wire down the main airport highway past the American position. No novelty. No sweat. One or two Marines watched the yellow Mercedes turn left at the end of the wire by the corner of the south parking lot.

There was no special movement around the Marine headquarters building, just the yellow truck. As for the building, the old shell scars and pock-marks indicated that the building was sound, a survivor. Most of the buildings in Beirut, as far as the Marines could see, were shell-blasted and bullet-pocked, if they had not been tumbled into the streets, great unexcavated mounds of masonry eroded by time and further shelling. For years the television news had brought in the images of almost ritual fireworks displays, the splash and thump of grenades and mortars, the flash of heavy artillery, and seemingly millions of rounds aimlessly fired for the benefit of cameras, out of boredom. The visible result was ruins.

In 1982 the Israelis had brought a real professional touch to the destruction, pounding down much of what had managed to remain standing. The Marines' own airport building had proved solid and sound. The Israeli army had used the building, and for most Americans the Israelis had a handle on Lebanon. They had smashed their way into Beirut, one more Mideast victory, knew all the local players, had been in the game forever. Not the Marines. They went where they were sent, one beach as dangerous as the next. For most Americans Lebanon was far away, mysterious, one more stop on the war tour, one more distant crisis with an alien cast and no plot. This remained largely true in the Oval Office at the center of the circle.

The Marine commanders had been told they were being sent into Beirut to "establish a presence"—whatever that might mean—and the airport administration building seemed as good a site as any. The Marines put observation posts on the roof, wired off compounds and parking lots on the ground, and added a few guard posts. A year later in October 1983, nearly 350 out of 1,250 of the American MNF were operating out of the building. Whatever the politicians in Washington had in mind, the Marines had become a "presence." They had found little peace to keep or conventional signs of a government to support. The first year had moved on through mine explosions, sniping, artillery exchanges, and news of violent incidents just out of sight. During much of the previous month of September 1983, because of an undeclared war on the fringes of the Marine positions by Shi'ite Amal militia men and a Druze sect private army, the men felt increasingly like targets. The Marines certainly had not been abandoned. A steady stream of brass arrived, inspected, conferred, and departed leaving most no wiser, and the Marines no better prepared. The brass seemed content. Their own officers seemed content if wary. The locals obviously were not content but that was another matter, a political matter not amenable to Marine doctrine.

Beirut 05:05 23 October 1983

Soon after five the yellow Mercedes turned into the south parking lot and began bumping along the perimeter at speed. At Post 7, on the other side of the wire, Lance Corporal Eddie DiFranco watched the truck without much curiosity. There were lots of stake-bed trucks around, all colors, mostly Mercedes, making deliveries to Marine head-

quarters or driving down to the terminal, lots in and out of the compound. No big deal. No deal at all. There were a lot of things stranger in Beirut.

The Marines had actually wandered into an undeclared war during a quiet period. During much of 1982 and 1983, the hard men were waiting to see what the Israelis would do, to see what the United States would do. So for some time the Americans were exposed only to the steady muzak of Lebanese low-level violence. The Marines sent out their laundry and played with the kids. Even in October 1983, they still carried unloaded weapons on guard duty despite the renewed fighting. The Pentagon press officers still discounted any real danger, any real war.

Not everyone in Washington was so sanguine. Many at the center of the military circle had long held second thoughts about a Marine presence. Secretary of Defense Caspar Weinberger, expressing the opinions of his senior military establishment, had recommended to the National Security Council meeting on October 18 that the Marines be withdrawn immediately. The American military under the shadow of Vietnam wanted no unconventional task, no war that could not be won, no assignment except the most popular. The Pentagon feared that the men had become more hostage than presence. Certainly, the appearance of regular Syrian troops and gunmen of the Iranian Revolutionary Guardsmen on the Marine perimeter could not be construed as a hopeful sign.

The day after Weinberger urged withdrawal, the Marines had their own local problem—again and still. On October 19, within the Italian MNF zone, a white Mercedes car bomb detonated in front of the Kuwaiti Embassy. The explosion nearly took out the jeep of the Marine commander Colonel Timothy J. Geraghty. A Marine truck returning from carrying a meal to the Bravo Company contingent in the American Embassy was not so lucky. No one knew who set the bomb. Yet no one reviewed security or took special precautions.

By October 1983, the Marines had become for many a tempting prey. They were the visible symbol of the West, the Great Satan, imperialists. They certainly had a long list of proclaimed enemies: Islamic fanatics (Shi'ites mostly), Palestinian guerrillas, private militias like the Druze often with imposing names and little discipline, single assassins, itinerant gunmen, terrorists of various deeply-held convictions, and even the bored ready to shoot any convenient target. The Marines—and most Americans—did not understand why they should be targets. Lebanon was not their affair nor their presence meant for any purpose but good. They did not even know the names of the players. "You feel helpless... What are we doing here?" Why were they risking their lives? Why had the undeclared September War been hushed up? No one on the ground seemed to have any answers. Word kept coming down the chain of command insisting on no American provocation, no offensive action, no loaded weapons. It was difficult in Washington or in Beirut for Americans to perceive themselves as evil imperialists and guilty of all manner of crime. They felt fair and decent. They were fair and decent.

Beirut 05:05 23 October 1983

Corporal DiFranco watched the yellow Mercedes truck turn and drive out of the south parking lot. It moved away on down the airport highway and out of sight. His gun was not locked or loaded. That would have been provocative. The guard posts around the huge administration building out on the airport road were not prepared for trouble—why should there be trouble?

And there had been no serious trouble at headquarters. Shelling, sniping, fire-fights, and now a car bomb had spread out from the city; but the militias and warlords had not sought to attack the MNF positions, to engage in any sort of regular battle. Everything in Beirut was irregular anyway, unorthodox. Nothing stayed the same or was as anticipated.

At least the first days had been happy days, passing out bubble gum to kids, making friends with the locals, sending out orders for this or that with the various Hello-Joes who had thronged the area eager to sell all sorts of wares and services. Then, suddenly, there were kids yelling "Khomeini good! America bad!" and no more friendly visitors. A Hello-Joe kept the laundry and sold the uniforms to a terrorist militia. Other Hello-Joes were just as likely to be in a terrorist militia. No one came down to the Marine perimeter any more. It was dangerous hanging around the Marine perimeter, even for Marines.

Beirut 06:02 23 October 1983

On Sunday morning Geraghty's 24th Marine Amphibious Unit mostly slept quietly at the edge of the airport, at the edge of Lebanese events. Lebanon was a crisis seemingly played at half-speed. Even if Lebanon was no longer good duty, at least the airport was quiet. In the hour since the yellow Mercedes had passed down the highway, nothing had happened. It was past six and quiet. One Marine was jogging, a violation in regulation. Nothing else moved in the quickening light. The Marines had an extra hour of sleep as usual on Sunday.

All cities have golden years, the perfect decade of the lost generation, forever lost. Mostly perfection is found in reflection. Beirut in the 1960s was a golden, glittering Levantine city, an elegant Arab *entrepôt*. French was spoken or English or any language would do. Anything could be found, bought. Correspondents at the bar of the St. George Hotel, epicenter of a spinning world of rumor, could send out for anything and be assured a reasonable facsimile would appear. It was a grand time.

As the entrepreneurs arrived on European flights, as new banks opened and new hotels and new opportunities emerged, time began to run a little faster. Beirut was still golden, could still ignore the misery on the city rim but not the ambitions of the Palestinians or the fall-out of the Six Day War. By the end of the 1960s, there were great black char marks burned into the tarmac of the international airport where the Israeli commandos had come in December 1968 seeking vengeance for terrorist attacks. To the south Fatahland of the Palestinian fedayeen was beyond the reach of Lebanese law. In Lebanon illusion had been everything, a country run with mirrors and wires that by ruse presented a nation where none existed, by agreement gave each political faction something if not enough, by habit allowed business as usual. Time ran out on the illusion.

For a generation Lebanon had been ruled by disguised barons and warlords, chieftains possessed of private armies and special accounts. All the greed of the sullen tribes, all the sacred egotism had been reined not by self-interest or good feeling but rather by a momentary correlation of forces that imposed stability. The Palestinian gunmen arrived in mass in 1970. Then the multiplying, unrepresented Shi'ite poor grew restless, became a new player with old grievances. Oil money kept for a time stability and allowed the machine to keep running. To the amazement of a great many in Beirut, money was alone insufficient to underwrite stability, but soon created more problems than solutions, more corruption in the region than could easily be tolerated, more pretensions of grandeur than could be met. As time trickled away without consensus, without general loyalty, without real governance or nationality, the ambitious sought their own advantage.

For the many in Lebanon, power was to be deployed not only to acquire but also to punish, to humiliate, to shame. Power was limited by capacity not restraint. And power was to be *used*, flaunted, a triumph every week insufficient. Loyalties were narrow, self-interested, parochial, and extended no further than gun shot and the farther hillside or the last village. All the spite had been held in place for decades simply by special conditions—a lack of opportunity. One day in April 1975, the long drift to slaughter had passed the fail-safe point: the year zero had arrived.

On Sunday, April 13, in the Beirut suburb of Ain el-Roumaneh, Christian Phalangist gunmen ambushed a bus filled with Palestinians and killed twenty-seven of the passengers. Later there would be reasoned explanations. Their spokesmen claimed provocation. In Lebanon no one ever, ever, is responsible—atrocity is always provoked, excusable, inevitable, necessary. Whoever was responsible the Palestinians were dead, sprawled bloody amid the flies and cordite fumes at the bottom of the bus. This time all the pipers wanted to be paid. The Palestinians had put up with enough. The Moslems had tolerated the Phalangists too long. Arrogance and brutality had run up debts now due. The next morning, Monday, April 14, the barricades had gone up in Beirut and the fighting began. It never really ended.

Everyone became involved, the decent and deadly, all the sects and warrior gangs, the private armies and paramilitary parties, every village, all the denominations, powers, and thrones. And in time strangers came to kill, the Syrians, and then the Israelis, next Iranians. So the most recent foreigners, Western peacekeepers, found only truce no peace. Peace in Lebanon, the golden years, had been an illusion drawn over an unplayed zero-sum game that once begun guaranteed only that there would be no winners only players. Beirut became a killing ground, an arena for all sorts of unconventional structures, loyalties and agendas, few found in academic texts and all potentially lethal to the involved.

> *"We built Lebanon and we will burn it."*

The Marines arrived well into chaos. There had long been a broad spectrum of violence that included the car bomb detonated before a supermarket or a magazine emptied into a car window as well as artillery box-barrages and rocket volleys loosely aimed. There had been and still were atrocities committed out of boredom, random murder on any lazy afternoon. The killing went on and on. Some continued to kill for a great cause, the Palestine nation or the triumph of the workers, others for the survival of the Druze or the defense of Islam. The simple and vicious soldiers of the slum simply shot at dogs and taxi cabs and anyone who crossed the far road. The proper and the decent and unlucky were all victims, all murdered as pastime or as symbols in a great struggle.

For those who would make realistic policy the peculiar horror of Lebanon had to be factored into any consideration, into every explanation. How to explain elsewhere, explain in New York or Los Angeles, murder as target practice or gunmen who shoot children because they would just grow up and have to be killed in less convenient circumstances? Families were shot, father, mother, children, the family dog. No one, of course, would take responsibility. No one in Lebanon was ever responsible. No one in Washington would find any explanation easy. Americans especially assumed all were like them and violence aberration. In Lebanon violence seemed convention, first-choice, cruelty and savagery the norm.

In Lebanon a different reason seemed to run, men acted like spoiled children possessed by wild tantrums and automatic weapons. Horror was convention and no one apologized or explained. Even for the specialists on the spot, even for those intimate with the sordid side of Lebanon, some of the horrors were unexpected, monstrous beyond ready explanation. Yet, much of the Lebanese conflict arose from real and historic grievance not simply the flaws in individual character or a people run amok. Those denied fair share, denied justice, ignored and humiliated sought redress even if others simply deployed wanton violence or sought private advantage.

Those armed and dangerous varied across a wide analytical spectrum. The Palestinians sought rational if improbable goals by distasteful means. They were, terrorists or no, revolutionaries. They had a national dream and an armed struggle. And their dream was pursued by a dozen competing movements and secret armies.

Out of sight in south Lebanon others were haltingly constructing another vision, another world even more alien, perhaps more dangerous. The Shi'ites, the largest Lebanese group, almost without leverage, had been increasingly touched by the militancy of Islamic fundamentalism. The appealing doctrine with Islam as the answer to anything, to everything, had been preached by the devout Iranians, often like Ayatollah Khomeini, and their own Mussa Sadr who "rose from the East" in 1961 and preached to the poor and the forgotten. His Shi'ites were largely forgotten once the Lebanese civil conflict began, their Amal militia ignored. Mussa Sadr disappeared on a visit to Libya in the summer of 1978, apparently murdered in an excess of enthusiasm by Colonel Muammar el-Qaddafi's military secretary. By then the faithful has established another world where the answer was Islam. The Shi'ites now had a fundamentalist identity, a friend in the new Islamic Republic of Iran, a growing private army in Amal, and a martyr, their Imam, taken from them. It was the West, however, the home of the Great Satan who had sent the Marines, they feared and hated. And now there was an answer: "Islam is the religion of agitation, revolution, blood, liberation and martyrdom."

In Lebanon there were others with general aspirations: Maoists and secular Arab nationalists, orthodox communists, friends of Nasser, those dedicated to the Syrian Ba'ath movement or to the Iraqi variant, those who trusted in Marx and Lenin. And many of the most parochial gunmen and unsavory terrorist for reasons of fashion flew such ideological flags. At the far end of the violence spectrum were those shaped by the most local considerations. Many Lebanese players, the Druze or the Lebanese Forces, or the Beirut neighborhood militias were neither crusaders nor revolutionaries nor proxy governments, not counterstates as much as tribal armies. Some had a religious base. Some had ethnic unity. Many had both.

Few had a commitment to real ideas or contemporary institutions. Some were village defenders or ghetto warlords. And some were criminal. Yet, each fashioned rationales for violence and proclaimed political reasons for traditional vengeance and violence. There were subsets eager for marginal gains and righting old wrongs, but not revolutionaries nor committed to vast change sought by the Islamic Jihad or the Hezbollah. Confusing the mix were the demented and the wanton, local killers eager for action, prominence, and profit, gunmen who liked to kill, charismatic chieftains of the moment. They were opportunists, free range killers added to the lethal orthodoxies and ancient agendas.

The Marines thus came to a Lebanon crowded with the frantic and desperate, those with visions and those with unsavory aspirations. Some were integrated into a galaxy of the faith and others killed alone. None were easy to explain for those eager for simple clarifications. Lebanon had for years been filled with simple answers, absolute truth, the laws of the tribe, the uncompromising faith of the insecure. Each established a safe haven for the faithful. Explanations of those who did understand Lebanon seem to fail—anarchy is a label, not an insight. Political violence implied politics, a reasoned violence that often seemed lacking along the Beirut Green Line. Even, perhaps especially, the old categories appeared ineffectual. Right and wrong, proper and improper, justice and tyranny had different meanings to killers who were often articulate, arrogant, always unrepentant. These gunmen often had a professional education, were well-spoken, and superficially sensible. Yet many were possessed with raw desire unrestricted by the possible. Moderates existed, could be found but had little influence and less leverage. Many "moderates"were moderate only at times and mainly in demeanor: shallow men in passé suits with gangsters on their payroll or brave new leaders with razor cuts and chic Paris tailors who deployed assassins. Others were seldom reasoned or rational, regularly brutal, angry if not mad—mostly beyond conventional categories, beyond understanding, beyond compromise, conciliation, or accommodation. So they were alien to the West in general and the Americans in particular who could hardly credit the degree of depravity and duplicity.

Across the world in Washington, in another time as well as place, sense had to be made of Lebanon. An American policy of sorts had to be fashioned, good to be done, harm to be limited, opportunities sought. To do nothing might open the Middle East to Moscow. To do nothing might lead to greater anarchy. So those Americans responsible for policy needed an understanding of Lebanon.

The reality of Lebanon had to be explained by the conventional to the responsible. Lebanon was neither conventional nor the Lebanese responsible. How to portray these strange players? Kamal Jumblatt had long been one of the crucial notables in the Lebanese game, a major player, the feudal leader of the Druze sect, dug into the mountains of southern Lebanon from the Shouf down to the Golan Heights and into Israel. The Druze were a secretive, post-Islamic sect distrusted by most Moslems, distrusting most Moslems. Many were citizens of Israel or Syria as well as Lebanon but true only to their own. The Druze formed the core of the National Movement, a complex of Greek Orthodox gunmen, fascist Rightists, local Moslems—and the Druze. Their Progressive Socialist Party was neither progressive nor socialist and not even a party. Mostly the Movement was an alliance of private armies built on the Druze base run by Jumblatt and his people. No one quite knew how to explain the Druze in the West. The Palestinians were classical nationalist revolutionaries, divided by ideology, inclination, and personality, by Arab habit, but united on an armed struggle. Their universe made some sense. The warlords of the village or the ghetto were criminals or vigilantes or defenders. They had some sort of parallel in the West. Not so those who found Allah as answer to all and preached a Holy War. There were too many gunmen possessed of the revealed truth. Yet something had to be done to ensure at least regional stability if not justice for Lebanon.

In the United Nations, the Arab capitals, and the Arab League, in Washington and the West, there were initiatives, proposals, suggestions, and pleas churned out, advocated, printed, distributed, read, discarded, all irrelevant. One massacre faded into the next. Sometimes life ran along with deliberate calm for some, water skiing to the background thud of mortar barrages, dinner out on near misses, never missed village festivals, and regular vacations in Europe. There was no center to hold but life went on. The villagers in the Bekaa or on the slopes of the Shouf harvested their crops. The businessman opened the store each day, paid extortion, avoided arson, hoped for the best. The media came and the foundations, the scholars and the agents of Egypt or Israel, the Syrian army on leave, drifters from Pakistan and Ulster, spies and agitators and self-declared revolutionaries on student fares. There was room for all. Beirut became a long-running television news serial. So there was ample exposure. No one was truly innocent of Lebanese matters even if ill-informed.

Year after year, the Lebanese acted out fantasy and grievance over and over: no remorse, nothing won, nothing learned, nothing forgotten. Even the Israelis, no stranger to arrogance, could not really understand the Lebanese, could only read the challenges and opportunities amenable to logic and thus misread the pulse at the heart of darkness. All the premises for anyone's intervention, for involvement and amelioration of wrong, for accommodation would prove dangerous for those who would tinker for advantage or in innocent good-will. There were many worlds in the land marked Lebanon, but most were well beyond the experience and expectations of the Marines, the Pentagon, or the Reagan administration in Washington, beyond the understanding of nearly all Americans.

* * *

In the global village there are always village explainers, often briefed by their betters, anchor men, columnists, leader writers, spokesmen, and seminar chairmen. And beyond them are the experts and experienced, steeped in history and statistics and years on the spot. The government in Washington employed such experts and analysts. The Lebanese may have no explanations about their world nor the responsible but the experts inside the Washington beltway do.

All over Washington vast mounds of paper had been churned out on Lebanon, reviewed, filed, and often over the years forgotten or regularly circulated, often across important desks. There were millions of words, often indexed, all on call, often relevant. This is almost always the case when crisis comes, nothing at first and then the flood on Laos or Somalia or Antigua. Anyone with proper clearance can get all sorts of intelligence, good, bad, indifferent, can find evidence for any predisposition, can find out answers to most questions. Neither the quality nor quantity of available intelligence is a problem. Filtering out the relevant is a problem. Asking the right question is a problem. And especially in Lebanon understanding the alien was an almost insurmountable problem.

Not all the Washington paper arises from academic concerns, pure analytical interest of clerks with higher degrees wearing green eyeshades. Some of the reports come from those who know the gunmen, have smelled the cordite and tasted the flavors of hate. those who have been eyewitness accounts of horror, regular Americans with special insight. These Americans, travelers in a strange land, knew gentlemen who take tea before ordering a car bomb for the afternoon or sullen young men who shoot live targets between football matches. It is an experience not easy to convey. In Washington priorities at the top were those of the moment and were often unrelated to Middle Eastern reality. Along the Potomac the nature of the arena must be filtered out into reasoned policy options by midmorning: great power interests, the oil companies, campaign promises, personal recollection jumbled on the table. At the peak of the power pyramid, most detail is necessarily filtered away. No time. No need. Often no interest. The specialists are curbed to a presentation of minutes.

There is no opportunity at the center of the circle for subtle distinctions, only for decision or delay. There is no way to introduce to official Washington the real Lebanese warlord with the tang of burnt powder still on his jacket. How to explain the inexplicable to the innocent—to those officials traveled and competent, experienced and capable, to the sensible, reasoned managers of American policy? The problem with the unconventional is just that: it is beyond the conventional. War shaped by the dynamics of the unconventional is beyond orthodox experience. Even to explain by analogy and example fails because the conventional are attuned to other assumptions. No matter how harrowing and awesome the presentation, most in power in Washington cannot imagine the terror and drama of a world without center—or grasp the imperatives of those not at all like themselves.

Those in power in Reagan Washington, more orthodox than most, more parochial, wanted to know only what was necessary. The men in Washington were not only innocent of complex international matters but also uninterested in fine-tuning American foreign policies. Israel was a friend. No one cared much for the Arabs beyond keeping the oil rich friendly and punishing the terrorist looney-tunes. In Middle Eastern affairs Reagan Washington focused almost solely on Soviet intentions: the evil empire's ambitions, the big picture and not on the narrow details of the region.

For an administration seized on Soviet intentions, the major factor in the area was Israel—Israel as ally against Russia and Russian-proxies. Israel was a friend, effective in war, even if in 1982 the Israeli invasion of Lebanon had proved neither elegant nor effective. It was a war shaped by the ambitions and mistaken assumptions of Defense Minister General Arik Sharon, ruthless, reckless, rude, arrogant, and bold. And in 1983 the Israelis were still in Lebanon and so Lebanon, was on the American and Western agenda. Lebanon proved a small black hole sucking in concern, assets, time, and interest few beyond the region wanted to spend.

The Lebanese remained quite beyond American military experience. This inability to grasp, much less explore, the alien meant that the Marines and the Pentagon had only a suspicion of the unconventional as doctrine. The American revolution was long ago, lethal political dreams, ethnic wars, faction fighting, heresy unknown. Vietnam had taught no lesson greater than the dangers of irregular war. The system wanted only one such exposure to limited war without popular support or attainable goals. Yet Lebanon was obviously not Vietnam. Lebanon simply did not fit any military text.

Lebanon was a free-fire zone for all sorts and conditions, some covert, most armed, and few covered in the manuals of war or the experience of American life. Told that monsters were in charge of the country did not mean that the responsible officers could imagine them. And few knew the capacities and intentions of the locals. The Lebanese arena, however, was real enough, had weight, substance and location, was filled with those driven by spite and idealism, malice and universal dreams, and the needs of the village.

There were power centers with shifting but real bounds—the Druze Shouf or villages in the hands of the Party of God. There were bounds and lines on the map and zones of control, but much more important were the minds that imagined alternative futures. Each dream gave center to a special universe of the faithful. These perceptions spun out from each center, converting the unwary, contesting other dreams, assuring volunteers and victims, were truly confounding to pragmatic Americans. What was wanted often seemed beyond reach or beneath concern. Many flew false flags of belief but many believed in impossible dreams. All felt justified and few could easily be contained. The Palestinians wanted vengeance, the Shi'ites salvation, the meek wanted justice and the rich vindication; this warlord wanted to prove a point and that gunmen merely a vulnerable target.

In a revolutionary world most governments must cope with but a single lethal dream, perhaps with variant readings, organized as an *altiplano foco* or a European terrorist net. In the everyday world a government may have a bandit problem, tribal insurgents or even an exile elite, but usually there is a center to hold, a visible banner, legitimacy of a sort. So in most revolutionary endeavors there is a government, a power system to oppose and to impose order. In Lebanon there was no government, no center, and so power was available to the bold. The dreamers often could operate above ground, but they still lacked legitimacy. Ambitions were contradictory. Zero-sum games abounded. There were those who would prefer to give pain than receive pleasure, those would cause turmoil for the sake of turmoil. And no one dream, no single coercive institutions could prevail and thus create a new reality. Lebanon existed without a center, without effective general institutions. Everyone saw what they saw and took what they could. The moderates, few on the ground, had been driven out. The ordinary was in hiding. The violent created their own imaginary worlds.

In Washington these Lebanese were a strange people in a strange land, far away, different. Let Lebanon adjust to the West, to Washington. Peace had seemed possible in 1982, during much of 1983. The West's MNF had landed and all seemed to bode well. There seemed no great risk. The worst was over. The fighting had finished. The months went by with only local incidents. The Israelis would surely leave. The Lebanese exhausted would accept a settlement. Syria lacked the capacity to resist peace. And all this at no great American cost, no serious risk—perhaps another 1958 ice cream war with low cost and no losses. Why, then, bother unduly about the minutiae of the arena or the character of the bit players?

Everyone in Lebanon was assumed by the Lebanese to have hidden and often vile motives. So for many, for most, America, the Great Satan, seemed intent on imposing hegemony. The West were imperialists. Their friends in Lebanon were the enemies of all others. So many of the involved perceived the Marines as a symbol, but not the one dispatched by President Reagan rather the one that fit their own interests. Some saw Satan in the MNF and others saviors.

Beirut 06:22 23 October 1983

The Marine sergeant of the guard at the base of the administration building for October 22—23 was Sgt. Steve Russell, a slender, taut, young man of twenty-eight, who had served in the Corps from 1974 to 1977 and returned from the reserves in 1982. His station was a sandbagged, wooden shed—rather like a movie box office—just at the southern entry into the four-story inner court. Russell had been in Beirut for five months, learned more than most about the street patterns but had as few ideas as most about the meaning of his presence in-country. The trouble had been getting worse since summer. Early in October 1983, Ayatollah Khomeini, increasing a dominate factor for Lebanese Shi'ites, had called on the faithful to "put an end to the shameful occupation of Lebanon" by the Americans and French of the MNF.

And there was soon evidence that MNF was a prime target. At least Sgt. Russell had been most impressed by the Marine reaction to the carbombing on October 18, that had just missed their commander, Colonel Geraghty. He did not contemplate the lack of security but only the quick reaction to provocation. The Marines had looked very professional, clearing the area and setting up a security perimeter. It had almost looked like offensive action instead of the usual defensive response demanded from on high. In any case, Russell had less than two hours until his duty ended at 0800 and he had other matters to contemplate.

In fact it was just coming up 06:22 when Russell heard a crackling sound from the south parking lot. Corporal Eddie DiFranco at Post 7 looked up to see a yellow Mercedes stake-bed truck picking up speed out in the lot. Perhaps it was the 0500 yellow truck. Its civilian Lebanese license plate number was 508292. It meant nothing to DiFranco. Anyway, this time the driver was putting on more speed for no good purpose. Russell turned around in his box in time to see the Mercedes truck zip through the gate in the chain-link fence into the parking area. No real surprise. The lot was often used by delivery trucks, most of which seemed to be Mercedes. This truck, however, was accelerating. And, suddenly, the pick-up smashed through the barbed-wire barriers, banged out of the south public parking lot and into the headquarters compound between Guard Posts 6 and 7 right in front of the guard shack and Russell.

The Marines at the airport were not the only ones unprepared for the unconventional. Seldom have those at the center of the circle charged with American security been as innocent of the world as was the case early in Reagan's first term. The new men were largely businessmen, political friends, wealthy travelers accustomed to executive suites, hired translators, rings of lawyers, and the like-minded. They were capable executives, capable negotiators, but committed to slogans and simple visions. They were justly proud of their own talents, their skills of the marketplace that had paid for their political pursuits. They were now, suddenly, rewarded with enormous power. At home they had in common a detailed agenda and real experience. They wanted to curtail intrusive government, reduce taxes, end woolly liberal policies. Abroad they felt that America faced an aggressive Russia, uncertain allies, eroded capacity; there was a need for military strength and an end to detente. Money spent on military systems, on technological wonders, would serve American security interests. All else was seen as distractions from their central domestic interests. What was needed were those dedicated to the new vision, not specialists or quibblers. America was to be strong, communism contained—but the prime ambition of the administration was to dismantle big government at home.

Most of the newly appointed had to learn on the job, an American tradition. In time some would become more sophisticated, move beyond the platform of the Republican Party, the network ten-minute interview, and the habits of the boardroom. In the meantime a very few competent appointees seized control from secondary positions and as for other problems and other priorities everyone made do—none missed the experts or felt a lack of skills. It could hardly be denied that the skilled academics, area specialists, and experts of the previous Carter administration had left a record massively rejected by the American people. Experience had paid scant political dividends. So the new men had no time or interest in the alien or unconventional. And Lebanon was both. And from that all manner of troubles flowed. The most devastating descriptions of Lebanese reality were adjusted not only for the Oval Office but in admirals' offices, filtered clean for Assistant Secretaries at the Pentagon, made reasonable in a reasonable world where history is ignored and management rewarded. Thus and not surprisingly for nearly everyone from Reagan to Colonel Geraghty at the airport, convention was imposed on Lebanese reality. Americans saw what they saw, wanted to see, did see.

Many in Washington knew the worst of Lebanon but could not share the burden. Many had no access to the powerful. The few who did never had the words. Americans generally choose to believe all men are much alike, much like us, seek peace, can reason together. And to begin to explain Beirut was to expose to view another different world with skewed priorities, where cruelty became virtue and time ran backward. The meek have not inherited Beirut but the frantic. How could a man in power, elected, respected, a national leader, who had taken courses at SMU in Dallas and worked in a Washington law firm, order death at dinner, request an atrocity for the weekend? And how could his father, an ancient wizened pharmacist, be a warlord deploying gangsters to shoot down yesterday's friends? Improbable. Impossible. Impossible to explain. Most adjusted "Lebanon" not their perceptions. Few really understood Israeli and fewer the Arab states. Few cared.

Still Lebanon reality had proven difficult to ignore and had engendered the Marines presence. It had been a long year. On Tuesday, September 14, 1982, a huge explosion had shattered the three-story building in East Beirut that housed the main branch of the Phalange. At midnight Prime Minister Shafik Wazzan announced Bashir Gemayel's death. For this special death Beirut Radio played classical music, a dirge for not only Bashir Gemayel, the President-elect, but also for the Lebanese. The Christian Maronite dream of hegemony had been bombed away just nine days before Bashir was to take office. Bashir, friend of the Israelis, friend of the Americans, man of the West, was dead. Lebanon was stunned but not surprised.

The Israelis were appalled—all the gains of the invasion might evaporate and terrorism revive in West Beirut. Ignoring all previous promises, on September 15 Israel moved into West Beirut, long General Sharon's goal, to prevent violence. Neither the Christians in East Beirut nor the Moslems in the West were guilty of planning violence—half the country was in despair and the other half in fear; but the Israelis had lost their last fragile grasp of Lebanese reality.

The PLO remained for Sharon the one enemy. Destroy the PLO and destroy Palestine. Sharon did not understand dreams. The Israelis built a state on a dream but seemingly had exhausted their empathy. Sharon believed that power could crush belief, that tangible assets could win a real victory. He knew the PLO forces were in the camps and so left them to the Lebanese Christian militiamen—the Lebanese Forces. For thirty-eight hours in the Sabra and Shatila Camps, they wrinkled out and slaughtered the Palestinians, almost all civilians, women, children, old men, babies, using whatever came to hand, grenades, automatic rifles, hatchets, revolvers, and knives.

It was a traditional, although especially bloody, Lebanese atrocity. There were at least five hundred dead and perhaps—certainly the Palestinians so believed—many times that number bulldozed into common graves. For the killers there was not guilt, only delight at duty done, vengeance taken. What had been done was to create martyrs not kill the dream, outrage world opinion, and raise questions. There was a crisis, both moral and political, within Israel. Not only in Israel but also elsewhere the long seething doubts about the Lebanese war boiled over in horror. Something had to be done.

The MNF came back to restore order, to reassure the vulnerable, to bolster the new president, Bashir's brother Amin. The appearance of the Multinational Force brought enormous relief to many Maronite Christians who saw it as an outward and visible sign of Western support for the fragile new government and for their own old, intractable ambitions. Moslems and Palestinians, on the other hand, saw the new government of the Gemayels as ruthless and deadly, a creature of the West. Israel assumed the West to be allies and the Syrians and Iranians the reverse. Everyone saw what they wanted to see. No Lebanese player was neutral and America was now a player. Everyone automatically had friends and enemies, an assumed and a real agenda. Thus the Thirty-Second Marine Amphibious Unit was perceived by many as the tip of a great imperialist lance. And the others, Bashir's heirs, the still ambitious Christians, their various associates, saw America as a player on their side of the Lebanese board. Everyone saw what they saw, what they wanted to see, what they expected to see, what they hoped to see, what they feared to see.

All these hidden hopes and fears focused on the Marine Amphibious Unit—1,200 men with 100 additional specialists ordered off the fleet to give America a military presence. These Marines did not understand any of this, did not understand their mission or their role. They simply followed orders and made hardly any preparations to defend themselves. They too saw what they expected to see: a strange city, mostly ruins, filled with everyday people who were irrelevant to the routine of the day. For nearly a year there was no pressing need for a reappraisal.

Mostly everyone in Lebanon waited to see what the Americans intended, what the Marines would do. All assumed there was hidden purpose, special stratagems, particular interests that Washington would pursue under cover of the MNF. Even the Russians saw imperialist plots. "The appearance of Marines on foreign soil has always in the past indicated the beginning of dangerous military adventures." But what sort of adventures? And dangerous to whom? How could the irregulars, the ambitious, the recently mauled and presently powerful benefit? What did America intend? What did President Reagan truly mean, really mean, when he had announced his peace initiative on September 17?

> The Lebanon war, tragic as it was, has left us with a new opportunity for peace. We must seize it now and bring peace to this troubled area while there is still time.

The months passed and Beirut seethed with rumor, speculation, and anxiety. And the curious trooped down to the International Airport to examine the Marines. They were at peace in a war zone, vulnerable in a world they could not see—a warp in a complex space filled with dreams, hidden agendas, attitudes and perceptions beyond all conventional experience. The Marines, the Americans saw the facade not the reality.

Beirut 06:23 23 October 1983

Corporal DiFranco in Guard Post 7 was too surprised to act as the yellow Mercedes truck bounded past him picking up speed again. He caught sight, just a brief glimpse, of the driver, a swarthy man in a dark blue or maybe green shirt. A man with a beard. Then the truck was gone. DiFranco slipped his M-16 off his shoulder automatically; but he had to lock and load. By then the yellow Mercedes was gone. He could not get a shot off. There was not even any point in calling up Russell since the truck had nearly reached the headquarters building. Across the way to the west at Post 6 Lance Corporal Henry Linkkila had his back turned and missed even a glimpse of the truck. He heard the roar—definitely not normal for a Sunday morning—and automatically dove behind his bunker. The yellow Mercedes roared on straight at Russell's shack, past the officers and staff NCO tents.
At Post 5, further in the compound, close to the airport road, Lance Corporal Berthiaume managed to get his M-16 off his shoulder but like DiFranco could not lock and load in time. He could not get a shot off either. Russell did not have time to worry about shooting. He was on his feet facing a truck bearing down at twenty-five or thirty miles an hour. The Mercedes motor was still revving up, the driver a blur. The truck simply plowed straight ahead toward Russell and the shack and the door into the headquarter's central court.

By October 1983, the Marines had for all purposes had been at war for months, especially the last month September, when the Phlangists, once more ambitious beyond reason, had sought the long elusive victory over the others. The Israelis had planned to move south and have the Phalange replace them. In August Pierre Gemayel, aging and vindictive patriarch of Beirut, said, "Let the war take place and let the strongest win." And it did but the Christians did not win. Walid Jumblatt and his Moslem allies were not as feeble as anticipated.

During the confrontation, the Americans, as part of the MNF supporting the government and hence the Lebanese army, became all but indistinguishable from the Phalangists, Gemayel's gunmen. The Americans sent officers to help the Lebanese army defend the besieged resort town of Suq al-Gharb in the Shouf foothills beyond Beirut. The United States Navy bombarded Druze and Syrian positions, thus alienating the Druze, the Syrians, and the Shi'ites. Yet the United States did not really feel involved, and a truce at the end of the September eased any potential threat. Despite all, the Americans still assumed that their presence was disinterested and took themselves at their own valuation. Besides, by October the Lebanese notables were planning, again, once more, a peace conference, this time in Switzerland. The Americans little noted the Ayatollah's threats. Iran was a long way off. They hardly noticed local threats, security at the embassy and at Marine facilities was cursory. None seemed to grasp that Dragonwars often begin without formalities, without visible armies.

These Dragonwars have been the wars of the times, a constant. Ideological fanatics, rebels, the very few, civilians in terror campaigns, rise up against government. Conventional wars between Iraq and Iran or Egypt and Israel, the Korean war, the Gulf war, even the British expedition to the Falklands, are all symmetrical combat, orthodox wars easy to detail afterward on scale maps and in academy classrooms. For fifty years most wars have been irregular, minor, insurrections or rebellions, guerrilla conflicts, revolts against the center, or gunmen in the streets. These did not seem *real* wars in the military academies, not wars like the Israelis and Arabs fought with main battle tanks. Yet for fifty years neither America nor Russia, the super-powers, could meet in conventional war. They were limited to regional proxies or to the dangerous edges where irregular campaigns would not endanger the balance of nuclear terror. And many of the ambitious or desperate simply deployed the gun in pursuit of denied aspirations without a bipolar conflict, sought a united Ireland or a free Eritrea, a Basque state or to save the Ibo, the Kachens, the Kurds. After the collapse of the Soviet Union, the arena for such wars was simply larger and their ideological classification more various. There were new sponsors and new conflicts but the tactics and techniques were easily recognizable by those who had studied Algeria or Vietnam or much of military history. Such conflicts were inevitably asymmetrical, will against tangible assets, irregulars against the orthodox, the frantic seeking to change history or maintain the faith, defend the old or open the way to the future. Most small wars were short and brutal, most won by the state, by real armies, but a few were prolonged as was the case in Vietnam. At times the revolutionary will did win over tangible assets as in the case of Vietnam.

Americans after Vietnam, long, long after Vietnam, sought to avoid the unconventional. The Pentagon especially wanted no limited wars without assurance of popular support and pressing national purpose. So the Pentagon preferred not to prepare for the unconventional. The unconventional was shaped to military texts, to techniques and conventional tactics. As long as the responsible in Washington refused to contemplate seriously the Dragonworld, then everyone was at risk if the county became seriously involved in an unconventional quagmire. Americans and the Pentagon chose to believe that orthodox power would be compelling in unorthodox situations: that the big battalions would win. The American way of war would serve. Almost all Americans, too, insisted that their world was everyone's world, that all problems could be solved by reason, by investment, by techniques and technology, and with dispatch and by use of conventional military assets, if it came to that. So the Marines had been sent into harm's way by those who would not recognize dangers that were alien because the alien was outside experience.

In Lebanon no one carefully calculated the risks to the Marines, to the military, to regional policy options, or to American prestige. Great powers have responsibilities. Presidents have always deployed force if national purpose required. Quiescence has great costs. Appeasement is capital wasted. The Pentagon may have feared intervention but not the Oval Office. Not for the first time America entered an arena lumbered with misconceptions, old habits, inapplicable doctrine, and hapless innocence. Vietnam had taught the lessons those involved wanted to learn, but in Lebanon that meant the military did not want to be involved and the administration did not want to be restricted. Once involved, however, nothing novel was done: the system kept ticking over. The Pentagon had feared the unconventional but had not prepared for the prospect. The Marines acted as if they were simply stationed abroad, no special tactics or techniques were in place, no special intelligence sought or offered. Security even—especially—at the embassy indicated the military was not alone in responding to perceived other than Lebanese reality.

Washington did not recognize that there was a war in Lebanon in 1982 and 1983, that camping out next to the airport was a risky venture, that unloaded rifles might sound prudent along the Potomac but not in Beirut, that very little things—one Mercedes truck—could matter inordinately in a Dragonworld. The administration and those responsible seemingly assumed Lebanon could be made American by treating it as if it were.

Beirut 06:22:31 23 October 1983

With the Mercedes driving straight at him, Sergeant Russell yelled, "Get the fuck outa here," to the stray Marine jogger and turned to run across the atrium lobby toward the north door in the opposite wall. He shouted, "HIT THE DECK," two or three times. He could hear the Mercedes getting closer. Bounding out of the north door he began to run faster. He shouted to a Marine walking post nearby to "Get down!" He cut across into the open north parking lot at an angle slightly away from the door and for the first time looked back. The yellow Mercedes had just smashed into the guard shack sending the demolished remains forward "like a wave of water rushing into the lobby, filling every space." The debris tumbled to the floor.

Russell ran on for thirty feet and then looked back again. The yellow Mercedes had come to stop almost exactly in the middle of the atrium court. The windshield was smashed and starred. The cab roof had been crumbled on the driver's side of the truck. The impact with the shack had sufficiently mangled the truck for Russell to wonder, ever so briefly, if the driver was still intact.

There was no movement in the cab. The truck simply sat in the center of the court in the midst of the smashed shack. There was no sound. Russell kept running but now with his head turned, his eyes on the crumpled yellow Mercedes. The unseen driver, a young man of special merit, Hassan Ali Talbakaran, smiled. A second passed. Another. Nothing.

Then Russell caught a glimpse of "a bright orange-yellow flash at the grille of the truck." The driver, the man with a beard, the man in the dark shirt, Hassan Ali Talbakaran, nearly twenty, had detonated his twelve-thousand-pound car bomb exactly at ground zero, the dead center of the courtyard of the 24th Marine Amphibious Unit headquarters housing 350 Americans

Washington had never really recognized the structure of the Lebanese unconventional conflict. Most at the center of the circle had been forced to learn of the Middle East while making policy. They had regularly reinvented the wheel or generalized from a single experience. The next secretary of state George Shultz, who replaced Haig in 1982, Weinberger at Defense, Vice President George Bush nor William Casey at the CIA, none of the friends of the President—Ed Meese or Donald Regan or the rest—knew the Middle East. Everything tended to be cut-outs, silhouettes scissored from afar, black and white. In Lebanon there were no recognizable insurgents. There was shooting now and again but no combat. There was killing but no battle. America wanted only peace this time. How could anyone thing otherwise? Surely such the American initiative was not irresponsible, not beyond the bounds of reason, and not dependent on a few Beirut gunmen?

At State under Shultz, the assumption remained that a logical accommodation could be imposed when the time arrived. As for the Pentagon, the military had been asked to make a simple opening move to avert anarchy and establish an American presence. The Marines had been meant as a small sign that the Lebanese government was worth a penny on the scales. Except there was no Lebanese government, only a husk inhabited by the gunmen with titles and office. Anyway, the Marines had been sent into Lebanon, had set up a perimeter, and were visited by emissaries and officials, were still but not charged with any special duty or warned of any potential catastrophe.

Washington had persisted in the belief that there was a Lebanese nation, a Lebanese army, that governance was real, that the leaders led through example and program. In such a world the MNF and the Marines could have a role, a mission. The Pentagon grew ever more dubious, all too familiar with the reality of Vietnamese "institutions." In time the unease of the military reached the top of the power pyramid. And Secretary of Defense Weinberger had at last began carrying those doubts to the National Security Council meetings in the spring of 1983.

Beirut 0:6:22:35 23 October 1983

As the bright flash blinked and grew at the Mercedes' grille, Russell was still making time, trying to get away from the headquarters building. He was barely aware that he had just seen what few living men have—the detonator flash on a car bomb: the very first moment of an enormous explosion. Then time ran out. All at once he felt "a wave of intense heat and a powerful concussion." For a long time that would be all he could feel. The car bomb's great shock wave lifted him unconscious into the air and then tossed him forward and onto the ground fifteen or twenty feet away, grinding his body along, face down in the parking lot. His body was torn and lacerated. In a second he was a maze of cuts, gouges, and deep bruises. His left leg was torn open from knee to ankle and a huge gash in his left hand would later require thirty-five stitches. His left femur snapped. His left ankle was broken and his pelvis cracked. Bones in his left hand were crushed. He was covered in dust and blood, a soggy, speckled body in a tattered uniform ground into the macadam. And he was very, very lucky. He was still alive. In that blinding flash that had blossomed out bringing down the building into a low crumble of smoking rubble, 241 Marines had been killed or mortally wounded.

By September 1983, the American honeymoon in Lebanon had already come to an end, except in certain persistent Washington hearts. The various efforts to construct a Middle East settlement on top of the Lebanese turmoil had all aborted. Nothing had gone very well. Ambassador Philip Habib expended his energy and American capital and was edged out by Colonel Robert McFarlane, who by the autumn of 1983 was edged out as Chairman of the National Security Council by Judge William Clarke, the most innocent of all—the family doctor sent in trust to perform a transplant with the instructions written on his cuff.

In September 1983, Amin Gemayel, who had tried to win everything, to humiliate, to cancel the "no victor, no vanquished" rule, had instead lost. The Gemayel militia had come apart to the delight of many. The Lebanese president had rushed to the United States Marines with pleas for ammunition, an outward and visible sign that he still had powerful friends, still had a future. The Druze and the Syrians had fired on the Marine perimeter when the fighting moved down out of the Shouf into Beirut. The Marines fired back. On September 9, the destroyer *Bowen* had fired to suppress Druze artillery, and a debate opened in the House of Representatives on the War Powers Act.

Even as the navy lobbed shells into Druze positions, no one in Washington wanted "war"—did they ever? Marine Corps Commandant General Paul X. Kelly told a congressional panel on September 13 that there was "not a significant danger at this time to our Marines." He did not feel that there was evidence that rocket or artillery fire had been directed at Marines. Yet on September 16, three American warships dropped 338 rounds on Moslem positions around Suq-al-Gharb. Two days later Congress authorized the presence of the Marines in Lebanon for eighteen months. On September 23, the American battleship *New Jersey* was reported sailing past Gibraltar on the way to the eastern Mediterranean. On the next day the French flew air strikes against Moslem positions. On September 25, the *New Jersey* arrived off the Lebanese coast. Washington felt the MNF supported government, aided the center to hold against faction.

For most in Lebanon, the MNF had merely been one more Lebanese faction, a force cobbled together to be used as a Western crusade. Worse, the MNF had not proven an effective faction, more vulnerable than intimidating. The West had done little for Amin Gemayel, little to awe: a lot of naval rounds dumped in the hills, a single airstrike, some return fire along the Beirut perimeter, and the dispatch of an obsolete battleship. The Great Satan, the West in Lebanon seemed more vulnerable than awesome. And so in October, the chosen, the Party of God, gathered beyond Beirut to watch a yellow pick-up truck transformed into an instrument of death. There would be Islamic vengeance over history, vengeance dispatched in a Mercedes, vengeance paid for by Teheran and in a device constructed under Syrian instruction, vengeance driven by an eager local volunteer, known variously as Abu Mazin or Abu Sijon, but born Hassan Ali Talbakaran in 1964.

Beirut 06:22:36 23 October 1983

As the concussion that had smashed Sergeant Russell rolled on across Beirut and up into the mountains, an Iranian photographer had released the shutter of his 35 mm camera set up before sunrise. Four miles from ground zero at Beirut International Airport, the lens snapped open at 06:22:31 seconds on a huge, eerie cloud—one part bulging black mass, boiling and roiling over the airport, and far above was a long, gray-white tail blossoming into another cloud, a second mushroom that had shot up from the initial blast. Some experts would later contend that it was the largest non-nuclear explosion ever, the largest explosion since Nagasaki. All witnesses would agree that the concus-

sion and cloud were awesome, even more so than the similar car bomb explosion that demolished French headquarters at approximately the same time. At the airport the administration building had been turned into an appalling ruin. The yellow Mercedes vaporized in the central courtyard. The explosion lifted the entire building up so that the top floors collapsed inward into a huge masonry heap trapping almost all the Marines. The concrete floors smashed together. The walls snapped off and collapsed into heaps and the rubble tumbled into the atrium court. Marines rushing to the rescue were disoriented—their headquarters had apparently disappeared, leaving only swirling gray dust and a heap of reinforced concrete one-story high. Above, still rising, the mushroom cloud twisted thousands of feet above in a bright, blue Beirut sky.

Those trapped in the ruins were symbolic victims, Marines sent into the Levantine to implement directives based on slogans. The military was lumbered with inappropriate structures, devices long in the building and sealed shut with priorities and concerns unrelated to the avowed Lebanese purpose. The establishment of an effective "presence" in Lebanon was quite beyond the system's competence. No one knew the nature of that "presence": an expedition, an invasion, a parade, a visitation by an ally, or a force on a hostile beach. None gave great thought to the nature of Lebanon or the prospects of the morrow, although at the top in the Pentagon there was increasing unease. The Marines *were* outside tradition, outside Marine practice.

Pentagon practice had evolved from reality. The Department of Defense had become a machine preparing for a great war none really anticipated and so not actually prepared to fight. The Pentagon had no effective doctrine for the unconventional, no proper units for the irregular in a world of high-tech weapons systems and nuclear theology. In addition since there were seldom enough wars of any sort to go around, in Lebanon too many commanded too few. Authority tended to be shared out until none was left on the ground—the system in effect had provided too many opportunities to command but had no doctrine to shape those who might command. Not sure what the Marines were tasked to do, the Pentagon was even less sure as to how to respond. The Marines, quite prepared to replay Iwo Jima, instead had been assigned to be a "presence" and so landed on a beach to keep the peace, dug in so as to provoke no one.

In 1983, even after a long Lebanese year, the Pentagon and the Marines were almost as unprepared as on day one for an unconventional car bomb attack. The Americans at the airport, in the embassy, in Washington, were unprepared even after repeated exposure to the irregular, unprepared even knowing that September events proved Americans were targets, and unprepared even after exposure to other car bombs. For some time Lebanon had made the Pentagon uneasy, but the answer in Washington seemed to be in asking for withdrawal rather than in improving the Marine's defensive potential.

Beirut: 23 October 1983

Sergeant Russell regained consciousness in a daze. He had no clear idea of what had happened. He knew he was in intense pain. "Oh, my God, I can't believe it." His left leg hurt. When he glanced down he could see that it was badly twisted and his left foot turned around pointing at the ground. In front of him, his hand looked tattered. There seemed to be a gash on the left side of his head. The air stank of the acrid smell of cordite. He was very cold. And when someone reached him, he first wanted to be helped up to see the headquarters building. There was no headquarters building only an enormous heap of twisted masonry, reinforced concrete ruins and billowing gray dust—no sign of the massive, squat building.

Russell was moved to a casualty collection point and then flown out to the Iwo Jima helo-carrier for morphine and more emergency aid. Almost immediately he was helicoptered back to Beirut International Airport and flown out again at noon on a white Royal Air Force C-130 transport, the first medevac flight to leave the airport. By early afternoon he was in the British military hospital at Atkrotiri in Cyprus undergoing treatment. The system seemed capable of coping with disaster, if not preventing it.

The arrival of the Marines should have imposed certain responsibilities on the military but in fact did not. The system did not work for irregular cases. There was no single command, no effective control, inappropriate intelligence, and all the communications were blurred with noise. And the Marines on the beach—the instrument of power—too proved faulty. They responded to the muddled tasking by rote as if they alone were immune to the Lebanese dangers. They wanted to be

Marines and taken as decent and disinterested, not players in an unconventional war requiring undesirable disciplines. So as ordered, they did not lock and load, neither did they secure their positions nor deploy to repeal terrorists. No one told Colonel Geraghty to do so and he told no one to do so and so nothing was done. The admirals and generals came and went, innocent of security, innocent of Lebanon, responsible but content.

When military actions was at last ordered, the response was conventional, ineffectual, and divorced from Lebanese reality. Those involved still felt disinterested, responding only to provocation, not part of the local equation. The navy arrayed battleships against snipers. And since naval bombardment was ineffectual, all the snipers were encouraged to continue sniping. No one in authority in Lebanon or Washington seemed to focus on Lebanese reality. The Pentagon allowed each level of command to structure and restructure operations from standard operating procedure, agreed doctrine, and formal experience learned far from Shouf and the Beirut suburbs. Just as Reagan's cabinet insisted on imposing American perceptions on the Middle East, so did the Pentagon insist on maintaining conventional military procedures. No Americans saw fit to imagine how such actions might be perceived in the Middle East. The conventional warplanes and ships displayed in Lebanon were no more than a display of the American way of war.

The American way of war failed, produced 241 dead Marines, and untold fall-out that hovered over American purpose long after the gray-brown clouds had dissolved over Beirut. Not only were Marines taken as vulnerable so too was the Great Satan. Will had clearly triumphed over tangible assets, one man over 241, a yellow Mercedes over all the hi-tech Western weapons.

Beirut 23 October 1983

> *Hour after hour, the stunned Marines dug into the great, grimy pile of crushed and twisted concrete, sifted through the ruins for the living and the dead. And the appalling total of 241 dead had been in exchange for one bearded Iranian in a dark blue, off-green shirt, a member of Huessein Musawi's Hezbollah, the Party of God, the hand of the Ayatollah reaching into Beirut to kill not only the Marines but with another bomb fifty-nine French infidels as well.*

The French, friend of the Iraqis, were as much the enemy as the Great Satan—the West is not neutral in the eyes of the ayatollahs. And even the experienced Israelis were vulnerable to car bombs. There might not be an endless supply of martyrs eager for salvation or reluctant to refuse glory; but in the autumn of 1983, a few were sufficient. So the Marines despite their *elan*, despite the carriers and frigates, despite the arrival of the *New Jersey* and the aura of a great power, like the French, like the Israelis, even at times like the Syrians, had been vulnerable. The enormous car bomb suddenly and irrevocably concentrated many minds. The dead dragged from the dusty heap of rubble beside the airport road under the mushroom cloud were for many too great a price to pay for whatever purpose America had pursued in Lebanon.

No one had been able to cope with the Lebanese Dragonwars because the American system was not so geared. The American way of war was a system to prepare for conventional, grand wars not to fight them, not even to fight small conventional wars; but then simple, brute power could impose American realities and fashion victory. In time the military shaped power effectively. Desert Storm in 1992 would be the prefect Pentagon war, fought slowly, conventionally with global support, against an incompetent and monstrous enemy and on to an assured victory. Most other Pentagon ventures were not such a nifty match to desire and capacity.

The conventional response in Lebanon was not simply inappropriate or ineffectual but assured escalating disaster. American practice sowed dragon seeds—the "learned" lessons in Lebanon were added to the rosary of previous misunderstandings: unconventional wars were to be avoided or ignored. In reality unconventional wars were best forgotten, discredited by diminution of concern. If one did not have an umbrella it would not rain. Preparation would simply encourage the unconventional and reduce conventional military capacity. The posture was ignored by all administrations. The Pentagon was regularly tasked with unconventional assignments. Each new unconventional challenge proposed by the shifting political agenda—transnational terror, the international drug trade, peacekeeping—was unwelcome but had to be accepted as inevitable, especially after the Soviet threat had gone. At least and at last the war against Iraq had reinforced all the old Pentagon priorities: American needed real armies, a great navy, high tech airplanes for real wars.

For a few the prospects of deploying irregular tactics, terror against terror, attracted; but an actual understanding of the unconventional agenda and perceptions remained elusive. Even the American "pragmatists" of unconventional war mistook the techniques for the core—missed the costs, missed the limits, missed the power of the dream and assumed secrecy meant license. In this the Americans were little different than other Western military powers who saw the techniques of armed struggles as useful in a harsh world and ideology or revealed truth as marginal to operational matters. Many of the orthodox do not grasp their own real power arises not from things or techniques but history and legitimacy. Few realized that to deploy terror against terror, defeat the monster by being monstrous, erodes the greatest asset of the state: legitimacy. Recourse to the techniques of a dirty war inevitably taint the orthodox, undermine the legitimacy of the state. The Basques or the Irish or the Corsican underground is not the great threat—state terror is the danger to legitimacy. And it is this legitimacy that the gunman and the martyr seek, to have their dream accepted.

It is inordinately difficult even after long exposure for the conventional, those at the center, the decent and legitimate, to understand the power of a revolutionary dream to shape reality, to foster an armed struggle, to fuel irregular war, or to seek legitimacy by rewriting history. And so the dynamics of the underground are in much of the West a mystery. The conventional depend upon conventions, the deployment of existing assets and the big battalions, and rely on the inertia of history. And mostly this has been ample. The Irish may still pursue the Republic but the gunmen of Italy and German are gone and so too their apocalyptic dream. The Basques persist and the clans of the Balkans but most of the Latin American guerrillas are history. The Corsican bombers are mere irritant. History mostly runs to schedule despite the grievances and aspirations of the underground: but not always as the Marines discovered.

In Lebanon all this—the nature of the underground, the limits on the terrorists, the costs of the covert, the power of the dream—were obscured by conventional assumption and experience. In Lebanon America tried to impose conventional assumptions on a confusing reality, make Lebanon normal, reasoned, predictable. And predictably the result was disappointment and disaster. It was not the first such exercise nor the last. Each experience was repeated not as farce but tragedy, some grand, some not.

> "... We came upon a man-of-war anchored off the coast. There wasn't even a shed there, and she was shelling the bush... In the empty immensity of earth, sky, and water, there she was, incomprehensible, firing into a continent. Pop would go one of the six-inch guns: a small flame would dart and vanish, a little white smoke would disappear, a tiny projectile would give a feeble screech—and nothing happened. Nothing could happen. There was a touch of insanity in the proceedings... and it was not dissipated by somebody on board assuring me earnestly there was a camp of natives—he called them enemies!—hidden out of sight somewhere."
> —Joseph Conrad
> *The Heart of Darkness*

Immediately after the explosion on October 23, the Americans appeared too stunned to respond to the terrorists. Once the rescue was over, the injured and dead removed, the replacements flown into the city, the local commanders concentrated on improving the perimeter defense. There were lots of visitors, important people rushed through Beirut to indicate support, off-shore commanders made hurried tours. All had a kind word or encouragement, a few had advice, none apparently gave orders, none seemed responsible for Marine security or very knowing on anti-terrorist matters.

A new replacement unit arrived on November 18 and found the defenses still unsatisfactory. More commanders came and went. The Commandant General Francis X. Kelly made the tour and still more admirals and generals. Around the Marines the Lebanese troubles sputtered on and on. Truces came and went. The Marines dug deeper. A formal investigation was announced. Eventually the investigatory commission chaired by retired Admiral Robert L. J. Long arrived on the spot and noted not only the ineffectual command structure but also the need to remedy the perimeter defense. They moved back to Washington to publish on why the Marines had perished.

By then the situation in Lebanon had deteriorated still further. At the end of October, irregular fighting had spread rapidly around the edges of Beirut. One more car bomb got through the supposedly experienced Israelis and detonated in a military headquarters. The Israelis immediately bombed the headquarters of Iranian irregulars in Baalbeck without waiting for guilt to be established. The Middle East specialists agreed that both Iran and Syria shared complicity in the car bomb attacks—funded the effort, offered aid and comfort and skill, encouraged those who needed no encouragement. Proof was another matter.

Then United States Navy F-14 Tomcat fighter bombers on reconnaissance missions over Lebanon attracted fire from Syrian SAM-5 missiles. At last America had a specific target—the anti-aircraft sites and a large, white building housing a radar complex: a visible dragon. On December 4, President Reagan authorized an airstrike fashioned by the Joint Chiefs of Staff. The appeal for retaliation had drifted up through the elaborate chain of command with the carrier advocates pushing for an airstrike. Vice Admiral Staser Holcomb, deputy commander in the European theater, had been adamantly opposed to risking men and expensive machines when the *New Jersey* was available offshore and in range. The carrier admirals won. Missiles had been fired at Navy planes, so Navy planes would strike back. The *New Jersey*, refitted at a cost of $326,000,000 could not, according to General John W. Vessey, Jr., chairman of the Joint Chiefs, be relied on for accuracy. There were no Navy spotters ashore. The *New Jersey* had been designed for other uses. So had Navy doctrine. Everyone, however, felt something should be done, someone punished. So retaliation was ordered from the very top with the recommendation of those responsible in the Pentagon.

The air attack orders had been prepared by various hands, drifting down through command levels, each attaching operational directives and restrictions. By the time the Navy pilots were set to fly on December 4, there was no interest in battle arena feedback: the system in all its complexity had produced the orders. Anyway, neither the Israelis nor the French had run into any trouble against the locals or the Syrians. The operational plan guaranteed maximum advantage to the defenders from time to height. The planes would, however, be without the benefit of electronic countermeasures, never purchased because the Navy preferred to have more units of existing systems than make effective the existing ones once they were operational.

The airstrike was a disaster. The ill-trained irregulars and their Syrian allies shot down two of the twenty-eight Navy planes: an A-6 light bomber, which cost over fifty million dollars, and an A-7. Another A-7 was badly damaged. Two pilots were killed, one of the ground, and a navigator, Lieutenant Robert Goodman, was captured and ended up in Syria. The strike destroyed nothing but American military credibility. The Navy complained about inadequate intelligence and the heavy anti-aircraft fire. No one mentioned that Lebanese conditions were ideal for an airstrike, as the Israelis had regularly demonstrated against far more difficult targets. Few in Washington wanted to admit that the local commanders had been restricted by the tinkering all down the command ladder by officers who, having read the book, knew best. No one seemed interested in local conditions, irregular capacities, special factors that made Lebanon—just as none of the stream of commanders visiting the Marines at the Beirut airport had recognized the vulnerabilities of the Marine presence either before or after the car bomb. Orthodoxy—the system at work—continued to pay bitter dividends.

On the same day as the airstrike, a Druze artillery shell hit Marine Combat Post 76 and killed eight Marines and wounded two. There was an opportunity to reply with naval gunfire. The battleship *New Jersey's* big guns were opened up on the unseen Druze positions in the Shouf. The famed fourteen-inch guns were, of course, not firing against fixed fortification or at static targets like bridges, bunkers, or artillery emplacement. In fact the *New Jersey* fired on map coordinates without concurrent target finding or correlations: the gunners did not know where or what their shells hit. The Navy had no spotters ashore. No one wanted to order Marine spotters into the dangerous areas beyond the airport perimeter. There were no minidrone aerial spotters because the Navy procurement process was still seeking the ideal American spotter drone after years of research and development while fending off as unneeded foreign models that were both cheaper and more effective than American prototypes. The *New Jersey* opened fire without visible targets, without spotters, without feedback, and without hitting anything but the fields of the Shouf offering little more than a photo-opportunity for the world's journalists. American power seemingly could not be used effectively. The system was not attuned to any effective use against the irregular and unconventional. Something had to be done so a battleship and hi-tech aircraft were deployed as if an irregular war could thus be made orthodox.

* * *

The December demonstration of American military incompetence, whatever the Pentagon explainers might say, made further adventures in Lebanon a high risk operation politically. The nation's military commanders, who did not feel incompetent or at fault, had felt their Lebanese mission inappropriate even as they defended the record. For months Defense Secretary Weinberger had argued against the whole wretched exercise. He wanted the Marines out whatever reasons marshaled by Secretary Shultz of State, wanted them out, perhaps, especially because Shultz at State had defended the involvement. A weak President listening to his Secretaries squabble over vital interests might have opted to be strong: to escalate.

Reagan was in a strong position. He was enormously popular despite his subordinates' failures and had been greatly bolstered by the inept but successful military intervention in Grenada in October. No matter how clumsy the execution or feeble the opposition, America took the island, stamped out subversion, stood tall. Thus he could on December 15 give the first hint that the American presence in Lebanon was not everlasting. On December 21, he indicated that, of course, the Marines would stay unless "the Lebanese government no longer wants them there." America was on the way out. On cue at the end of the month, the Lebanese Ambassador in Washington indicated that his government wished it had never invited the Marines.

In Congress General Kelly's earlier explanations on Marine base security appeared increasingly shabby: the gate had been open to allow the yellow Mercedes to drive through and the vaunted sewer pipe barrier that Kelly had cited as a deterrent appeared on closer examination to be a couple of pipes so far apart as to allow the truck to drive through. Reagan defended Kelly on December 20, but three days later the publication of much of the Long report opened up the matter again. Finally, on December 27, Reagan accepted full responsibility, "If there is to be blame, it properly rests here in this office and with this President. I accept responsibility for the bad as well as the good." He did not believe that his commanders should be punished for failing to understand fully the terrorist threat. And they were not. The buck stopped in the Oval Office and few reprimands trickled down and no major systemic changes were suggested.

Nothing had changed except the Marine perimeter was finally tighter. They would be withdrawn—sooner or later—when face had been saved and George Shultz could be placated. The House Democrats called for a prompt and orderly withdrawal. The White House announced Reagan would ignore the resolution. On February 4, the President said that if the United States were to cut and run it would be a signal to terrorists everywhere to wage war against innocent people. By then the Lebanese Army had disintegrated. Shi'ite and Druze militia moved deeper into West Beirut. On February 6, the United States flew jet strikes against the militia positions and Navy warships fired at Moslem positions. By the next day the Moslem militias had almost cleared West Beirut and at last Washington stopped pretending. On February 7, President Reagan announced that the 1,400–man Marine detachment would be evacuated to ships off Beirut where they would remain a seaborne "presence" safe from car bombs and the continued faction fighting.

The British and Italians announced they would go. Foreign nationals, including 884 Americans, were gone by February 12, partly under cover of American airstrikes. A State Department spokesman in Washington felt that the Pentagon and the administration had panicked and withdrawn in haste unnecessarily: "... a capitulation to terrorism." The Lebanese Marines were withdrawn between February 21 and 23, most in the lull negotiated by Saudi Arabia. All the Lebanese factions and notables were still at odds. Nothing had changed. The temporary winners were bloated with pride. The losers kissed the hand they could not cut off. And eventually, the Sixth Fleet and the Marines sailed off. On September 20, 1984, Islamic terrorists drove another car bomb into the United States Embassy annex in Yardze—the explosion killed eight and injured several, including the American and British ambassadors to Lebanon.

As the months turned into years, nothing good would happen. The endless rounds of war, random slaughter, truces and pauses, conferences and killing went on and on. Those who felt Islam the answer would grow more radical, difficult to lead, hard to manipulate, generating new generations of gunmen. Parties split and fronts collapsed. Hostages were taken or released. The founding fathers of hate lived on joined by a new dreamers desperate to die as well as to kill. Yet, in time the violence ebbed except in the south where the Israelis were engaged in a low-intensity holding operation. Elsewhere exhaustion muted malice and murder. Lebanon would have a government of sorts. A center of sorts would emerge. By the end of the decade much of Lebanon, much of the time was if not at peace no longer a killing zone.

The American establishment sought to rationalize the Lebanese experience, concentrate on minor errors, ignore the implications of Dragonwars. On February 22, 1984, President Reagan revealed for the first time that the Marines had been part of an effort to prevent a war between Syria and Israel; news to all, not the least to Syria and Israel. By then and long afterward the surviving Marines felt, perhaps, they had bought time for Lebanon; but no one knew for what purpose. Most, however, felt the entire exercise had been for nothing. Secretary Weinberger agreed that it had been a dirty, disagreeable, and miserable job; but he told the Marines to blame it on "people working on the political and diplomatic and all those other sides." At the beginning of March 1984, his cabinet colleague Secretary Shultz was no less forthcoming publicly about military enthusiasm: "Remind me never to ask for the Marines again. I've decided next time there is something like this, I'm going to ask for the opposite of what I want because I know the Pentagon will always be against me."

This seemed the only lesson learned: the military wanted no part of unconventional tasking and one should act accordingly. This was, indeed, a lesson but one that the responsible did not incorporate into doctrine or folk wisdom. Schultz's successors would have to repeat the past. There were a few little flaws to tidy up. The Long Commission found that the chain of command "did not initiate actions to ensure the security of the USMNF in light of the deteriorating political-military situation in Lebanon." And Colonel Geraghty, later commander of the Marine Corps barracks at the Norfolk, Virginia, Naval Station, felt, "We were taking just about every precaution that I thought was prudent to address the threat as we knew it."

With considerable magnanimity, President Reagan had already publicly announced that if there were to be blame, he would accept it—the fault was his, not the strategic planners of the NSC, not the Pentagon, not the chain of command, not the Marine commander on the beach, and most important not at all the system. So there had been no systemic faults, no one at the center except the President need seek blame. And none blamed the President, who was too decent, too popular.

On the Lebanese record, no one could predict that the administration or the next one would do much better since none would admit to systemic error. The Pentagon had not wanted to go into Lebanon in the first place, never again wanted to be drawn into a protracted unconventional conflict. So Weinberger felt his Marines should have been withdrawn sooner while Shultz insisted that they should have stayed longer. The Long Commission found special and particular faults, reported it, and saw the report published and forgotten. Most of all, no one from the Oval Office down to the field command had understood the nature of the Lebanese arena and the capacity of the unconventional to reach up and out to change the tides of history. Efficient, ruthless, and brutal terror can have an escalating impact when deployed—achieving at least prominence and at times, as in Lebanon, determining in part the future, changing the perceptions of the involved.

The American disaster was read by the region as a triumph for Islam, a triumph for all those oppose to the pretensions and power of the West, a triumph for terror as a means, The asymmetrical confrontation was perceived a victory of will over tangible assets It revealed for those so concerned the failure of Washington to address the perceived reality of Lebanon. The Americans wanted reality to be congenial. American military procedures arose not only from such assumptions but also from a system that could not cope with Lebanese reality. The cost in contrast to Vietnam was low but not high enough to warrant systemic adjustment. And so Americans continued to think like Americans, respond to events like Americans, maintain their American institutions—national characteristics are not immutable but real change is usually incremental or achieved at enormous cost. So Americans did not change.

More than all else, Americans lacked empathy with others, with other nations and peoples, with the alien. And so Americans failed to understand those driven by faith, those who hated history and were determined to control the future by means of an armed struggle. They could not imagine the returns of malice, the bandits' agenda, the warlord's priorities. Americans, not surprisingly, persisted in responding to any unconventional challenge as Americans and not unexpectedly sought to impose perceived reality on the arena. As had been the case in Vietnam, mere coercive power, mere tangible assets, and especially the assumptions that all grievance was amenable to repair by right reason, by investment and enterprise all combined to engender ruin. Still, America was rich in resources and resiliency, could pay the cost of innocence.

So little was learned that imposed a shift in perception or in priorities. Americans chose to remain American, retain their assumptions and ideals and attitudes, manipulate the Lebanese experience so as to avoid any serious change in national assumptions. Just as they had done in Vietnam, Washington found the fault in details and procedures not in the system nor the national ethos. That ethos arose from American history and habit and national characteristics not easily shifted by an act of will. Americans were apt to stay as Americans not for advantage or as a matter of choice but because this was a constant, the end product of time. And America as inherited tended to shape the procedures of the Pentagon as well as the attitudes of the public. The American way of war, the American perception of the postmodern world, the deployment of battleships or Marine expeditionary forces, the general response to provocative terror or Islam were molded by the perceptions of the nation. And change had not been imposed and was not perceived as necessary.

With the expeditionary force withdrawn, the administration went on to other matters, priorities were the same and perceptions. The small adjustments of form made by the Department of Defense and the reinforcement of popular ideas about the Middle East, Islam, the risks of intervention and the responsibilities of a great power with military power meant that American was ill prepared for the next Dragonwar. Those who did understand such unconventional phenomena remained on the margins of power and those who did not—most Americans, most of the responsible—pursued policy and interests innocent as always of Dragonworlds and Dragonwars.[1]

---

1. The chronology, public statements and many of the quotations can readily be found in the press of the day. Those quotes directly related to the events of 23 October 1983 can largely be found in the highly detailed Eric Hammel, *The Root: The Marines in Beirut August 1982–February 1984* (New York: Harcourt Brace Jovanovich, 1985).

   There is a wealth of material, often by those present, a sampling of the available might include works such as:

   Anthony McDermott & Kjell Skjelsbaek, eds., *The Multinational Force in Beirut 1982–1984* (Miami: Florida International University Press, 1991)

   Michael Petit, *Peacekeepers at War, A Marine's Account of the Beirut Catastrophe* (Boston: Faber and Faber, 1986)

   Larry Pintak, *Beirut Outtakes, A TV Correspondent's Portrait of America's Encounter with Terror* (Lexington, MA: Lexington Books, 1988)

   and the official report—the Long Report—Department of Defense, *Report of the Department of Defense Commission on Beirut International Airport Terrorist Act, October 23, 1983*, (The Long Report), 20 December 1983.

   There is less on those who placed the bombs, some material not always published, from the Israelis, some public and private statements out of the revolutionary Shi'ite community over the years, nothing relevant from official Arab sources.

On Lebanon during this period there are in many languages a library or works on every aspect of the conflict, by representatives of nearly every faction or group involved—gunmen, politicians, women, and soldiers, officials and journalists.

There is, as might be expected, a vast literature of recrimination and analysis out of Washington, some in book form, that reaches no consensus except the incident was a psychological and policy disaster.

Part II

The Dynamics of the Underground

3

The Dragonworlds: Revolutionary Ecosystems

What the West in general and the Americans in particular found in Lebanon was exposure to the unknown and unpalatable, a world filled with monsters, a Dragonworld. There were in Lebanon more than one such world, ecosystems that protected those dedicated to imposing change through violence. Such underground systems, covert and illicit, are focused by the compelling need to transform history. The movement at the core of the revolutionary underground is seized on the quest for power over events, power over history, power to imposed a transcendental dream. Others have forms, attitudes, and tactics similar to revolutionary underground but seek not to change history rather to pursue grievance or interests or tradition. These irregulars are empowered not by the energy of a dream but by the lure of advantage.

A classic revolutionary underground is not tangible but a perception, a state of mind arising in a special place at the end of a long history of grievances perceived at first only by the few. These who resist the present are possessed of a need to change history's direction with a paucity of resources. Some want to return to past glory, others to rearrange the present, many to destroy what exists in order to posses the future. And some others—tribes, factions, defenders, opportunists, the weak, the greedy—are attracted to the prospects offered by irregular violence and so by necessity adapt many of the aspects of a revolutionary ecosystem. They are part of history not seeking to escape into the future. All, however, those with a dream and those without, must use unconventional means, must evade power, must find an effective irregular strategy.

All underground have special motives but those with incandescent ideals have a revolutionary ecosystem and a sense of purpose that gives a more special impetus. With greater aspirations they are a greater risk to stability and the center. Few enter such an underground as the first choice. During turmoil, however, many find a vocation in the irregular or at least opportunity in chaos. Escalation of the campaign attracts the less committed, escalation may even reveal not volunteers both those drafted into the crusade. The greater the intensity of the campaign, the more conventional the underground. In any case all the irregulars, idealists or warlords, seek power from which all else flows. With the idealists there are at first so few that leaders and followers are apt to be one until tumult and confusion has generated immediate prospects. A great war or a collapse of order may open the way. Lebanon had examples of all, the dreamers, the traditionalists, the ideological gunmen, and the ethnic separatists: a wilderness of dragons partly visible and largely covert. And beyond were the others in Lebanon who sought more unsavory advantages, the pleasures of power, vengeance, a quick return on risk, or merely to survive in troubled time. Some were not unmindful of the power of an ideal and flew false flags over interest. Because these, the criminal and the demented, had recourse to unconventional violence there were often mistaken for the true believers when merely similar in techniques.

The classical case, the model of a revolutionary underground, deploys the dream as engine to drive the armed struggle and so change history. The others find irregular means valid and are encouraged by the arena, by opportunity, by wont, by lack of options and at times conviction, necessity, and practice. What gives impetus and coherence to revolutionary activity under any banners is commitment to a dream. In the beginning the conflict must be unconventional for there is no other way forward. The faith is always at risk, opposes the orthodoxies of the times, opposes the center, the power of the state, the assets of the conventional, requires great risks while the other irregulars seek less revolutionary results. None are pure, some villages have dreams and dreamers have interests, but the revolutionary underground runs to a different dynamic.

The rebel to effect events must have a safe haven. They shape a special world that nourishes their dream, inspires their secret army, protects their cadres, and determines their history. In the cause of the dream, they go into the underground not a first choice but often as last,

The Dragonworlds: Revolutionary Ecosytems   63

go not because attracted by the unconventional but to become conventional, legitimate. It is a world created by perception, a terrain not found on orthodox maps, where victory is won in the mind and death comes when the dream dies. Their underground exists only to make possible an armed struggle.

America's exposure to the unconventional in Lebanon was painful, imposed imperatives on policy, and indicated once again how conventional and orthodox were the assumptions of Washington. Few Americans could readily image lethal dreams—American society is content, grievances few, prospects grand. The nature of the Lebanese arena, the forces at play, the dynamics of the violence remained mysterious—and unpalatable. The competing and mingled ecosystems of perception were largely ignored except by the experts. These specialists found no ready market for their analysis nor easy means to explain the irregular. America, the professionals and the people, preferred to find terrorists and gunmen both awful and inexplicable. Movements dedicated to compelling dreams were alien—and with them most worlds shaped by the perception of the dedicated and deadly. Americans sought to see what they expected to see not a dark complex of injustice, salvation, hope, all shaped as fact and thus logic transformed into rationale for atrocity.

For a generation Lebanon had been an arena filled with the unconventional: the irregular, terrorists, secret armies, tribal levies, and conspiracies. Lebanon as a nation was illusion. The state had been at best facade, an accepted fiction, and more often a movable asset of one faction. There was no Lebanese nation, no Lebanese people, no state, no center, only agreed fictions, postage stamps, flags and ministerial offices, police to direct traffic, licenses to buy and politicians to sell. What did exist was the lethal mix of families and cliques, sects, factions, movements, parties, institutionalized conspiracies, and ethnic structures. None in Lebanon had access to compelling force, real legitimacy, or at times even reason or decency, so none could impose on all. What evolved was a lethal competition for independence of action by those organized through various loyalties and varied dreams. Lebanon became a petrie dish of unconventional violence.

Within Lebanon, as elsewhere, those with limited resources by necessity had been forced to resort to unconventional means. For a generation all the variants of the unconventional, the dreamers and the desperate, the revolutionary and the mad dog killers, were readily to hand. In the crevices and crannies the factions found for a time, now and again, room to maneuver, prosper, and persist. Others pursued the future in secret, emboldened by the truth but limited by the competition.

In this arena the Islamic fundamentalists of Hezbollah fueled by the faith and Iran, tolerated or encouraged by Syria, opposed the Zionists and the West, opposed much of the post-modern world and so operated out of an ecosystem of belief that inspired an armed struggle against all odds and even and often against the conventional interests of the avowed constituency. In this Hezbollah were engaged in one more armed struggle, one that seeks to go back to the future by using the means of every such war. And so too were the varied Palestinian movements and less so those dedicated to narrow ambitions—a Maronite hegemony or Druze autonomy.

When the Marines came, they came largely in ignorance and innocence: to do good. Americans are apt to imagine everyone American, capable of reasoning together. Not in Lebanon. There no one had a similar agenda and few seemed moved even by the returns of peace and quiet that accommodation offered. What was unique in the arena was not so much the variety and intensity of the various dreams or the wide spectrum or irregular actors but rather the lack of any general authority. There is no center in Lebanon. There was no effective state, no real center to hold divergent interests in equilibrium. Thus the various aspirations, universal goals and village priorities, had inspired both special ecosystems that permit the faithful to function and also room for irregulars to operate.

This had often been done in contradiction of the legitimacy of others or the system at large. Even the most radical and ambitious had been able to find secure zones beyond the reach of the others. In Lebanon everyone, each group, the dreamers, the bandit gunmen, the tribes, and ethnic factions were limited by their own lack of assets, the power of their competitors, and the orthodox intervention of external force. To compound matters, there had been the covert interference from external forces: intelligence agencies, transnational terrorists, drug cartels, commercial interests, media priorities, criminals who needed safe haven, and simple drifters seeking opportunity. The avaricious had too rushed into the vacuum, the empty center.

The result was that for much of the time the arena was in a state of institutionalized anarchy: no law but necessity and no restraint but that imposed by weakness. There was free play for the random and contingent. The various perceptual ecosystems shaped by the Lebanese allowed the old dreams and the new, the old ambitions and great universal ideologies to flourish. There were real gunmen living in imaginary gardens.

While those in Washington were apt to expect—all but demand—that others be as Americans were assumed to be, few in the West fully understood the priorities and imperative driving those in the arena. Many, the French and the Italians and even the Israelis, were often as mistaken and mystified by the forces at play. The tiny Lebanese enclave caught between the sea and the Syrians, once prosperous and open, had collapsed into horror. Lebanon was a free fire zone that even after a generation and general exhaustion remained unstable. Almost all that mattered took place in frantic hearts, in matters of perceptions. The result was visible violence without reasoned explanation.

For many in the West, the arena repelled rather than fascinated and so was relegated to the margins of consideration: future wars would not be like Lebanon and Lebanon was no prologue to the post-modern world. Lebanon was Lebanon, a horrible singularity. When such singularity reappeared in Africa or Latin America, those in control, those responsible or in Washington were again appalled and ill prepared. The armed struggles of the Lebanese factions would remained enigmatic and forgotten, irrelevant. The only lesson that they had learned was that there was no logic to the unconventional, no preparation possible. The American system was not adjusted to allow an effective response to an armed struggle but rather the reverse. The national security establishment in power and out focused on conventional missions and assumptions. Armed struggles were defined so as to force no adjustment in perception or mission. Those who advocated special responses, special missions, special units were a small minority. Even with the end of the Cold War and the search for novel missions, there was no enthusiasm at shaping great power to respond to the gunmen. There was no real institutional enthusiasm for the irregular even within the various intelligence establishments. Yet, over and over, the White House saw gain or necessity in committing force to unconventional tasks no matter the Pentagon's reluctance. Somehow even Lebanon had not dissuaded those in control in Washington.

When American withdrew from Lebanon, they did so largely with their illusions still intact—despite all. The Lebanese were ungrateful, the initiative was valid, the system had coped. No one was at fault because of the car bombs and no lessons needed to be learned from Lebanese events. Unconventional operations were the conventions of orthodox policy, regulars on special missions. Americans persisted in believing in the practical: the real, numbers and assets and the tangibles, and in the universality of man's nature. Evil was a biblical concept. History could best be changed by programs and investment and construction not dreams. If something special was needed send not special forces except as a last resort but power displayed—battleships and airstrikes. Few rarely adjusted to the unconventionality of the rebel heart. The Pentagon might issue texts of tactics on anti-insurgency to military academies, within the curriculum but hardly on the agenda of the system. Unconventional war might have certain attractions to special operations buffs but was not a major consideration in a world undergoing radial transformation, vast technological change, a shift in the basic balance of power, extensive development, the elimination of time and distance. For admirals and generals the shadow of the gunman was neither long nor dark. The armed struggle, once a fashion in radial politics, lost even the cachet of trendy. Why focus on the weak? They did not count and their dream could not be counted.

Armed struggles deploy will not weapons. And weapons usually win. And America in particular did not take revolutionary will seriously. Revolutionary dreams had ended in America with the failure of reconstruction. After that America and the West built on reality, built by recourse to pragmatism, enterprise, patriotism. Ideological appeals found few American converts: communism or fascism were threats because they were un-American and backed by state power. America had the Sixth Fleet and stealth aircraft, a prosperous people and command of the media, had freedom and history to hand and no need of dreams. Dreams, however, may in special circumstances permit the will to win over the big battalions, may permit those without, those without a triumphant history or stealth aircraft to win. Even if the predominance of force usually triumphs there have been exceptions: Hanoi insisted that conviction and the direction of history overcame American power. For many in the Pentagon it had simply been that insufficient power had been deployed. Power wins over will. And it did not in Lebanon where a battleship shelling the hillside revealed the limitations of tangible power deployed against a dream. For two centuries the armed struggle as a means to change history has been the weapon of the desperate. At times the armed struggle has been a means to power, as Giap and Ho claimed in Vietnam. Often it has been a means to great power as in China in 1949 or to the confusion of great power during the anti-imperial struggles against the European presence after 1944.

Dreams can become real. Who would have believed a small man sitting at a cafe table in Switzerland would replace the Tsar or armed agitators harried into wilderness would capture China? Castro started with a dozen men and the British Empire faced independence movements composed of a few discontent radicals armed with revolves and resolve in Cyprus or Palestine or Aden. Who, in fact, could imagine that one man in a truck could transform reality, American priorities, and the mind of the President? The frantic and faithful in Lebanon might not have seized history but they denied all others, corrupted the center, imposed chaos rather than control.

Each armed struggles is different and yet each is the same. Each comes at the end of a long and special historical process, proclaims particular aspirations and grievances, and yet is shaped by objective reality in similar ways. Each must pursue the cause in similar ways. These are the classics, the FLN in Algeria or the Sandinistas in Nicaragua. Those without a dream, the irregulars, those with special loyalties or particular grievances or aspirations, also must deploy unconventional means: ethnic militias, tribal warriors, pirates, and separatists. They arise from different ecosystems. Some hidden worlds are structured by the purchased loyalty of a mafia and others based on parish custom. These undergrounds are not unlike that of the revolutionary but with other priorities and so at times more open to cooption or coercion at the core: the mafia mainly shaped to economic interests or the village militia to security. The center can hire the mafia with stipend and legitimacy or turn the militia into a regiment with banners. Revolutionaries have greater aspirations.

The Lebanese arena revealed a vast menu of aspirations and dreams and of enterprises, ventures, and movements. Some "armed struggles" were little more than village vengeance, murder without even political explanation, the politics of atrocity. Other struggles were driven by universal religions or by dynastic ambition, the great faiths of the times deployed by both bandits and the idealist. Each struggle deploys varied assets for varies purpose and to mixed ends. And all of these were and remained beyond the American agenda.

* * *

The archetype armed struggle arises from a community of the faithful who, seeking to change the future, resort to violence. They imagine the future as the dream dictates. The campaign may be more effective if there is a social history shaped to resistance, a suspect legitimacy at the center, a tangible agenda of grievances—real and actual, tangible injustice institutionalized—and, of course, prospect of success. Into such an arena the faithful take the assurance of what is wrong, what is the goal, and what is the necessary means—have answers unequivocally, absolutely, validated by history and observation.

The contemporary world is filled with injustice and so grievance. The world perceived has never been perfect. There are at present extremes of wealth, the clash of cultures, people oppressed, and faiths denied. Mostly the miserable suffer the day and take comfort in small matters. Mostly the rebellious heart is sullen and the prospects of change daunting. As modern society has grown more complex, some dream of real change, necessary change, history written to an ideal. They may even find the past perfect, seek to return rather than transform. In each case the present systems is obstacle to aspiration.

For two centuries the global system increasingly has been organized by states, increasingly nation-states, and these increasingly have incorporated the mass of the people in a maze of complex, central, often authoritarian institutions. Governance rests best on highly structured polities dedicated to the general good, but governments may find simply force adequate to rule if not to build the nation. The nation-state is the key, the prime loyalty, the basis of world order, such as it is. Universals must fit the system. Great and general ideas are apt to be applied nationally: the Iranian Islamic Republic or the slow evolution of a European union. Small people are apt to find themselves citizens within a nation-state however artificial: Zaire or Singapore. And a small people may aspire to their own state: Slovakia, Macedonia, Kurdistan, even without a historical ancestor.

Those who find the present in whole or in part intolerable find the future in control of states, nation-states. So even those who assume a universal enemy—the Imperialist State of the Multinationals of the Italian *Brigate rosse* or the West of the Great Satan—usually find a particular nation-state as enemy. Most who seek radical change in the present find those in power, those at the center, intolerable. Many on the margins find those at the center an imposition, unnecessary, not an obstacle to a better future but an intrusion on cherished liberties. Some—

the criminal, the warlord, the pirate—find any order, any law, any limit restricting, necessary to evade or to reject. It is, however, the revolutionary who pursues an armed struggle not to limit the center but to replace it, to change everything. They begin simply by saying "no." The archetype revolutionary core consists of the few and the faithful who will not tolerate the direction of history and cannot find appropriate means to act except through recourse to force. And so their dream is contradicted by the reality of state power. They know the enemy and the goal and are, therefore, driven to the means: the armed struggle.

If there is sufficient force or the center weak, then there need be no armed struggle. Even if the center does not fall, there can be conventional war, a civil war, an traditional war. There is no need for secret haven but only audacity. Then there need be no underground. Most African officers may take power in the palace in the course of the morning with luck, a few friends, and sufficient brutality, and so broadcast the proclamation of triumph on the evening news. This has been the way to power for many in Latin America with the way of the revolutionary guerrilla left to those whose dream is more general. Real wars or quick coups, democratic procedures or efficient brutality obviate the need for an armed struggle. Yugoslavia collapsed not into competing undergrounds but into an irregular war between national factions. Ethnic cleansing cleared the potential constituency for an underground leaving a mosaic of defenders and unconventional forces.

For an armed struggle there must be compelling power at the center and yet room for an armed conspiracy. At any given time, however, for generations, there have been, still are, hundreds of conspiracies pursuing power from the underground. Some arenas like Lebanon had a wide variety of types seeking various measures of power for various purposes. Some arenas have institutionalized struggles like the Troubles in Ireland or the rural guerrillas of Columbia. At times the state cannot controls the margins where both dreamers and recalcitrant provincials may deploy force. Most who attempt to attack the center directly have a record of futility. Coups fail. Tribal revolts fail. Ethnic resistance to amalgamation fails. Most armed struggles are crushed, usually swiftly. Most revolutionaries end in shallow graves, in exile, or in despair. Unlike Che Guevara, few revolutionaries are remembered, few even leave example or find a place in unread texts, family memories. Some may leave a guerrilla heritage but most do not. So the *Brigate rosse* is history and so too the Huks in the Philippines, the Montoneros of Uruguay and the Black Hand of Macedonia. The failed may have as legacy an enriched arena: certainly any Irish nationalist builds on failure and the Italians on the revolutionary tradition and the rural insurgents of Luzon on previous experience. But the Black Hand of the last century is seemingly a matter of records and books, not really relevant in the uncertain turmoil of the Balkans at the end of the century where ethnic bases permitted overt militias and inchoate regulars—lethal, ill-disciplined but rarely covert, authorized terrorists with new banners.

Often the few who succeed have been those opposed to institutions that lack the will to persist—the empires of the West or the successor governments to the chaos of general war—the Russia of Lenin and the Germany of Hitler, Italy of the fascists, all offered a ruined center. Lenin, Hitler, and Mussolini left the armed conspiracy behind and seized the state. Most states can be defended. Even the fragile governments of Latin America for a generation resisted the Cuban model—until Nicaragua and that adventure came to confusion. Democracies, especially those without a nationality problem, have proven stable. And so too authoritarian states if both brutal and efficient. Yet, much of the world is not democratic, not free of nationality problems, not simultaneously brutal and efficient. And tomorrow is not today. So there are always new armed struggles to go with the old bandit lords and tribal gunmen, the cults and the criminals.

Each such struggle is special, but all such struggles configured by a state enemy, by the imperatives of the covert and illicit, by the lack of tangible assets, and by the nature of the dream. It is not the content of the dream that matters but the existence of the faith. That faith must fuel an armed struggle in certain ways because the arena is uniform, state-centered, and so each armed struggle is determined by the same imperatives. If there is to be a campaign and an underground, the intensity and sophistication of the dream is what matters most. Once the dream exists all evolves within a particular context, at a special time, open to the contingent and unforeseen as well as habit, custom, and wont.

In fact, custom or opportunity often are sufficient to allow gunmen to operate, irregulars to appear even without a dream. And at times the dream is faint, habit more that revelation. So the ideal for those opposed to the center may range from rationalization and self-deception through the habits of the tribe to an incandescent and universal revelation. And so too may the capacity of those who oppose the center by irregular means range from the assets of a single individual to those of a counter-state capable of waging conventional war. Obviously, the more effective the military struggle, the more representative, the more likely the campaign is to be recognized as valid. The closer to conventional power the more conventional recognition: partisans can no longer be hanged and guerrillas discounted as bandits or defenders with proper flags be treated as criminals. Insurrection becomes war if those at the center do not flee but persist, resist.

The legitimacy of the struggle depends upon a spectrum that begins with a historically recognized ideal and ends with the delusions of a few fanatics, the slogans of self-appointed defenders or local warlords. What is sought by a revolutionary underground is legitimacy: a righteousness proclaimed by the involved and recognized by the avowed constituency, by the international community, and even by the opponent. Arafat began with a few friends in a cafe and ended speaking before the United Nations, a long trek to recognition that in time made possible negotiations with the Zionists if not proper concessions or the Palestine dream a fact.

Legitimacy comes in various ways including the barrel of a gun. The IRA inherits legitimacy from two centuries of struggle, a legitimacy denied by London and Dublin but still real—Irish gunmen are not criminals or mad, not unreasonable, may be brutal and ruthless but share a dream with most nationalists. For generations the Irish thus begin each campaign with recognition while most others must find legitimacy the faithful claim during the struggle. The Palestinian terrorists, pariahs and fanatics, evolved into a movement recognized by the international community and at last by the Israelis. In Lebanon various groups possessed degrees of legitimacy as well as various military capacities. These groups arose in part from dreams, some great and universal but many focused on the most parochial advantage. Some few were archetype armed struggles and others merely criminal factions. The arena was filled with the real and imaginary, haven for the unsavory and idealistic galaxies of the faithful, some underground and some all too visible.

Generally, anywhere within the global arena at any given time, there is an arc of examples, various campaigns. There are archetype armed struggles with historical and ideological justification for pursuing a protracted campaign. There may be counter-states pursuing all but conventional war with liberated zones and ambassadors in major capitals, or there may be nomads of bandit zones pursuing an ancient vocation. There are titled warlords with pretensions of grandeur like those who appeared in Lebanon or later in West Africa. There are the hereditary bandits of Ethiopia or the Yemen. To the innocent all these gunmen share tactics and so definition, but among them only the idealists seek to capture history, seek such power with a gun, seek a classical armed struggle. In Lebanon there were those traditional Lebanese groups who deployed violence for traditional reasons when an opportunity arose to seek advantage, give pain, wreck vengeance, acquire assets, or display power. Few even wanted to win completely and so deprive the future of victims and rivals but rather to persist in a village world of legitimate vengeance and constant defense. The Maronite Christian factions were incipient nationalist movements, sought ethnic domination, perhaps if need be ethnic separation, avowed a special tradition, a special religion, a history unlike that of their Arab neighbors, looked to France or the West as mentor. The most typical nationalist movement was actually that of the Palestinians, illicit, covert, in exile, engaged in a transnational armed struggle against the Israeli center but in Lebanon distracted by the benefits of local power and the opposition of local factions and forces. Indeed, the Palestinians offered an entire spectrum of nationalist ideological variants, the special dream made more special: the Ba'athists, communists, the Islamic, the democratic, the purely national, and the friends of those elsewhere in power. And within each such current there were other smaller riptides. Each was ideologically special but inevitably each dominated by the charisma of a single individual. All competed for legitimacy. All sought to use the Lebanese arena not simply as base but also as power resource and so each had an agenda with Lebanese items unrelated to the national goal: a Palestine state for a Palestine nation. The national dream was quite different from the aspirations of the Druze: to be left alone amid weak and quarreling rivals. There were movements that sought to change the world not by the establishment of a nation or the protection of a people but by changing all history. The faithful sought an Islamic future, confusion to the West, an end to the Zionist enemy, and the ruin of the Great Satan. Allah would be the answer, the answer to everything, a global answer, a universal answer. The Hezbollah was in all ways a traditional revolutionary movement, dedicated to changing the existing world fundamentally. And so Hezbollah waged struggle against other faiths, other tribes, other rituals, the schismatic and heretical as well as the alien and oppressive, a traditional Jihad shaped as an armed struggle.

Analytically the local warlords displayed only the techniques of an armed struggle, not the dynamics, not the ambitions of even the Druze. They existed because there was neither center to impose order nor a return sufficient to attract those with power to eliminate disorder. Thus lacking military resources, ideology, a base beyond self-interest, consistent friends, long-term prospects, and disciplined ambitions, the warlords existed for momentary rewards, the delight in harm, the tolls of the check-points, the pleasures of power displayed, the joys of a gun. They had neither the traditional rituals of the tribe nor the dreams of the rebel, were without faith or heritage, without great ambition, were in all ways faction. These warlords fashioned their own small wars with the tools to hand: the sharp and lethal grit of greater factors ground down by anarchy.

Anarchy was not unique to Lebanon. In Somalia clans would give coherence to the collapse of government, but in Liberia countervailing power and no center let the contenders run free, pursue gain and excitement for a decade. Often there had been no center, no law but habit in the outback. Their like was to be found any place that the center failed and the options canceled out—once in China and Mexico and more lately in the Sudan or Zaire. Some set up small counter-sites disguised as provincial governance and some keep the outback for themselves.

Such dissent reveals capacity at command center; then the marginal may seek extended power, seek to govern the province as state or march on the capital. Mostly bandits remain with the trade and the warlords savage the land. These last are the most irregular of all, factional killers dressed in loot, addicted to cordite, involved for the slaughter of the moment, the power of the gun in a confused and barren life. Much of China was once ruled by warlords, much of the ruins of the Soviet Union has been prey to the species. All deploy unconventional force and the tactics of the irregular. On the evening television one can see headmen with Jeep Cherokees as command vehicles, combat jackets and aviator glasses as uniforms, overseeing pogroms or children in T-shirts and flip-flops shooting out windows with assault rifles. There are life-presidents amid tribal bodyguards giving television interviews to CNN while out of sight one tribe slaughters another. There is no heart in this darkness, no central core of tradition, no compelling dream but ample display of the techniques and tactics found in any armed struggle.

In any arc of orthodoxy, the tribesmen and traditionalists are in a sense most orthodox of all: they survived from the time that the center lacked the resources to shape the arena and the locals had to cope, rely on habit, experience, and force. The frontier ran to parochial rules, justice was imposed if at all by the rituals of the outback. Thus the map of the state, the nation, often included wilds beyond the law, beyond the interest of the capital. This has not everywhere changed. There are bandit zones and so village defenders: orthodox careers.

Warlords come and go but in some areas the traditional is stable if violent. Ambush, piracy, or cattle raids appear as the means of the guerrilla but are in many places apt to be no more than the usual: the way things have always been done to assure order, a steady income, the redistribution of wealth, the gods' delight, or the coming of age. The usual central respond to the irregular is to deploy the assets of the center in an anti-insurgency campaign, the future against the past until the point of diminishing returns. What is often daunting for the state is that the traditionalists find different definitions of winning and losing. For those on the margins to fight may be ample reward, to die may be a triumph, to steal may be more worthy than to accept bounty, to kill the innocent may be required. In fact, just as the most elegant, revolutionary anarchists claim none are innocent, so too may the roving warrior: all but his own are prey.

The refusal to play by the rules that so infuriates the orthodox in an irregular war—the Indians will not wear red coats or the terrorists fight fairly—arises from the assumption that the center rules OK. Right is made by might and possession of the flag and a seat in the United Nations. All opposition is defined as illegitimate. It is not acceptable for the irregulars in the outbreak to run to different rules, play another game, prove difficult. In fact the conformity demanded by the center often is seen as opportunity by the margins. The warlord who would rather give pain than receive pleasure presents problems in rewarding. Much more difficult is the tribal gunmen who seeks confrontation and risk as advantage, proof of capacity that can only be demonstrated by display. Many in the Horn of Africa were delighted with the arrival of the Italians—a new supply of enemies, new opportunities. The roads and clinics and new cities had little appeal to those who found triumph in war. Pacifying those who seek a bandit vocation, copying with parents who skimp and save so the son can be properly armed is far more complex than killing a revolution.

Fortunately for those at the center even warrior tribes have various priorities. Thus pacification by coercion need not mean genocide—may mean any end to tradition, turn the Masai into game wardens or the Bedouins into a tourist attraction. The Italians bought time and limited peace by co-opting the young into an armed militia: career bandits but still careers that offered fewer victims. Dreams, however, are not as easy to corrupt as warriors or warlords and so revolutionary armed struggles may not be easily appeased or cheaply opposed. Yet many such aspirations are difficult to define. Some dreams are muddled. Some idealists are criminal. And most armed resistance in any case must be countered with the tactics and techniques used against bandits or criminals.

* * *

Many struggles between the orthodox center and the irregular at the margin reveal a conventional confrontation, an asymmetrical conflict between those with state resources and legitimacy and those who rely on cunning and the imperatives of habit. These are wars of modernization, unconventional in tactics and techniques but after two centuries merely an aspect of the evolution of the nation-state. In time it is assumed the Zulus will be pacified and the jungle refugee only for animals. What has become unconventional is that the new state is often merely a tribe with a banner, a faction with a flag. Often the state is more irregular than the bandit. The state's unpaid, untrained army can be dressed in stolen mufti, wild, drunken, feckless, armed and dangerous, part of the problem not the solution. Some of those who pursue an armed struggle like the Viet Cong or even the IRA may in fact be orthodox in structure or tactics, as disciplined as their state opponents. Such undergrounds are tightly organized, committed to a dream not advantage, and determined to be worthy of legitimacy.

Not only are there various kinds of traditional irregulars and revolutionary conspiracies but there are as well other unconventional challenges to the state. There are cults mimicking the revolutionary. Thus the center may find paranoid dissent within the state, the more complex and intrusive the state then more paranoid the reaction. Such dissent arise from anxiety, lost of place, a fear of the new. The converts are charmed by the charisma or message of a special visionary, resent the complexity of life, the lack of explanation. Seeking certainty, seeking a devil as well as heaven, the fearful create their own special faith. They are born again not into revolutionary fervor but a cult. Some cults are composed of those whose commitment to a compelling dream is absolute. This is despite and because the dream is viewed as deviant, debased, and peripheral. Were not once Mormons a cult or Christians deviates? Some few cults may find expression in political analysis—find an enemy in the state and the system—and so organize resistance. If society is sufficiently open, they can find room to organize and so mimic traditional revolutionary cadres. Even one or two such pilgrims can in a hi-tech society bring down a building, shoot a world leader. Yet only the few organize violence much less to seize the center since most assume they already are the center. And none are merely the zealous defenders of great faiths, assassins for this god or that, but rather convinced advocates of a new reality beyond reach of the unenlightened, the everyday, and often beyond reason.

Some within such cults are not the simple by-products of sophisticated ideas and societies but the response of the traditional to chaotic times. In Uganda the Lord's Resistance Army pursues an armed struggle of a sort: regional bandits made bold by a new sect, armed refugees turned rural guerrilla, dangerous, distracting to a weak center, not cult alone and nor really an armed struggle but haltingly aware of the advantages of the guerrilla vocation. When Zaire collapsed, some of those who emerged armed and involved had no real agenda and others had appetites and uniforms. Africa has repeatedly generated tribal cults that modernize with automatic weapons, with borrowed slogans and camouflage jackets—tradition adapting to change rather than simple bandits or real guerrillas or just a new cult.

Cults in developed societies, in modern states can be far more sophisticated if no less bizarre. Thus Shoko Asahara in Japan created *Aum Shinrikyo* with a mix of ideas and programs and convinced the faithful a real defense was necessary: an armory was established, nerve gas devised, biological weapons stockpiled, equipment to produce machine-guns acquired, and even a Russian helicopter imported. In March 1995, *Aum Shinrikyo* released gas in the Tokyo subway system killing eleven and injuring thousands and released nerve gas in Matsumoto. The purpose in large part was to defend an inchoate dream about reality—unconventional attacks not unlike those pursued by others under different banners but parallel to not identical with an armed struggle.

The Japanese state, the center had to respond to a cult, dangerous, lethal, and bizarre but not to an underground army—not the Japanese Red Army, *Sekigun-ha*, in many ways also a cult but one differing from *Aum Shinrikyo* in proclaiming a radical, secular dream similar to that espoused by other university revolutionaries in the late 1960s and early 1970s and exhibiting less capacity to do harm. As is everywhere the case with tiny movements, *Sekigun-ha* exhibited characteristics of the cult. Intense, militant revolutionary ideas may attracted the deviant. The paranoid know everything and have grandiose expectations and so too *Aum Shinrikyo* not because the group posed a terminal threat to Japanese society but because the means were grandiose—chemical and biological weapons that had engaged the concern of Western analysts for a generation. In fact most such centers of belief converted few and often collapse in murderous encounters—and so too *Aum Shinrikyo*, frightened a great many, killed few, and disappeared.

The frantic dreams of the cult are not always without recourse to reality and just grievance. The dreams of many revolutionaries seem to the conventional to be just as fanatical, beyond reason. How could Che expect to overturn a government by marching a dozen men through a Bolivian jungle? Yet, Castro had landed with a dozen. How could Menachem Begin take on the British Empire with a few volunteers armed with mismatched revolters? Yet, in time he became prime minister of Israel, a national leader as had Kenyatta and Makarios and Mandela and the rest. Slotting one set of fanatics into a cult and others into legitimate conspiracies is not simple—and often for those under threat not relevant. The motive of the gunman may not matter greatly while there is shooting. So cults can produce lethal gunmen just as can revolutionary cabals and such cabals share much with cults. And both once the shooting begins, guerrillas or gurus, appear to the targets no different than the tribal killer or the village defender firing from ambush.

Some revolutionary movements seem more easily amenable to psychological explanation that political. In Germany the *Rote Armee Fraktion* (RAF) attracted a special variant of marginal middle class drifters but offered as well a political agenda. The convinced functioned best underground so, in effect, they sought within the armed struggle both change and therapy. Some groups like the Symbionese Liberation Front that captured and "converted" Patricia Hearst or the bizarre Charles Manson with his hope for a race war sparked by murder—"Helter-Skelter"—are proper pathological studies for those concerned with aberrant behavior, were lethal deviants under false flags.

It is easy to see all assaults against the system as deviant—the Soviets had recourse to the asylum, drugs, and therapy for those so demented as to dissent from socialist norms. To the conventional especially in America, any recourse to violence seems alien, aberrant. The warlords of Bosnia, the tribes of Sierra Leone, the urban guerrillas in Rome or Berlin are all weird, abnormal, unreasonable and unreasoning. The few always seem unreasonable, engaged in a futile exercise in violence: movements of the mad whether militias besieged in Montana or bombing abortion clinics in the Middle West or the gunmen of the secret armies in Ireland and Ceylon. They appeared "cults" of one sort or another whatever the professed agenda. And there are religious and secular cults who find evil in the times, often attract those drifting in desperation, free floaters in the ideological shallows. Conspiracy offers meaning not only to those unsatisfied with conventional religion but also those unable to cope with the everyday. These zealots are not without parallel in the revolutionary world.

The Peoples Will or the Stern Group, small, desperate, appropriately paranoid and lethal, were engaged in armed struggle, propaganda of the deed, campaigns with seemingly unreasonable political aims. Who could imagine a few intellectuals replacing the Tsar or modern Zionist zealots seizing the promised land from the British Empire? In time such aspirations became reality. An unreasonable commitment to the efficacy of the movements' capacity to impose change and to change history that fueled an irregular campaign indicated that the faithful were a revolutionary underground not some mad visionary secular cult. Time and operations indicated a true armed struggle not a delusion, indicated that the gunmen were not mad but driven and like all of the ilk criminally optimistic.

The aberrant converts like the religious zealots may disturb the center, threaten order, frighten and dismay, but they do not propose a valid alternative reality, offer an old and cherished way of life or a new and revolutionary future. The center may at times treat such challenges as if the conspiracy were serious but the limits of the cult, the reality of delusion is mostly apparent. The dreams of the cult are narrow, black, intense, often bitter and always innocent of reality, the imagining of the paranoid and desperate, the marginal and the monomaniacal. Most often even their political dreams are merely patina over fantasies—and mostly those fantasies are clearly religious in origin if adjusted to the day. They need a great a powerful enemy, a savior, a revealed truth that explains all, a way of life that eliminates decision, an isolation from a world too complex. Many seek salvation through destruction, often their own. Most seek redemption but not to change history. They find everything is wrong and seek perfection in violence as an end not a means: salvation is an end not a beginning.

A real armed struggle arises from a different, more practical dream, a viable analysis of history and a general aspirations beyond tribe and village and fantasy. The converts to the Jihad may seek redemption and the triumph of Allah but also a real state in Egypt or Algeria. The sacrifice of the carbomber may be alien to Western practice but is reasoned: individual salvation is assured and a victory for the collective dream. The Holy Jihad wanted history changed not their anxieties eased—that would come with victory but would not be victory. The willing death of one pilgrim was not an end but part of the beginning. Such volunteers share much with any religious movement, much with the cults, with the conspiracies of the arena, share with all undergrounds a dream and so a campaign for power to impose that dream on history. In Egypt the Moslem Brothers, like the Holy Jihad, found in Islam the answer, in sacrifice a means, in the armed struggle a strategy. And like most the Moslem Brothers failed, ended in Nasser's jails, executed, disgraced, and in most cases unrepentant with the dream hidden but still valid and available to another more ruthless generation. And Holy Jihad was that generation and so far sacrifice has not imposed the dream on history—but for the faithful the struggle is far from over and the Marines no longer in Lebanon.

* * *

Every armed struggle, especially the classics of the last two centuries, the major revolutions, the national liberation campaigns, the risings and insurrections against imperial power, the urban guerrillas and transnational terrorists, has been the same; and yet each is special, different. Those who seek Allah as the answer in Egypt today differ substantially from those of the Moslem Brothers that did so a generation ago. The dream is similar but the arena transformed, the appeal novel, the components rearranged.

The faithful always assume themselves special. Each knows, however, that the dream is unique. To the involved, those once underground or still engaged, their own struggle may be announced as a front in a general war, a special front; but it is more and more likely in their heart of heart, unique with special grievances and particular aspirations, the rebel is more parochial. Some may find allies and examples in ideologically similar movements; but few accept that the very nature of the struggle imposes uniformity no matter the ideal, the peculiar historical circumstances and the special cultural climate. Most underground inevitably assume that only those with similar grievances, similar aspirations are truly engaged in an armed struggle: the others are denied the legitimacy of revolution. The fascists in Italy felt nothing in common with the *Brigate rosse* or the Zionists descendants of the Irgun nothing in common with the Palestine fedayeen. For the dedicated the content of the dream makes the difference—and in this they are in error for it is the actuality of the dream that shapes the armed struggle, the resistance of the center, and the objective reality of the arena.

Armed struggles oppose, adjust to received reality, make do with what is available and so are apt to appear in varied configurations. These structures are not inconsequential, only necessary, but disguise the similarities of all—even and especially from those underground who see their struggle as unique, their dream as special, their struggle as singular if within the announced tradition. Yet each struggle arises from the energy supplied by the dream—any dream. And every struggle is opposed by a state. It is the state that imposes the necessity for secrecy and all that this implies. It is the state at the center that imposes scarcity and all that this means for the underground. And it is the recognized state at the center that opposes legitimacy so sought by the faithful who seek the title deed to the future by violent means. Even the most feeble and dubious state has the assets of resistance—if not then the rebels would sweep to power overnight. And each state is apt to respond to conspiracy, to threat, to an armed struggle regardless of the dream in similar ways: the bandit is countered, the village defender defeated, the tribal insurrection quelled by means also deployed against gunmen and terrorists, the lethal dreamers.

The assets of orthodoxy over the last two centuries have been sufficiently similar whether in the hands of the Tsar or the Life-President to impose continuity and form of the functioning of the dream. The campaign requires the same techniques and tactics of the bandit and warlord but is driven by quite different aspirations and so conviction. As always it is the working of the dream, the perception of the faithful, the energy and impetus supplied by the vision that is opposed to the power of the center.

Every Dragonworld may be special but is the form given by a dream that is endangered by existing power. The world exists within other worlds, exists in the perception of the faithful. It is a world that is not structured by traditional values or by personal consideration, by the opportunities of the moment, often by objected conditions but by an ideal that cannot be achieved except by recourse to violence. Some movements hover on the edge of that violence, seeking other ways forward. These are revolutionary parties, seminars for inchoate gunmen, front groups, support groups, organizations that are in process of discarding options, moving imperceptibly underground. Most who go underground only emerge transformed, disappointed and denied, sitting in prison repentant or defiant but residue of past hopes. Some, of course, like Che never emerge. Most failures are simply forgotten, bodies dumped in ditches, the faithful lost along jungle trails or buried in prison yards. Sometimes there are survivors, dream intact and legitimacy available as legacy. The Irish militants have spent a long history waiting for the few opportunities to act—a two-hundred-year institutional history. Others, as in Italy, can rely on a society that anticipates, even tolerates, recourse to violence as a means. This may be a wasting asset but as the violence of Bosnia indicated old dreams, ancient grievances, desperate ambitions do not easily disappear. The post-industrial world has been filled with irregular war, many traditional, a few freshly made, but most enduring are those fueled and formed by the dreams of the faithful. In Lebanon there were all sorts of dreams and irregularities—and no real state, for it is the state that gives coherence to the armed struggle, shapes the campaign in an armed confrontation between the ideal and the real, the responsibility to the present and to the future.

The dialogue of the dream and the state, asymmetrical at one remove, most visible in the armed struggle, has been a constant in modern times. For the underground this has meant a special world not simply a special struggle. The armed struggle is the purpose and most visible aspect of the Dragonworld but not the ecosystem itself, rather the product. The ecosystem has its own special dynamics even if all is dominated by the struggle being waged—or about to be waged. There are within the postmodern world a great archipelago, clumps of the faithful, hidden, desperate, armed, and dangerous: the real gunmen in imaginary gardens. It is composed of perceived reality, imaginary entities, unreal systems that generate most harsh reality, tangible killers, and visible armed struggles.

4

The Design and Determinants of the Dragonworld

The postmodern world is filled with irregular war, militant undergrounds, and violence at the margins. All share a similarity of tactics, a lethal dialogue between the conventional center and those without tangible assets. Some of the violence is shaped as war or revolution, as conspiracy or insurrection. The definition often focuses not on the dynamics of the challenge but the visible means, the tactics of the guerrilla or the stratagem of terror. There is no single satisfactory, all-encompassing model possible for the unconventional. The tactics of armed dissent have always been similar, the recourses of the weak. The responses of the center to rebellion and conspiracy have for centuries tactically been similar. It is the quite various motives of dissent that are apt to shape violence. The strategic purposes may vary enormously but the similar means, regardless of motive, tend to engender similar responses from the challenged. Yet all armed dissent is not alike nor amenable to similar counters simply because the tactics are so universal. Some irregulars have no more than bandit ambitions and others are enlisted in a crusade to transform the world. As far as the security forces are concerned, bandits or visionaries present similar tactical problems and similar immediate problems.

The classical revolutionary armed struggle is the means that those possessed of a dream discharge their responsibility to history: their dream supplies the energy that drives the campaign as well as defines the agenda. This means that the challenge presented to the center is apt to be protracted. Still revolutionary dreams are not everywhere the engine of struggle, for others seek not to capture the future but to reinstate history. And many engaged in irregular war do so not for any incandescent dream, any desire to go back to the future or forward to salvation, but merely because the times, the habits, the opportunities so demand. The village is endangered, there is profit to be made, the clan is at risk. These are the traditionalists, the gunmen of the moment, who focus on tactics, who are underground by necessity. When the times change, so too does their commitment to rebellion, the village is safe or no profit is worth the risk. When those who denied the past can be persuaded to take advantage of the present they, if not co-opted, may be sullenly content. The center may deploy coercive power or compromise more readily when there is no dream involved, when the threat to the center is not terminal. On the other hand, as long as the dream is vital, as long as history and the system must be changed, then the classical rebel persists.

The cults, the greedy warlords, the bandits of the outback, the factions with flags, and the tribes seeking to seize the nation are more amenable to the reality of the arena than those driven solely by a transcendental dream that requires the faithful to change history, rewrite the past, and control the future. The devotee of a cult can be manipulated, converted, co-opted; he proposes an aspiration that has limited appeal to the many. The bandit on the hill or the village defender have personal interests as primary. The clan or the tribe shape their aspirations to a narrow agenda that is often within the capacity of the center to accommodate. The absolute revolutionary deploying the same tactics, hunted in the same way by the same state, on the other hand, lives in a different world, one both visionary and realistic, and one not easily amenable to coercion or co-option or accommodation.

Such a revolutionary ecosystem is created almost solely to pursue an armed struggle: the only appropriate way to determine the future. As a result a special Dragonworld is both reflection and template of an unconventional pursuit of power. It rarely exists without acting as haven for an armed struggle. If the armed struggle evolved into a civil war, an orthodox war, no such ecosystem is needed. If after a protracted campaign the arena changes, the underground may emerge during a end game as victor or vanished, as exiled cult, or perhaps in a more conventional configuration. When the protection supplied by the underground is no longer needed the ecosystem evaporates—the Dragonworld is gone.

The Design and Determinants of the Dragonworld 85

This world is fashioned by the perceptions of the faithful and is not a matter of formal membership, boards and operational commands, not easily charted. Such a creation, like romantic love or a mystical experience, is easy to recognize but difficult to define, subject of similes and metaphors, poetry and commentary. The underground offers protection, a haven for faith and the galaxy of the true, and a base of operations. Entrance is only by conviction that transforms perception. Within the Dragonworld the committed, no matter how various their special faith, may pursue diverse goals in various ways, create various forms for the armed struggle, but all dwell within a similar world. Each knows what is wrong, what is wanted, and what must be done. Each fits a model just as do lovers and mystics. The French in love are not so different from the Italians in love and the Baptist born-again-into-Christ is much of one with St. John of Cross emerging from the dark night of the soul. Each is changed and empowered. For the rebel life is transformed, enhanced, charged with meaning. And all rebels are apt to face very similar obstacles in the pursuit of the armed struggle: the assets of the state-center, a reluctant constituency, the lack of legitimacy, and the penalties required to persist underground.

Each Dragonworld is a galaxy of belief. Thus when the Marines arrived in Lebanon, there were many such perceptual worlds in the visible arena, many who sought to achieve a Palestine nation or to impose Islam as answer. There were as well those with less compelling ideals. Those who wanted the best for the Druze or vengeance against the Sunnis or a dominant role for the Maronite Christians. And as well there were those who defended the village or achieved advantage. Some deployed the means of the armed struggle but did so without the energy of the dream. Others were possessed of dreams quite alien to the Americans. And none, the lethal dreamers or the village defenders, needed to cope with a coherent center. Instead they had to cope with each other. Mostly, elsewhere, the center tends to give coherence to the lethal dialogue with the underground, mostly elsewhere only a few undergrounds can emerge in one arena.

In Ireland there have for years been only tiny factions to compete with the dream of the orthodox Irish Republican movement—and all of these differ only on capacity and tactical choice. Always the IRA has been the core of the militant republican movement in contrast to the Irish nationalists in power in Dublin, in control of the electorate and everywhere recognized as legitimate. For the faithful it is not votes and governments that give legitimacy but history and the dream—and this they claim. In Lebanon a Palestinian fedayeen or a volunteer in the service of Allah would recognize the logic. In Northern Ireland there have been various defender militias of Protestant paramilitaries who seek not to transform the future but to protect the present union or to return to an imaginary past. These factions are not unlike those found in Lebanon. Some of the world's underground aspirants are like the IRA, classical, engaged in protracted armed struggles. Some like Hezbollah in Lebanon are a mix of Islamic revolutionaries, Lebanese Shi'ite defenders, transnational zealots and agents of Teheran and Damascus. Elsewhere in Egypt or Algeria the fundamentalist underground is less complex but as dedicated to the same dream. National dreams engender irregular war in Bosnia and religious ones in Afghanistan. Defenders can be found not only in Ulster but also in Yugoslavia and the Caucasus. Rural guerrillas in Mexico mimic not the dream of redemption but the commitment to the armed struggle, act out political theater at low cost. Elsewhere, as in Lebanon, there are all sorts of armed struggles, all kinds of irregular war, but only one category of revolutionary underground that is generated by an ideal.

Even then, any model underground must be adjusted for local conditions, for the times. None is really perfect example unless stripped to the very few faithful, often at the very beginning—the *foco* as a tiny universe of a dozen faithful. In Germany the underground of the *Rote Armee Fraktion* was in theory a classic neo-Marxist struggle but in practice was explicable to purely psychological explanation—a secular cults of the marginal in a single student generation. A group like *Brigate rosse* during the years of lead in Italy was classic, largely isolated in one country, competing not only with other Italian cores with similar dreams but also with a neo-fascist underground dedicated to a very different and lesser dream but one pursued in remarkably similar ways. In any case each Dragonworld spins out from a central few, those there at the creation, those who went underground and opened the first campaign of the armed struggle.

In each instance the center of the Dragonword was tiny, dedicated, absolute, possessed of a vision that energized all. Any protracted struggle allows movement toward the center, new and at times different recruits to leadership arriving at the center of the faith. Often those in the beginning last the course: most struggles are not as protracted as they seem, Begin like Yasser Arafat there at the end as well as the beginning. Some arenas have for the taking a revolutionary legacy but rarely an existing movement as base. Only the Irish seemingly go on forever and even then after a generation many of the names are the same. In any case, at the core of each ecosystem—the galaxy of the faithful—is a command center, ground zero of the Dragonworld.

Close to the command center are those fully dedicated to the faith and beyond them sweeping about that magnetic commitment are the faithful, the followers. Some of the most faithful cannot be deployed, some can often not be reached once the struggle has begun. Those who can be deployed are found in the support groups, the committees and fronts. There are as well the publicists, those who parade and vote, buy the newspapers and clap at rallies. And further out still those who agree and those who tolerate violence and far, far out on the margins are those who may in time glow too reflecting the central conviction, attracted by the spectacular or the truth as dispersed.

Like a real galaxy the motion is constant, the whole throbbing and expanding with successes, thinning and narrowing under coercive pressure. Operational deeds may light up the whole universe. Then those otherwise quiescent may sparkle in support. And effective oppression, internal blunder, or simply exhaustion may in turn darken much of the enthusiasm, narrow the galaxy, snuff out all but the faithful few at the center. As long as the core retains the faith, their conviction that the future can be transformed by the campaign, then the galaxy threatens the stability of the recognized center. The center can tolerate only the dead black star-hulks of the failed or low-intensity, low-cost persistence near the margins of control. The responsible want no living heritage of rebellion, no live rebels. The ruthless and efficient find comfort in killing all who dream and most who might do so and clearing out the badlands. Often, however, the center eschewing final destruction in consideration of costs and other priorities and so wait on time, co-option, reason, and the secret police.

Always, always no matter how flawed the rebel faith, dubious the cause, restricted the power of the dedicated, it is the pretensions to legitimacy that is regarded as intolerable by the responsible, by the government, by the system. Often the challenge to the center has in reality acknowledged historical roots, a viable appeal to the avowed constituency, even if defined as misguided and unduly uncompromising by the center. The rebel is labeled alien, deluded, evil, or criminal—and the more likely to be so labeled by a center with dubious credentials. The center seeks to deny any validity to the dream. At times this is easy. Who did Lenin represent? At times less so, since Castro seemingly represented all the decent and democratic. No matter the center denies the dream. Those underground are zealots or fanatics, separatists, alien or bandits or demented. The IRA cannot win an election, the Sandinistas nationalists were communists, and the UNITA rebels merely an Angolan tribe with ambitions. Underground legitimacy even if arising from real factors often must remain underground. Thus their dream is crucial and too much reality dangerous: the IRA can win votes but not a national election, the Sandinistas were Marxists but also Nicaraguan nationalists, UNITA is tribal as are their opponents.

The dream is a threat to stability, to the center because of the claim of legitimacy: this is the true path, the one way, the promised future. And dreams can promise anything and everything while the state must be responsible, collect taxes, repair the roads, enact rules and regulations, allot entitlements. The degree of threat, the nature of the future proposed in large part determine the degree of toleration and the intensity of coercion provoked at the center. The more legitimate the dream and the greater the aspirations of the underground the more intense the struggle.

In Northern Ireland, even when Protestant defenders erode legality and just governance, they are no threat to the system, none to the regime, rarely even to law enforcement. No one can go back to the future in Ulster or slow time so the loyalist paramilitaries use limited and symbolic violence to counter the threat of the IRA. And the IRA avows very great institutional change: the withdrawal of the British, sovereignty for a new Ireland, a republic of thirty-two counties, and so an end to all partition regimes including that in Dublin. The aspiration is grand but the capacity of the IRA to pursue these aims has always been limited, tolerable if unpleasant. London is apt to balance other priorities and the limited capacities of the IRA with their own perceived responsibilities in the United Kingdom, in Northern Ireland. And the IRA does not challenge the existence of the United Kingdom much less the future of British society only the existing arrangements in Northern Ireland. The Irish dream limited by capacity is tolerable.

The Design and Determinants of the Dragonworld 89

An acceptable level of threat is not always possible. At times the dream engenders great power—or at times the capacity of the margins with or without an ideal simply endangers stability. In Algeria the Islamic fundamentalists want to change the regime, the system, and society and so leave those at the center little choice but to persist or flee, to counter the insurgency and seek to co-opt the more pragmatic dreamers. And the more effective counter-insurgency, the more brutal and ruthless and desperate fundamentalist tactics. The Islamic dream also in Egypt or Turkey is taken as a terminal threat to the center—and the more so because the dream is seemingly legitimate. In the Sudan the dream in power in Khartoum does not guarantee security for much of the country is untouched, beyond Islam and beyond easy rule: the margins generate a disorder beyond the capacity of the state to control. So Egypt is at risk because a dream may generate capacity and the Sudan merely because the center even with a dream in place, even with recognition and a flag, still lacks assets to impose on the margins.

The more marginal the dream the less threatening. Those opposed to Khartoum seek limited goals—not even at times a separatist state. Their aspirations may be open to co-option as well as denied. In most cases a state can easily tolerate bandits or a village defender, even the Ulster Volunteer Force killing Catholics or tribes at the margins running up separatist flags over old habits. In such irregular conflicts the center has more assets or at least enough to persist if not rule with grace and dispatch.

As the perceived aspirations of the underground escalate so too does the adamancy of the opposition—more stand to lose more. The greater the underground dream the more dangerous as long as the aspirations are practical. Thus a new regime need not threaten the entire establishment, did not seem to do so in Cuba. So when the tide turned and the Cuban underground became visible only those close to the palace fled. If Castro had come under red flags, more would have been threatened and more would have resisted. Those who would rid Algeria not simply of the old regime and the old system would also discard the old assumptions, the old reality. The underground wants to replace all with Islam as the Answer, and so resistance has been fierce. Each campaign is different, a muddle of capacity and aspirations, a special lethal dialogue with the center over the future.

The violent dreamer must offer a legitimate future as did Castro, as does the IRA and the Islamic fundamentalists. The radicals in Germany envisioned a completely transformed Germany, a new communist society, a new socialist people as part of a world revolution. No other Germans could imagine such a prospect. Of course, few Russians had imagined Communist Russia but in Moscow and Petrograd in 1918 the dream shaped capacity in an arena where few had compelling power—few had faith and fewer trained cadres. Lenin's dream imposed capacity and so shaped a legitimate contender. In 1968–1969 in Germany as in Italy, the radical dream was impractical, lacked sufficient legitimacy and so lacked the power to convert many beyond the central core. The state had assets, unlike the Russian provisional government. Baader-Meinhof or *Brigate rosse* wanted too much, had the capacity to do harm but not to offer a valid future or destroy the center. Instead they were destroyed. In Ireland, legitimacy eroded but intact, the secret army can persist but not win, can threaten but not succeed, wants a viable dream but one deemed not practical by most. So in Ireland the dream does not die, even during the peace process grievance remains, a base exists, and so tolerance permits if need be an armed struggle despite the flaws of the ideal and the limits of the movement.

What few at the center of a revolutionary underground chose to accept is that almost every dream is flawed: the constituency is imaginary, the world as analyzed is fantasy, the cause is special pleading. Islam is not the answer for everyone. Not even every Irish nationalist wants unity nor did the radicals of Italy and Germany offer a viable revolution to the masses. Revenge on the satisfied is not a dream. Not enough workers in Turin or Essen cared, felt need of vengeance or radical change. The dedicated turned to terror in the streets and the workers watched on their new television sets. The Zionists overlooked the Arabs in Palestine and the Holy Jihad in Iran ignored the attractions of the secular. Dreams seldom reach out to everyone, to all the people, to the whole nation, even to the proletariat or those already converted to the faith. Most dreams evade complexity, alternative aspirations, the intricacies of real life. This is the nature of dreams. The lover's idol is perfect. All imperfections are consumed in the fire of love and especially so at the command center: all Irish Troubles arise from the British connection, the answer is Islam, the creation of Zion will usher Judaism into a new world, evil is institutionalized by the Imperialist State of the Multinationals, the Great Satan denies all glory and justice. No matter the limits of the dream, the command center believes—and spends much of the armed struggle extrapolating and enhancing the basic commandments, the aspiration arising on grievance, the faith.

With only the faith as central asset, most inchoate undergrounds would seem little threat even to the inefficient. The world outside certain authoritarian states is filled with movements and factions and parties that advocate the most radial change, appear to shape their agenda as an armed struggle, and they all look like losers. The faith is espoused by frantic hearts, peripheral orators passing out smudged tracts, by cranks and romantics and the self-educated, by students and clerks and lawyers without clients.

In April 1902, after one more split in the socialist ranks, V. I. Lenin arrived in London to edit *Iskra*. His mean office at 37A Clerkenwell Green was soon filled with proclamations, galleys, heavy books in various languages, visiting radicals of all the more extreme varieties, tea glasses, unanswered post, unpaid bills. Such offices are easily found this century as they have been the last. The revolutionary office is a permanent artifact of radical life, often the only visible indication of an underground. Wherever there is grievance and ideals, the necessity for an armed struggle, the faithful open an office.

In the 1960s and 1970s on the edge of Lusaka in Zambia in a meager compound of low, shoddily built huts, dusty, dry in the heat, a set of liberation movements kept similar offices. Each front room was filled with the same smudged journals, out-of-date newspapers, dirty dishes, and persistent hopes. Beyond the offices lay an almost quiescent African underground. For years ANC or SWAPO could be found in Lusaka-exile offices. Their triumph to come. Then, unexpectedly tomorrow was not like yesterday. The correlation of forces shifted and the aspiration for an effective armed struggle became irrelevant. The limited armed struggle haltingly fueled from exile headquarters had after generations eroded the will of White Africa. The seedy offices were abandoned for real power as the end game played out elsewhere.

Any tour of the active underground often leads to offices, often the only visible sign of an underground. In an unfashionable piazza in old Rome overlooking discarded market garbage and dubious trattoria there were those who spoke of the faith, but the gunmen had gone so far underground that the office had become no more than relic—a symbol of the faithful left behind. In Ireland the secret armies have always been less secret and militant Irish republican more likely to be tolerated within reason. So all the tiny offices on the Northern edge of central Dublin with the phone bill unpaid and the banners furled in the back office are within the compass of the underground. The police parked outside know that at times not simply the faithful and their friends may appear but also the wanted and dangerous. In Cairo Islam redeemed is to be found in the midst of the shoddy and dust of a Cairo office, a building protected by a decaying blast wall and jammed with too many people, unpleasant odors, and a constant din, lit by a flickering fluorescent bar light. The wanted, those with guns are elsewhere, not even in Cairo, but the faith is thick in the office and often so too the police outside. Everywhere in these seedy rooms, Lenin would have found the outward form of the dream readily recognized.

The office is a universal icon. Seemingly visible evidence of futility. The rebels always look like losers, often are, seldom display conventional assets or appear to be winning anything, even time. And all such offices look alike: 37A Clerkenwell Green gave not a hint of what was to come even to sympathetic eyes. And no one but the most optimistic could imagine that from the grime of 37A Clerkenwell Green would come the new Tsar. Few now can imagine that the president of this front or the chairman of that movement could ever be more than a querulous ideologue seized on a fool's cause. Mao was once a wanted man and Begin hid in a closet disguised as a rabbi. Grivas lived in a dugout and Mandela spent a lifetime in prison. Most dreamers may end like Che in Bolivia but all do not—Che in Cuba did not. Sometimes the dreamer has capacities as well as aspirations.

Every revolutionary cause, claiming possession of invisible assets, hiding underground, looks hopeless—and not only at the beginning. Even when Lenin got off the train at the Finland Station over a decade later in 1917, he did not look a winner nor did the Petrograd offices of his faction appear much different from those in London or those of other radical Russian contenders for power. Lenin was a small bald man who spoke in garble dialects about a world none but the faithful could see or imagine, reading, writing, analyzing: no successor to the Tsar, no threat to the professional soldiers and the sophisticated politicians. His shoddy offices were all too typical of radical failures everywhere. And some movements are too small or too far underground even to open offices.

Lenin's armed struggle was largely a rush pass abandoned barricades, audacity triumphant, not a true campaign. It became an armed struggle in retrospect. The real armed struggle, the archetype, was left to Mao. He did not even have an office at the beginning—his cadres were hunted and killed and persistence was achieved only by flight. Given all this—the shoddy offices, the impenetrable analysis, the cheap newspapers, dense, dull house organs, the bills unpaid—the sensible would assume any threat arising from such dreamers is minimal. When there is a threat, it is from gunmen, bandits, guerrillas on the run. In fact even when the shooting begins the underground is often unable to do more than plant the odd bombs, dress up as guerrillas, distribute proclamations. The dreamers at best can deploy only a few men with guns to destroy an empire. What matters is that an alternative faith lives, exists underground, and has the capacity for enormous growth. Dreams can transform reality more easily than reality dreams.

Such dreams are often seen by the conventional, the powerful, and the threatened as nightmare. Those in the underground in the grip of the revealed truth simply discount such characterization as self-serving rationalization. The truth is so obvious, so blinding, that none can imagine any real flaws: after all they are engaged in sacrificing everything for the ideal. Oscar Wilde noted, "A thing is not necessarily true because a man dies for it." Underground such sacrifice is validation of revelation—and such revelation, however flawed, however dreadful to the center, is not simply zealous fantasies sold as analysis. The dream *is* the great asset—and sacrifice its due: that a man dies for the dream is vindication and validation.

Buttressing the ideal and parsed at great length are the grievances of the present. Each armed struggle that survives at all does so amid of arena rich with grievance as well as aspiration. These grievances are rarely imagined although before the underground generally considered unalterable or inevitable or tolerable. The British empire appeared constructed for the ages. The Tsars had ruled Russia for a millennium. Ireland as a prosperous part of the United Kingdom, beneficiary of the imperial mission, seemed a reasonable arrangement to many. The present always has great weight even when the present does not work well. Irish nationalism was always there, erected in part on long years of deprivation and injustice—especially for the Catholic Irish. Long before the British empire was fully constructed, the world map shaded in red, the rebels had been at work in Ireland for national cause. Russia did not work well, the poor and the weak were exploited, the state could not provide appropriate entitlements, manage the country properly, avoid the corruption of power, and most of all prosecute the war. Times changed in Russia as for the British Empire. Liberation fronts, secret armies, underground conspiracies emerged, persisted, often triumphed. Yet times in 1968 were misread by the radicals in Italy and Germany as they had been in 1848. Clues to the future are elusive. Still it is not merely the perception imposed by the dream that generates grievance but also agreed reality. Much of the world displays undeniable injustice, just grievance, offers a rich medium of complaint. The dream may not be the solution but there are a real problems. Italy had problems in 1968 as did the British in Ireland in 1916. Those who would find their opportunity in the dream are optimists. They are most apt to be wrong. Like the anarchist Michael Bakunin, they at best mistake the third month for the ninth. They are, however, not always wrong or entirely wrong: the old Tsar is gone and just as Dublin foresaw and there is a new parliament of sorts in Dublin, somewhat as many Fenians imagined, and still hope for unity.

In sum, revolutionary ecosystems are founded on the assumption that the ideal is, indeed, the solution, the necessary and singular solution to injustice. These perceptual systems emerge as incubators of faith and then are shaped to husband, nourish, and disperse the energy so generated in order to achieve power over history, over events, over the arena. Those involved in this process, the armed struggle, seek an unconventional campaign against the visible enemy, seek such a struggle in the name of the avowed constituency, in the light of reality analyzed by recourse to the dream. They want absolute victory for the faith for they are responsible not to tribe or interest, not to fear or greed, but to history.

Design

The Central Core

At point-zero of the galaxy are those who first and most fully grasp revelation. They have been redeemed by the revealed truth, by the power of analysis, by the power of the ideal. They are absolutely consumed by the need to act for those ideals and in time realize that there is only one such means to act: the armed struggle. Those most convinced, those mostly at the beginning, those to whom the truth has been revealed are cut off by revelation from the everyday world, by the need to adorn and to cherish the dream and very soon by the secrecy imposed by the underground. In many ways they seem the same, keep regular hours, talk about this or that, go about the day's work but within each all is utterly changed. Those who know what they know increasingly know little else.

They gravitate in serious matters to their own. They have their faith reinforced by the faithful, find no options, no valid alternatives, repeat themselves to themselves. As is the case with encapsulated truth in monasteries, ashrams and utopias, secular or religious, the faithful bond and replicate attract the innocent and the curious. Life is intense, purposeful and dangerous. Service requires the constant focus on the truth that nourishes, invigorates, and sustains. Such a revolutionary life entails sacrifice and risk—and this is an asset, a charm not a price.

This environment imposes on all undergrounds special standards and reactions that in turn strengthen the dynamics of the galaxy. Because the faithful believe, they believe still more and so accelerate the circle of conviction. The reality of risks to the ecosystem, the dangers of the campaign, the real noise of war added to the demands of the faith generate desperate conviction at the expense of normal experience and expectation. Such an increasing overload of conviction tends to narrow the capacity to respond in any matter but that expected. And this is increasingly a focus on the inevitability of the armed struggle as means to eradicate grievance, achieve justice, redeem the past, and secure the future. Those underground cannot spare the time or attention, cannot empathize or seek out alternative directions: their created reality is too pressing, too loud, the music of the struggle. The more that is shared the less the external world is needed: the galaxy supplies all, isolates the faithful from the outside and is renewed by the conviction from within. As the perceived world opens up, as history becomes real and the armed struggle offers action, the faithful close in about themselves, close in about the central dream, and become more like themselves. They see what they see—revealed truth shapes all to perception. The larger the core of the faith the greater the attraction and the more likely tangible power. And, of course, numbers may include the dubious; growth is not without problems.

Popular Support

Every revolutionary underground assumes that the universe of the faith is grand, that the constituency that is to benefit from the armed struggle is incorporated into that universe. If in practice this does not seem to be the case, if the people fail to rise or the masses remain indifferent, then the fault likes elsewhere not in the ideal. Conversion and conviction are assumed, merely a matter of time and proselytizing. The rebel assumes he struggles for his people, his class, his nation. The conventional claims that he does not; that they at the center—the people, the nation, the workers—possess both legitimacy and the claimed constituency. In fact the threatened regimes hold plebiscites, display election results, organize mass rallies, present over and over evidence that the dream has no power. And the orthodox are often right. A crowd in the street or at secret congress declaring independence does not guarantee majority support. The IRB did not ask the Irish nation to support the Easter Rising in 1916. They relied instead on the authority of Irish history, the patriot dead. Neither Lenin nor Mao could number a majority nor for that matter could George Washington. In Palestine Begin's Irgun Zvai Leumi was a minority within Revisionist Zionism that in turn was a minority even among the Jewish minority. The *Brigate rosse* in Italy spoke for a proletariat they had not polled. Allah may be in the answer in Egypt or Algeria but the faithful do not seek validation by the numbers.

If some rebels represent only themselves and the proper text, others have more general backing. Certainly most Greek Cypriots in 1955 sympathized with *Enosis*—union with Greece was an old dream. Obviously the Turkish Cypriots did not but this did not flaw *Enosis* for island Greeks nor did the ideological opposition of the communists or the reluctance of the conservative. Only when the real Greeks in Athens balked at sacrifice did the faithful readjust to a newly perceived reality. Cypriot numbers might not have mattered but Greek interests were paramount. Sometimes there are the numbers as well. The Algerian Moslems appeared ready to win a fair election. The regime shut down the polls. The rebels in Iran represented not only their own, the mullahs and bazaars, but also to some degree most Iranians: not all, not the discarded regime, not the radical Mojahedin, not everyone but certainly many and had the toleration of more. In Ireland or Cyprus or Iran those engaged underground knew what the Algerian Moslems could prove: the people were theirs.

The Design and Determinants of the Dragonworld

Rebels prefer to look into their own hearts to find the nation's desires. With considerable honesty the ideologues of the radicals in Germany felt that the people were misguided, did not know their own interests. The radicals did and so entered into an armed struggle bold with reveal truth. Some underground recognize that they must often strike simply so that the people will notice, be vitalized by the deed, rise if not on schedule at least in time. So in Ireland the Easter Rising was to be if not a blood sacrifice then a call to the legacy of the past. It is always easier to find validation in sacred texts and in history, or arising from martyrs' graves.

If the forces of the conventional are overwhelming, the predilections of the people hardly matter. A thousand flowers can bloom in China only if Peking chooses to sow such seeds. The armed struggle of the Palestinians before and after Gaza-Jericho is still narrowly limited by Israeli power. If authoritarian governments can cancel numbers with force, such force is often not needed, for dissent is a minority matter: a few middle class radicals at the university, clans outside the law, or the irresponsible aspirations of the young and foolish. And democracies that can adapt to real grievance, legitimate demands for radical change, have little to fear from armed rebels. Riotous assembly, violent strikes, symbolic bombs, civic disobedience, a few bank robberies by the fanatic are the muzak of an open society. There are always, often even in the most democratic and accommodating states, national separatists who may seek more than cultural autonomy: Puerto Ricans who seek to assassinate the American President, Austrians in the Italian Alps who want to be on the other side of the border or the more serious armed struggles like that of the Basques who persisted even without Franco as the ideal enemy. Political terrorism comes when democratic societies function poorly or cannot cope with an unrequited nationality problem. An armed struggle elsewhere arises when the center is inefficient, unrepresentative, and insufficiently brutal. Effective brutality can easily discount unpopularity, minority support, genuine grievance, and institutionalized injustice. Dissent is hard to mobilize, to maintain, vulnerable when visible, often ineffectual when hidden.

Mostly, the rebel appears where the people can be swayed, can be claimed as an effective asset, cannot deny them, appear when dissent cannot be swiftly crushed by the brutal. And so popular support matters, as it should, in people's wars but often less than the rebel dreams and more than the center would like. As long as the avowed rebel constituency tolerates the armed struggle, the campaign can continue and all involved can hope that the faith is valid. Even when denied the faithful may persist, their underground difficult to eradicate and their vision constant. As always it is the vision that empowers not the people.

The Armed Struggle

Forsaking all other means except as adjunct, the revolutionary underground exists so that the faithful can pursue power by an armed struggle. Few Dragonworlds arise and none continue without entry onto a campaign, an armed struggle, for this is the imperative of a dream. Mostly revolutionary ecosystems exist only during the struggle. This struggle is so compelling and so dominant that those involved are hardly aware that their world is not their struggle. Their environment is in large part shaped to pursue that struggle. Consequently, most of the involved are apt to ignore more general aspects of their condition. Whatever their ideological pretensions, their strategy has been determined: will triumphs over the assets of injustice. Tactics are imposed: persist and seek escalation. And so operational details, maintenance, propaganda, and the requirements of the day dominate the underground.

The dynamics of the struggle are often, in fact, all but identical to many of those of the ecosystem. In any case the anatomy of the universe of the faithful, the ecosystem of the armed struggle, reflects a variety of more conventional structures and movements. Secrecy, revealed truth, opposition, internal exile are not unique. Cults and radical parties, secret armies and conspiracies, crusades, tribes, extended families and clans of the saved share much. What the faith requires in the underground is good works: the armed struggle pursued assures salvation. The defining characteristic is the thrust for power to change history by violent and unconventional means. In so doing the faithful may find personal salvation but do not seek it, find position, authority, and tangible returns but do not seek them, find a vocation underground but do not want it—and in the end most likely find a broken heart, a ruined life, disappointment regardless of the direction of reality. Service in such a cause arises from perceptions that will not stand too much reality but permits great risk and sacrifice for the general good as defined by the ideal. The Dragonworld is a perceptual structure, an intangible but effective revolutionary ecosystem orchestrates an armed struggle that, however perceived by the involved, is lethal, real.

The Life Cycle

Most institutions, most faiths evolved as does a narrative, a beginning, growth and development, maturity, decline and decay. At times evolution seizes up and nothing happens, tomorrow is like yesterday, one Pharaonic Egyptian century is like the next, and at times process is ended not with decay but with dispatch: a civilization cut down by strange men on horses or a universe closed by the occupying power. A revolutionary underground is a construct of those sharing an incandescent faith. The faith is imagined more general, more pure, more irresistible, and so more a responsibility than any alternative creed. To inspire recourse to violence the faith must be intense. Those who are converted, usually in the beginning by communion with fellow pilgrims, come to the ideal slowly—later recruits move more swiftly to the same conclusions, a few transformed at once but most by exposure and by experience often at a propitious time in their lives and in history. The Tsar is falling or the Imperialists are vulnerable—the change is suddenly there and the volunteer is young and idealistic and ready for a cause. The initial motivation is a matter of conversion that persist not only as a sustaining motivation but persist during combat, may deepen, may adjust but arises form the truth revealed.

These recruits are as one on the necessity for a campaign; and so the underground matures or decays, triumphs or not as a result of an armed struggle. Unlike most institutions longevity is unlikely, stasis difficult—there is Liberty or Death and in the interim persistence against the odds. A Dragonworld is a sometimes thing, quick, bright and bleak, a means to an end, a passage that shapes all but only for a generation—after that the system comes to confusion or power. To persist after defeat requires special conditions, even to persist beyond a revolutionary generation or two requires great and rare flexibility. Mostly revolutionary ecosystems are site specific. Shifts in time and alignments, cultural assumptions, players and local conditions usually eliminate the applicability of the faith. All that is left is memory or a cult, exile, shallow graves. One destroys the regime at the national center or does not, Cuba becomes communist and Guatemala does not, Afghanistan becomes Moslem; but none have general answers. Those who fight with another's manual are apt to fail, anti-imperialism is a collective of special cases, the communist cause not achieved by one guerrilla text.

Most protracted armed struggles exist not only because the involved can adjust to the passage of time but also—far more important—can hide in isolation from real power. This was the case with the Irish, almost a special cases in perseverance, and has been the case with the guerrillas of Columbia, the Karens of Burma, or even the Eritrean nationalists, each beset by coercion badly deployed but still effective. Usually the rebel risks and loses. Sometimes the weak win but rarely can the failures find refugee to persist and so the dream dies. And without a dream there can be no struggle and so no need for haven.

The Beginning

At the beginning much is hidden, even the dream is haltingly grasped. There is, in fact, often a single moment when the founding fathers sit about the table, when a name is chosen or the first operation authorized, but that moment is the culmination of an evolving perception of reality. Revelation comes on the installment plan. Reality is seen as a long train of abuse and grievance, institutionalized as injustice by the system and the center. This is what is wrong—the ideal is what is sought and the delay focuses on the means. The armed struggle is rarely first choice but inevitably all that is left.

Three men may organize for a jihad over tea on an afternoon but they sit at the end of a millennium of confrontation with the infidel, with the West, with corruption and betrayal. The seven men of the Provisional Irish Republican Army Council in 1970 sitting in a cold room planning their campaign were laden with history, with examples and precedent and so with authority to act in history's name as had their predecessors and as would their successors. History as validation, history as inspiration, history as legacy plays an enormous role in the founding of an underground even when the involved seemingly have no history, arise from those denied history. The inchoate masses, the humiliated and wretched find systemic explanation for their condition, find history evolving wrongly and so assuring injustice. It is

The Design and Determinants of the Dragonworld 101

always the past that is at fault and the future that is at stake. Grievance and grace are heavy in the atmosphere of the underground.

Only a few armed struggles have emerged with long institutional roots. Even those parties who go underground often do so with new cadres and new analysis as did the faithful in Malaya in 1948—Mao as seen from Malaya by local Chinese. Mostly even secret parties seek political means, seek power through conversion and co-option, if for no other reason that the odds against violence. In Latin America only the Venezuelan Communist Party opted for revolution. The other communists sought instead to mobilized the masses. Elsewhere communists came to power with the support of orthodox armies in Eastern Europe or in the luggage of the central command flying other more general banners as was the case in Cuba and the African colonies. Only the very bold and the deeply dedicated see in the armed struggle an effective way forward, leave politics behind and go underground.

The faith, whatever faith, any faith, must impose a single strategic option to history's responsibility. The underground exists not to exist, not to keep the faith, not to prepare the people or the proletariat, not to maintain a dream but to allow the faithful to act, to act as soon as possible, to act by recourse to violence, almost certainly unconventional violence, to act in a campaign against the oppression at the center. That dream charges the underground with energy and impetus, imposes direction, transforms the convert, shapes a special world, dangerous, exhilarating, often brutal, exhausting, even addictive. Those who have lived underground can explain much, the routines and risks and responses, but none can explain the impact of the vision—only mystics, those romantically in love, those born again, those absorbed by religious ecstasy experience the power of the transcendental. It is this that makes the Dragonworld special, makes the risks of an armed struggle appealing, makes action certain, persistence likely, and victory possible.

The Central Campaign Structure

Neither simply bold nor simply brave but rather driven, the faithful have a singleness of purpose, a tunnel blindness to options, a mind closed to compromise that assures at the least that the armed struggle can begin. All else has for them failed except the vision. All other means have proven flawed. Some may seek ideological ancestors and some find them but increasingly what all want is action. A few like the Irish republicans need not begin at the beginning, but the others must begin at the beginning.

The ecosystem may pulse, expand and contract, during the campaign, but there is usually but one campaign, one beginning and so one end. At times more than one revolutionary generation emerges during the struggle but the campaign is continuous. Inside the movement time ran differently and more to the point so too the central core. During the struggle the entire underground is shaped solely for the intended purpose becomes one with the armed struggle. If there is an open society, if there are liberated zones, if the underground can remain in touch with the everyday, then the central core can be renewed with reality. If coercion is effective, if secrecy absolutely vital, the central command, the still point of the turning world, may even lose touch with the rest of the galaxy.

Those faithful may continue to spin to their perception of the center but, cut off, the universe changes, becomes one with the avowed constituency, speaks a different language, finds urgency eroding but not the faith. This isolation of the central core is a factor in many urban arenas where to function a secret army must be truly secret. A modern state is quite able in many cases to intern all who might be in a universe of support—intern, isolate, and intimidate the entire galaxy. The *Brigate rosse* could not operate in contemporary Italy as a galaxy of the faithful but only as a command center for terror. Thus the underground was separated from the galaxy by intrusive power. A powerful and brutal state will impose such separation with dispatch but even an open, representative society can isolate armed dissent.

Under most conditions the command-center, the secret army, those engaged in maintaining operations, bleed out into the galaxy of maintenance imperceptibly—the whole underground becoming less intense, less committed, less illuminated in proportion with the distance from service, sacrifice, and risk. Those further out may want to serve but cannot, may aspire to risk and sacrifice but be denied. Whatever the reasons those not at the center of a campaign, those not fully underground, those with other lives and responsibilities, are less touched by the perceptual dynamics. For them life is altered not transformed.

If there is protraction, if command center tires then a new generation arises to persist and are as driven and dedicated as their predecessors. The few at the beginning of the *Brigate rosse* struggle in Italy were long gone by the time the campaign had escalated and even their replacements were mostly gone by the time the campaign dribbled away in arrests, recrimination, and apostasy. Those few who kept their faith and freedom at the end, however, were as one on all crucial matters with those who had been the founding fathers—success or failure, time or circumstances do not matter especially in matters of faith. And for the few who lasted the faith was still valid even and particularly after years of isolation.

Those there in isolation at the center, the IRA Army Council or the Irgun High Command over time endured. They, if anything, became more committed, more ruthless, even more like their culture and clan—the Spanish more Spanish and the communist more communist. Time shapes what is there by excluding the irrelevant: elegance, compromise, toleration, humor, and empathy are not revolutionary virtues, do not assure survival or a victory over history.

Protraction

For there to be a campaign, for the secret army to survive and perhaps prosper the struggle must be protracted. None at the center anticipate victory at once, the *focos* are tiny, the men about the table few, the arms scant. All, however, anticipate escalation and sooner rather than later victory. Since time underground is intense, enormously aging, consumes energy and youth, fills the moment, protraction often appears indeterminable. Those relatively short anti-imperialist campaigns of Grivas or Begin, the march on Havana, the hidden wars of Latin America to the involved are not a matter of years but forever. Existence underground is a struggle, persistence is a struggle, any escalation hard won and often transitory. And most often the campaign evolves into an intense routine of risk and cunning and all the while time slips by. Thus command center in particular and the faithful in general seek to make a virtue of reality. Amid the operational demands and the search for a way forward, many ideologists are apt to find virtue in triumph deferred. Not winning is turned into an asset. The new nation is being born out of struggle. The new people are mobilized through struggle. Time spent underground is not time wasted.

Others find if not a vocation underground then a congenial way of life filled with excitement, motion, power, and risk. Even in prison or exile there is romance and mission, and victory fades as motive. The campaign like those in Columbia and Ireland becomes institutionalized. There may be escalation as well as action but mostly there is simple persistence. One joins a secret army for all the same reasons, pursues the same career, deploys dedication and sacrifice but anticipates no immediate triumph, persists in part because youth has been invested and normal prospects and conventional habits discarded. Some want nothing more—find a vocation and thus are apt to be eliminated if there is victory and the new center finds too many old gunmen to hand. Persistence is permitted, transformation often welcome but always the armed struggle is process not end, the underground is transition to be left behind: years of terror and intensity, habits never quite forgotten, but a beginning not an end to the dream, a necessary passage.

The End Game

The end in theory is easy. The goal imagined, the goal pursued is always victory, absolute, shimmering, transcendental: the dream as real, history in hand. The core, the community of the faithful, the galaxy, the constituency all persist and then history is vindicated, all redeemed. And the intrusion of reality can be shattering when there is no viable prospect except endless protraction. The secret army cannot win and will not lose and the dream does not die. More distressing is the prospect that such protraction must be pursued with declining assets, a shrinking galaxy, in the face of exhaustion and effective coercion. Such a reality may impose priorities on the absolutes that serve as foundation for all armed struggles. *Venceremos* does not leave much room for maneuver between Liberty or Death. Yet, at times the weak can win, not all as in Cuba but twenty-six Irish counties and a Free State government or a Zimbabwe with majority rule but the best land in settler hands. And always the dream was more than tangibles to be counted, more than a treaty could give, more than victory could bring. Thus all end games short of a triumphant march on the palace are difficult—even when there is much to be gained at the end. What had been sought was everything and everything had been risked.

Armed struggles do end because those involved accept, at last, triumph is beyond capacity. Such a realization can be very slow in coming. An investment of great commitment must be made to pay

The Design and Determinants of the Dragonworld 105

despite the odds, despite everything. Time must be on the side of truth. The last "generation" of the German terrorists did not end their campaign until after a sniper killed a government official Detlev Karsten Rohwedder in 1991—years after anyone but their own cared about a futile armed struggle. The few refugees from the Malayan campaign hiding out in the jungle after a generation of obscure isolation finally formally surrendered. The aged and innocent survivors returned to a world transformed and their dream irrelevant. Sometimes those in a secret army simply secretly go home and the war is over without documentation or visible recrimination. More telling, a few armed struggles are transformed by the prospect of power offered, power that may allow history to be changed without recourse to a gun, power promised. The IRA was willing to risk a peace process strategy—but only for seventeen months: after that an effective means forward proved elusive, bombing difficult to initiate and difficult to stop. So seventeen more months produced only a second sullen ceasefire. Sometimes simply waiting on events will bring about real prospects as Hanoi discovered: the Americans went home and Saigon was open. Protraction had worked. The faith had been vindicated. The campaigns ended with new banners over the center.

Mostly the post-war imperialists simply ran down the flag and left, often retaining control by other means and often giving up in the name of democracy or common sense or because times had changed. Thus many undergrounds looked to anti-imperialist struggles as models; but there is no happy end game model for mostly an underground must accept more reality and more swiftly than is comfortable. Mostly the center holds. The campaign ends and so too the underground. When victory is not at hand nor the armed struggle without reward, the prospect of an end game, negotiations for something between liberty and death, can be divisive.

An underground is not shaped for conventional tasks, does not impose conventional patterns or expectations or capacities. A secret army must remain secret to be an effective army. What works by necessity underground is unnecessary and often counterproductive above ground in a conventional world. The gunman is not a soldier whatever his banner says and cannot so fight, lacks the training, habits, and discipline of the professional soldier, is vulnerable to competence openly deployed. If there is chaos or turmoil the gunmen may find an effective military role but if there is real war, even real but irregular war, then the faithful are vulnerable. The Viet Cong above ground at Tet were decimated. In Palestine the Irgun and Stern Group deployed against the Arabs in defense of Israel were ineffectual. The IRA parades only at ceremonies and funerals, remains otherwise by necessity a secret army. And when the secret army emerges to wage war by political means, there again is a lack of skill, lack of practice. The underground is shaped but for one purpose: to wage unconventional war in a hostile environment not to deploy real soldiers and not to ready political cadres.

Unless the command center moves directly into power, one day the NLF engaged in an armed struggle in South Arabia the next day sitting about a table in Aden as the government, there are problems of transformation. If this transformation focuses on the dissection of the dream into the possible and the postponed, internal dissent is assured and almost inevitably internecine violence. Dreams are not divisible. Dreams cannot easily be negotiated. Who can divided justice or allot vengeance? During an armed struggle no one need contemplate such matters: *Venceremos*. During an armed struggle the contemplation of such prospects in fact assures a decline in coherence and capacity—the closer to real power the dream no longer answers everything no longer supplies the energy needed for the campaign.

The end game always reveals an underground in turmoil, values shifting, the unreconcilable certain that their responsibility to history requires absolute victory not merely some power, some part of the agenda. The pragmatic stress their pragmatism, the reality of the arena, the intractable forces of the present, the eventual prospects of accommodation. They offer compromise now and full gratification later. The faithful no longer have the proper answers to what is wrong, what is wanted or what is to be the means. Matters have grown too complicated, too subtle. The very discussion of now and later, the prospects of concession, the whole process inevitably, often swiftly, changes the underground—secrecy begins to go, variations in intensity emerge, options appear. Any risk is no longer acceptable. Some risk more, seek in action salvation and coincidentally a return to campaign days. Some want no risks at all, accept that the armed struggle is no longer a valid means to achieve the dream and become engaged in the conventional world of negotiations, accommodation, trade, and barter than the purity of the dream made impossible during the armed struggle.

Each underground engaged in an end game evolves differently but what is always the same is the transformation imposed by the dream. The dream is not longer identical with the ideal and so begins to dissipate. It is closely held by the few, the most faithful even if they cannot continue the struggle and simultaneously claimed as banner by the pragmatists. Inevitably the ideal is lost as a force that shapes the field, shapes the perceptions of all, enhances life, and authorizes the gun. The ideal is institutionalized as myth in the move to governance as love is by marriage and revelation by the monastery. If an end game must be played, it must be played out over the death of the old dream and the acceptance of a more practical vision, a conventional agenda with conventional rewards. What matters is power, power to achieve what has been promised, power to deploy from the center and so achieve position, prestige, control over history, over what the campaign meant, and what the dream sought. Power can write history and can domesticate the dream.

There may not be full power but enough. There can be enough power so that the African National Congress can effect events and write history. In South Africa there is no longer need for a dream, for an armed struggle—the struggle is over stability and development. There was in 1921 enough power so that the Irish Free State could drive the militant republicans underground, rule twenty-six counties, issue postage stamps, and soon attend the League of Nations. In a divided Ireland the dream did not die. Mostly this is not the case: the underground dies during an end game as surely as most do during a campaign of coercion. An end game is a death watch but with hope that the future will be fair if not as promised. And this promise is another matter, a conventional matter—and rare.

Perception as Reality

Dragonworlds on average have a short life cycle. Most armed struggles fail, most fail swiftly, chopped down at the beginning with the police rounding up the conspirators or the army waiting on the beach. Most that do begin fail to escalate, fail to persist very long, the students arrested in their safe house, the guerrillas of the *focos* starving and tattered, the gunmen in exile or in prison. Those that persist often fail as well, impetus dribbling away as the center holds and the faith fails to attract a new generation. If time can be captured or haven found, some movements persist, irrelevant, really bandits or impotent conspiracies, too costly to ferret out but not too dangerous to tolerate. Many parts of the world have always had bandits, wild zones dangerous to travel, host resident guerrillas with or without valid ideals. Some parts of the world lack even a valid center to hold, offer free fire zones. After 1960 central Africa became a patchwork of provincial powers monitored by a few state systems skimming profits for a rotating elite. Afghanistan or Lebanon or the *altiplano* of the Andes often are beyond reach of convention, haven to dissent and for warlords. Here the emigres from armed struggle may persist, the dream not yet dead but prospects gone. Mostly time kills the faith and the faithful.

Defeated in a rising against the ayatollahs in Iran in 1981, hunted down, executed, imprisoned, dispersed, the Mojahedin of Iran could neither sustain a conspiracy within the country nor seriously propose an armed struggle from abroad. Their contribution to deposing the Shah forgotten, denied, their constituency of the modern middle class too small and too intimidated to support insurrection or even conspiracy, the Mojahedin could exist only in exile. They created their own special world with martyrs and saints, honored families, with a private history and world view, with a private language of analysis, a life embellished with symbols and rituals, with icons of the truth and holidays of the struggle—and always the litany of grievance and promise. There were, as were so many before them, a vital but temporary world in exile, increasingly divorced from the arena, increasingly irrelevant to events but viable as movement, not as a revolutionary underground—the Dragonworld had died in exile.

The Mojahedin, the other exiles meeting in seedy offices, practicing with stolen rifles in the woods, parading out of sight and out of mind, the acolytes for dreams denied, all those once enhanced by the power of the vision were without prospects. For the dream to be appealing, an armed struggle must be a viable and compelling aspiration. If not the faithful become members of a cult, advocates of a secret order, party members in an illegal party, are easier defined as this or that than gunmen for the faith.The single purpose of a Dragonworld is the armed struggle, not to husband a dream but to deploy the faithful, not to seek power by other means, not to memorialize or even cherish, but to serve as haven and stage for a campaign. If there is no campaign, then the underground evolves into other forms with other dynamics. What the Irish did was to make waiting sufficient service, penury and risk kept the dream alive.

The Design and Determinants of the Dragonworld 109

If the campaign is merely reduced to ritual or a bandit career, if the volunteer finds a career in a secret army, then there is no living dream. The dream imposes vocation not career. Those within the dream are not empowered with a permanent revolutionary vocation but rather with a mission. The mission gives a role, an explanation for action. No action, no dream, no Dragonworld. The internal dynamics of such a Dragonworld rest on the perceptions of the revealed truth. The faith may be absolute at the beginning but always is sufficient during a struggle. An end game offers reality in the prospect of power over events that does not come directly from the barrel of a gun. Control over history is not divisible. Power over the people, over events is divisible. A dream is not easy to translate. Thus in an exile cult, in defeat, in the dry Irish years or in latent Iranian underground, the dream may be more pure, more intense more rewarding to some than an unarmed struggle over documents and clauses, constitutions and taxes. Every revolutionary underground has a beginning and an end. To try as Mao did to go back to the future by creating an internal revolution, to find again the intensity of the struggle, was futile. The underground arises from power denied not power disposed. The intensity of the dream is honed on denial and danger. And the dream is the constant underground. There perceptions create a special world that will ordinarily but last out the armed struggle, a transitional world offering haven shaped by desperation and a dream.

The Basic Assumptions Underground

The revolutionary underground is a nexus composed of the assumptions and aspirations of the faithful. None but the convinced may enter and none leave but the apostate, the exhausted, and the triumphant. The battlefield arena for those underground is truly not real but, as many insurgency texts have indicated, within the hearts and minds of people. The key people are not the mass, the nation, the peasants or the workers but those involved at command center, at the heart of the galaxy. These are crucial, the prime players, for these are those most shaped by the dream that shapes their ecosystem and who in turn most shape that world. And that world may defy reason, run to different time and rules but is the reality opposed to the center. Even the smallest underground is lethal, capable of growth. Feeding on the dream, on any congenial evidence, on the day, the faithful are not easily swayed by the mundane, losses or denials or reverses. Lucidity and prudence and right reason often do not sell well underground. Commitment ends only with death, despair, or when revelation destroys the assumptions imposed by the dream.

Undergrounds are real enough, not inventions. All undergrounds are of course underground, covert, illicit, but some are more underground than others, more illicit or more isolated, some are more classical examples—and a few barely fit any pattern. Among other factors what all share is a degree of secrecy imposed by the need for haven during the subversion of the center. And as well as the tangible secrecy of tradecraft, there is a more profound secrecy arising from the hidden perceptions of the involved that shape the underground. Ideals, ideas, assumptions all play an enormous role—and one not amenable to easy analysis. In fact the dynamics of the underground that make possible a campaign are seldom of great interest to those underground. Operations are performed but grievance and the ideal are almost the sole subjects of analytical concern. What can be done is what is done: this one killed or that bomb detonated. This is not so with their opponents for it is the product of the faith—the armed struggle that concerns all at the center, those with guns and those who watch, not the assumptions and conceptions of the faithful. Once truth is revealed, one need not be introspective. Once the campaign has begun skills may be acquired but the faith is the key not analysis, competence, or prospects.

In an underground campaign, size is not relevant nor tangibles, what counts, what makes a Dragonworld is a vision that compels recourse to unconventional operations. Some visions transform the believer into a crusader or a missionary, offer a vocation not simply a goal. Not so for a revolutionary underground: the vision demands action so that time need no longer be spent underground. The Dragonworld is filled with those desperate to leave, to achieved power. They cannot even be bothered to learn their trade, to acquire the craft to win. The volunteers focuses not on the means but rather on the end. If the campaign is protracted and survival possible, the killing trade is inevitably learned as a necessary requirement for service.

Intensity

What is most special about life underground is the enormous intensity of the experience. Once underground bonded and buttressed, dedi-

The Design and Determinants of the Dragonworld

cated to the mission, empowered and moving at a special pace the ecosystem satisfies most, all for a time, most for years, a few become addicted. The central command will continue if not destroyed until minds change. And with changed minds the magnetic core loses all power to attract. Then there is only gain on the installment plan, only self-interest, only private values. None of these have the commanding power of the incandescent ideal: the lure of everything.

To achieve everything there is the campaign, there is the whirl of events, the hope of triumph, the charms of the dangerous edge. Within this world all are dedicated to the ideal life, to total victory, dedicated with a fierce intensity that is beyond easy understanding for the conventional, for those with reasoned agendas, log books to fill, families to support, or payrolls to make. And conventional armies are just as confounded, for their orthodox deployment, their victories in numbers, the reality of casualty lists and suspects imprisoned and weapons captured often have only marginal impact on the will of the faithful. Faith can be blind to logic if not to the police trap or the range of an assault rifle.

Not all underground campaigns display the same characteristics, some are more orthodox, some very small, some protracted. The more conventional the struggle the less intense the exhilaration of those operationally involved: the peasant guerrilla hidden in the jungle is not unlike any rural irregular except in motivation and expectation. The revolutionary is different. The terrorists, the gunman, those who go home at night or can never really go home again, those who appear normal and are not, lead lives of quiet exultation, engaged, at risk. Their life is not so much one of tradecraft employed as a long-running and hidden existence within a dangerous medium, the campaign arena. That arena to the innocent is the everyday but to the hunted is filled with risk and with challenge. Often those so involved are hardly aware of the craft, the constant demands of survival, most have learned on the job and those who do not are seldom about for long.

The nature of the covert is crudely found in thrillers, a landscape of complex procedures, game rules and taught wiles. What is missing is that in the real underground the threat is ubiquitous, exists because of the game. Everything is a risk, going out or staying in, eating with strangers, mailing a letter. Anything can be a danger, an old friend, a sudden raid, a traffic jam. None but the involved can imagine how the presence of a bomb in a car concentrates the mind of the driver, creates an urgency that often defies procedure and order. None but the professional know how easy it is to follow the suspect, to monitor telephone and family, to detain, intimidate, imprison. Nothing is so hard as to do nothing without attracting attention, to appear as bored and normal when filled with exhilaration and purpose. To appear conventional is a constant task. To operate effectively is often a matter of compensating errors. To persist requires audacity, luck, and a natural sense of the aberrant. In the later case a sense of the arena, what is not present, what should not be present, the reality of the dog that does not bark, cannot easily be taught.

The practice of the underground, the survival of the sly, the challenge can be enormously attractive, addictive. Yet, those who find full satisfaction in an armed struggle, in operations, in risk and danger and action, are not assets but suspected. The dream demands not competence but conviction. Those who volunteer for a gun, for adventure, for profit, or for pleasure are usually discarded. The central command looks for the sound volunteer not one who is dedicated first to the action rather than the ideal. The Dragonworld can use only the faithful who accept the armed struggle as necessary, a means not an end. Thus the underground is not only secret and illicit but also transient and so inefficient.

At times, however fast perceptual time runs, the campaign neither collapses nor succeeds. Craft can be learned all but incidentally. Inefficiency can be eroded. Protracted campaigns encourage capacity. Mao could demand accuracy in marksmanship, obedience to orders, and skill in deployment without endangering the faith for his march was long. Most armed struggles are not. Most revolutionary ecosystems are quick and bright, filled with intensity and danger and devotion. And, win or lose, most evaporate into myth or ritual once the armed struggle ends.

For the faithful protraction is not so much an opportunity to acquire the skills needed to persist as an opportunity to allow the dream to institutionalize. A long march then generates almost as an aside technical skills the central core may use but suspect. The need for survival skills requires that the volunteer possess a delicate sense of the environment and the audacity to take risks, the assurance that this time like last the operation will succeed. Every operation occurs at the end of a long train of complex arrangements and adjustments. Every act underground rests on a sensitivity to danger coupled with a willingness to proceed: the contradictions that generate the exhilaration of

The Design and Determinants of the Dragonworld 113

the dangerous edge. It is an existence even more addictive that the killing game and one not unknown to the professional who no matter how far out in the cold remains an agent of orthodoxy. The professional is paid and can go home again while the rebel has no home but the underground.

In that underground home the faithful tend not only to become like each other but often intensify their existing characteristics. The Arabs become more Arab and the Irish more Irish just as each becomes more committed to the dream. The underground answers not to the world, the system, or the decent opinion of all but rather to the needs of the faith and the habits of the parochial. So what is acceptable as an unconventional means, a legitimate atrocity, in one culture may appall everyone but the most faithful. Thus Arab support may be engendered by a massacre that horrifies everyone except the movement just as the murder of Lord Mountbatten appalled both the British and Irish establishment but not those the IRA intended to influence. National and ethnic characteristics may or may not have advantage. Indominitable Irish persistence has paid. And in the Palestine Mandate in 1936, for example, the Arabs fashioned a most effective revolt by relying on traditional means of coercion: riots, arson, rural ambush, sabotage, and terror rather than attempt to fashion an underground army or a revolutionary movement. The introduction of Kikuyu oathing in the Mau Mau rebellion in Kenya in 1952, on the other hand, tended to divide the potential base of support so that the emergency was as much civil war as anti-colonial revolt. In these cases, the armed struggle tended to intensify cultural patterns. Within the IRA the volunteers become more Irish, within the *Brigate rosse* the terrorists were more Italian and the driver of the yellow Mercedes truck in Beirut was an Islamic solution to an Islamic challenge. Even a universalist strategy, language, and agenda cannot erase the local—or can do so at great risk. And this is a risk many rebel leaders take. The communists in Malaya ignored the parochial in the name of Mao—and did so to operational disadvantage. It is more effective to shape the formal structure of an underground to existing social patterns—clan habits or the street gang rather than what Marx suggests or Che assumed.

Whatever the requirements of the theorists and the intrusion of intensified cultural norms, every underground is shaped only for transition between history as imposed by others and history as envisioned by the dream. There underground out of sight the end is all, blinding, demanding, appealing, a great core magnetic that spins the faithful universe on a course of struggle. That struggle is shaped by the dynamics of the revolutionary ecosystem not the other way around. And that system is the outward and visible sign of the faith.

The medium of the galaxy is an intensity of emotion, life on the dangerous edge, enormous risk, and great certainty. All those involved are deeply involved—in conversation perhaps mundane, at home conventional, not visibly so different from the everyday while walking the street or even while driving to kill. Their perceptions are quite different from the everyday for not only do they live at risk but also must assume that the unseen is real. Sooner or later the masses will rise, the rules of history apply, or the spirit triumph. And if it be later, the rebel commander persists, waits to catch the inevitable tide. The campaign, the waiting, reflects the intensity of the underground system.

The gunman has no certificate, no authorization but the dream, no prospect of sanctuary and so must stay out in the cold warmed by the ideal, by the returns of service, by the realization that the present will stop and history will begin. And there is as well and worse the never fully articulated fear of betrayal that bonds those involved in the armed struggle. The truth does not set them free but opens the prospect that others will not be sufficiently redeemed. The truth found can be the truth lost. So not only is there the constant danger from the orthodox but also the constant anxiety of betrayal. Those underground are not free agents, not free to kill in any way, not free of guilt or restraint or moral suasion but are the hunted, cribbed and confined by greater power and the responsibility of revelation.

Some never again will live as intensely. None will escape untouched. Few can ever quite give up the practices and necessary habits of the hunted or the recollection of the unity of purpose and the anxiety of responsibility to revelation and so the high seriousness guaranteed by total commitment. In time the time spent underground will cast long shadows even for those, especially for those who win through to power and conventional prestige, sit in the palace or run for parliament. The great power of exaltation, the prospect of glory justly earned, the life given transcendental meaning is not easy to repeat. The goal of all revolutionaries may be power and power has pleasures and rewards but few will ever live as fully as they did underground—and a few, like Che Guevara, would in the end not live in any other way, could not die in any other way.

Time

For the orthodox the appeal of the underground is misconstrued as a world without laws instead of a closed world of the faith. That the faithful are often dedicated is apparent in their sacrifices if not their skill. That the faithful because of special perceptions lead special lives not simply secret ones is not apparent. For most of the orthodox, time is a constant, may run slower for the Irish than the Germans but is counted by a clock. Time is never a matter of perception. Underground time runs differently driven by commitment, by the intensity of existence, by need, and by the intrusion of the orthodox world. Most important, time is not a single constant but runs rather both too slowly and too quickly and not to conventional expectations.

Armed struggle are fought between those with asymmetrical perceptions as well as asymmetrical assets. In the flux of battle these may fuse, pressure molding a single reality to the vulnerable and desperate under the gun but even then not always. Often time runs quite differently—the whole underground can be alien not just time. The Shi'ite driver of the Mercedes looking out of the cab at the last moment shared nothing with Sergeant Steve Russell but space, one rebel dreamer, Hassan Ali Taibakaran, on the edge of martyrdom, one Marine on the way to the British military hospital at Atkrotiri in Cyprus. One was a volunteer in a Dragonworld, soldier in a secret army, and the other an American Marine keeping a peace no one could find, no one define. The Marine and the Marines knew nothing of the world of Hassan Ali Taibakaran and in a conventional war this would not have mattered. There compelling force would prevail. In Beirut the vision proved more powerful.

Time for the Marine, for most is clock time. Time for most in the West time runs to the clock—but not always. The child finds every day too short while the years stretch out, birthdays far, far apart, while the old discover that their days are long, often tedious, but the years short, speeding away back into an early past, often closer than yesterday. Neither can easily explain their perception to the other. Time is relative and no place more so than underground. How also to explain the wondrous moment before martyrdom, a few seconds as eternity? The prospect of execution not only concentrates the mind but also transforms the passage of the hours. The underground moves on with time perceived as short, intense, filled with terror, every hour long and dangerous, a conception unlike that of the everyday soldier who finds long stretches of boredom and training separates the few career moments of horror. Between mortar rounds the world stops, sprawled on the ground waiting, the seconds both whirl by and stop. Underground the volunteer lives under unseen mortar fire, waits for discovery, the police at the door. And yet time creeps along toward triumph, years do not matter as long as the faith persists. So underground time runs to perception, intense and extended.

Guerrilla campaigns seem to last forever to those underground, for the risks impose concentrated time. And yet there is all the time in the world to wait on victory, a moment just beyond reach. Time is so protracted and so concentrated that clocks are irrelevant but to set bombs and meet friends. The gunman is hardly aware of either cycle of underground time but both are currents in the ecosystem: the great now and then that allows operations and persistence. The mix means time is an intense factor in the dynamics of the underground: and is hardly noticed. Realization occurs only to some and only when the secret world evaporates. Then the intensity of the time underground is realized as some move on to power and others to a regular life. Man's time, of course, is always perceived differently. A peasant may remember clearly events two hundred years in the past as more compelling than the events of the afternoon. The emigrant's child may have no history, no family luggage to carry, no artificial memories, and so learn the nation's past, his past, her past, their past, from a school text. Americans share a history learned rather than one lived while the Irish assume, not always accurately, a heavy historical burden. And anyone knows that pleasure is fleeting, time in pain seems endless but is quickly forgotten, anxiety makes a clock creep.

Time spent underground ages all but a few rapidly, nearly as quickly as time spent in prison. Prison can keep one young and innocent, offer a vocation, or age the impatient and feckless. Prison can be mere time spent for the criminal but is incubator of the future for the revolutionary, although the end product is never certain. Some remain hardened and some evolve into the everyday. What the underground needs is energy and vitality, virtues of the young. Rebellion is a young man's game and ten years is a long revolutionary generation. So for those underground time is both swift and slow. Some few never age, stay the same, but most have only limited capacity for absolute commitment. Their energy cannot be endlessly replenished and they age so

The Design and Determinants of the Dragonworld 117

that normal time begins again. They have not lost the faith but rather the capacity to contribute everything. Their time has come and gone. Perception of time is so alien to right reason that few in the West give it much thought. Clock time is assumed real time but underground time is relative, runs differently and so too not just time but the direction of events, the pace of battle, the winning and losing. Each, all, are all a matter of interpretation. These form the ethos of action, the sphere of combat, the perception of reality. All those who are to enter the Dragonworld as recruits in a dangerous crusade feel an escalating sense of urgency, the need to act, a urgency unknown to the comfortable where one twenty-four hour day is seldom more pressing than the next. A moment in the dark at the top of the stairs or at the side of the road when at risk seems to last forever while progress in the armed struggle has not even the visible certainty of the slowest hourglass. And yet for the faithful the long cycle is as certain as the short is dangerous.

Even the rebel leaders, ever optimistic, soon accept the necessity, even at times the advantage of protraction, the rewards of waiting. Waiting on history is intense not frustrating, waiting is a requirement of the dream just as is action and so time underground is adjusted to both. They are compelled to wait forever until the Americans go home or the center fails to hold, wait until victory. The result is that reason does not seem to factor into many revolutionary strategies. The faithful, if cunning and elusive, ruthless and at times effective, still are not practical.

For the conventional with orthodox weapons and schedules, eager to win on the ground, find a military solution, a technological fix, an end to rebel arrogance, this sullen persistence, the protraction of an armed struggle without the skills or tools to win is beyond easy understanding. The gunman or the bomber takes enormous risks for little apparent gain, kills seemingly simply to kill, seemingly kills because it is habit or pleasure or arises from some unsavory aspiration. The conventional perceive conflict as a process of winning, moving forward, progress past measurable mile posts, time running along to the clock—one year in country or three, four years and out. Underground the volunteer may live in time little different from the orthodox—certainly the terror of action is the same but not the boredom. One is never bored underground. This means that time is an asset in the long run, persistence sufficient until conditions are ripe, until the will wins over the tangible. Time spent waiting is not spent idly but often intensely. So the war stretches out, is attenuated, speeds up in escalation, slows down in frustration, persists in protraction. Everything that occurs underground is shaped to allow the armed struggle to continue.

Time underground, then, for those at command center, for those near the core, for the active and operational runs too fast and too slow, runs within the mind to the march of perception not the clock. There is the long cycle that lets the cause flourish in the hearts of the many, that secretly constructs the nation or mobilizes the masses. And there is gunmen's time when the world stops with a revolver shot. Time underground never runs the same twenty-four hours each day. Time spent in this closed world, time of risk and mission, is time spent with the few and the faithful and often with few others. Even those who go home at night do not necessarily go home to the same intense intimacy. The closer to the central core, the closer the bonds and the more dreadful betrayal. So close to the core those fully involved, fearful of the fragility of the dream as much as the concrete power of the enemy, constantly seek renewal and reassurance. And within all undergrounds time is special: Irish revolutionary time is not that of the Arabs but both run too fast and too slow, both arise from the perceptions of the involved and the intensity of existence.

The orthodox approaches underground reality from the outside. Time is simply one more factor in the asymmetrical dialogue. The center seeks the direction of battle in matters that can be counted and read, the battleground as physical with morale but a factor and psychological consideration irrelevant: as would be the case in a confrontation of main battle tanks or great fleets. In a Dragonwar, however, the great battleship has no visible enemy, must shell the hillsides while the gunmen and the bombers leading lives of quiet desperation plan murder one death at a time. The two barely live in the same time much less on the same battleground.

The Battle Velocity Conceived

Much that occurs in the revolutionary ecosystem is not noted by the involved, the dynamics are neither very visible nor very interesting. What imposes on the aspiration and shapes the agenda is the course of the armed struggle. And in this the underground and the

The Design and Determinants of the Dragonworld 119

center see what they see and by so doing determine often to a great degree what is to be seen, what is real. If the will goes so do the values and priorities of the concerned—and will is in the perception of the involved. If the armed struggle is seen by the threatened as beyond compromise and beyond coercion, then this is soon the case—and the values and priorities of the center are so adjusted. Since much that concerns the center, the state, the government is truly tangible, can be counted if not always weighted as asset or liability, it is within the revolutionary ecosystem without many quantitative assets that, as always, perceptions matter most. And what often matters most is the velocity of the armed struggle: How close is vindication and redemption?

From within their ecosystem the rebels seek to shift the perceptions of others by violent means or by example. The spokesmen talk of actual victory and the optimistic are apt to assume such an eventuality exists—Did not Mao win?—but most accept that history will be changed when the center's perceptions change, when justice is conceded as inevitable, when events imposes a verdict. The proposition that victory will come when the will wins, when minds are concentrated at last and the dream accepted, is not without evidence. The national liberators saw the imperialists sail away, still better armed, better trained, better paid but without desire to remain. Castro saw the regime flee rather than attempt to defend the privileges of the palace. Even those who resist to the end do so without the necessary driving sense of purpose as Mao demonstrated. The dream wins—as well as the guerrilla column.

And since none can measure the will of the faithful or the persistence of the center, none can say for sure the direction of events. The state surrounded by symbols of authority, possessed of legitimacy, holding real assets, an army, a flag, tanks and guns, is inevitably confident and for good reasons for the center usually wins. If for whatever reason, incompetence, oversight, sloth or arrogance, the center does not crush opposition and subversion, then the future is not as certain. The great unknown is the effectiveness of the campaign—the velocity of war. Is the underground merely persisting for lack of option and a refusal of the center to invest in coercion or is there invisible momentum, a tide toward the triumph of the will?

The underground assumes the direction of the great tides of history has been grasped. This assumption may be managed through dialectic analysis or pursuing the heart's desire, relying on the old values or the example of others. In all cases those underground *know* that the faith will win even in a long war, especially in a long war. And the means to this end remains pursuit of the armed struggle. Knowing allows persistence but does not assure progress. Progress is measured both at the central core of the underground and at the center of the state by perception. And both see what they see.

What even the sympathetic are apt to see offers little comfort to the underground. All rebel causes appear hopeless. Before the first shot one finds the same dingy, rented rooms, the same dirty coffee cups, discarded posters, anguish at schism and dissent. Then comes time spent underground with the dedicated wandering ill-fed, ill-maintained, ill-led in the wilds or hunted at the margins in hi-tech societies. Ruthless, bold, persistent, and criminally optimistic, for the few surviving seems triumph enough. Only at the very end is there indication that the dream has led the committed to power. In protracted campaigns, therefore, acceptance that the armed struggle has been futile comes very slowly indeed. The last resort of the committed is the belief in the justice of history. When this goes, so does the capacity to persist and all that is left is inertia and in time exile. The faithful without a faith are no longer players, life is lived without intensity. Some may go into exile but all are already exiled from hope. The dream erodes in exile but not the logic of the original assumptions

End games are often as swift and unexpected as are the initial stages long and inconclusive. This is true when the center does collapse or accommodation becomes viable either through the inevitable rise of pragmatism within those who have watched a generation sacrificed and endless funerals or through an acceptance in agonized reappraisal by the center that an accommodation swiftly reached is better than the cost of persistence or the risk of collapse.

Thus those at the center as well as those underground prefer, often prefer for years, to imagine what they choose to see is what is important. The center refuses to imagine alternatives to the present: How can legitimacy compromise with terror? The threatened seldom can imagine that the gunmen can ever rule the nation. In East Africa as an interned nationalist, Kenyatta was the Force of Darkness and Death. In time, not much time, as a Kenyan President he was welcomed at the Commonwealth as the Grand Old Man of Africa. There was after the end game a shift in perception that served everyone by disguising the complex nature of the Kenyan national struggle and the British

The Design and Determinants of the Dragonworld 121

withdrawal. The Mau Mau ended hunted like game in the wilds and Kenyatta in charge and none would admit that perceptions had mattered, that will had changed history, that power deployed had not returned the rewards assumed by anyone.

In the end the velocity of battle cannot be readily measured by filled prison cells or liberated zones, cannot be discovered by numbers or by rebel claims. An armed struggle is not only asymmetrical but also perceptual: what matters is what is assumed to matter. What happens does, indeed, happen: the prisons are filled, the soldiers dead, the terrorists loose in the capital. What it means, what it all means, is another matter.

What the underground contends is true. The will is what matters once the underground is in place, once the faithful have shaped their own perceptual haven. The will of the underground is opposed not just to the assets of the state but to the will of the state—and that will is more complex, more visible, more general, and less apt to be wishful than those pursuing an armed struggle accept. Yet the will of the center is, too, a perceptual consideration often disguised with the language of law or politics or example. The British are in Northern Ireland in considerable part because such a presence pays the London establishment psychological benefits not declared and often not recognized. And those engaged in an armed struggle often do so under false flags. Some undergrounds seek vengeance not simply a dream. And as well all those involved in the asymmetrical armed struggle are realists so that what matters to everyone, to anyone, to the distant, and especially to those at risk, matters as well: numbers count and weapons and executions and ambushes. What counts *more* is what the involved perceive as the direction and velocity of events, of the armed struggle.

So, again, what matters is often what is assumed to matter. And the underground will to persist is structured not from things but perceptions. And like much else in an armed struggle will is not quantifiable and secondary evidence scant until the process of winning or losing is irreversible. Che was never going to engender a revolution in Bolivia but the Huks in the Philippines, the communists in Malaya, any of the urban guerrilla movements in Latin America appeared for a time terminal threat to systems that might not have staying power. The velocity of the battle indicated for some time underground capacity—the will of the center was seemingly in doubt. The battle velocity that is shaped by the will of the involved is also muted by a practicality. The slogans of the armed struggle are apt to ignore such complexities. In fact the faithful can in time be persuaded to make an accommodation: so the dream is more effective as banner when the cause is pure and the cause pure because accommodation is unlikely. Thus the darkest times are often the most intense and perceptions most shaped to desire and prospects of power imposes tangibles in lieu of ideals.

Summary

Dragonworlds are obviously not the only galaxies of the faithful, not the only systems that support violence. And an underground is not merely perceptions. Reality intrudes all to visibly. Assassins and terrorist are all too real. Those filled with fervor eager for sacrifice, zealots desperate for risk, do not go lightly into harm's way. Whatever the odds, it is a responsibility for each gunmen to survive, to matter, to serve the dream as well as embody it. Yet the dream has transformed the world as imagined—everything is utterly changed. Each such world is different but each generates the energy for the committed, husbands the dream, and shapes the armed struggle. Some Dragonworlds are classics like the Irgun Zvai Leumi's in the Palestine Mandate or the long war of the IRA or the short one of Castro in Cuba. Some centers of armed dissent are not true undergrounds but shadows projected from warlord ambition or habitual bandit habits. Some blend into my traditional resistance to authority, some slowly, some at first, some always. Some defenders—Lebanese Druze militia, Bosnian Serbs, South Sudanese insurgent factions—shape an environment that in part shares much with classic examples. There are all sorts who would seek the assumed legitimacy of the underground, run up the banners of an armed struggle to cover other priorities. Mostly it does not matter for most are transitory, brief flashes or revivals of old habit. Only a few undergrounds persist into history text or case-studies.

The center mostly has the assets and often a legitimate ideal as well. In fact at times, the authorized armies of recognized regimes arise from societies that often seem in the grip of a general dream. Trotsky on his armored train seeking world revolution does not seem so different from the Hezbollah engaged in a jihad against hundreds of years of humiliation by the West. Trotsky represented an unrecognized state and the Hezbollah of Lebanon acted for Islam with Iran as

The Design and Determinants of the Dragonworld 123

sponsor. Both sought to shape the future through force. Soviet Russia and Iran needed no underground but still sought to change history. Some states are frantic to change history and without real assets sponsor subversion. Such authorized galaxies of the zealous may dispatch the faithful illicitly but once there is a haven, a state, a visible font then the ecosystem is different and so, too, the faithful's expectations, agenda, and sensitivity. Failure is not as dangerous, persistence salaried even when the involved yearn to sacrifice. The faith may be as intense but not the risks and the intensity of existence.

What distinguishes a valid underground from other havens, from those supplied by states or geography, or a cult, habit is that the revolutionary faith so shapes reality: makes a place for the rebel heart. That place is no special place but rather the combined perceptions of the faithful. And their perceptions engender a course of action, a community, attitudes and assumptions, largely determine reality and so generates a lethal dialogue with the present, with the past, with the center. The faithful go underground solely to pursue an armed struggle—their only means of redemption.

Such an underground existing to foster an armed struggle must exists in secrecy, is usually illicit, and inevitably unrecognized. Some undergrounds are more secret than others, some more illicit, some even have patrons and recognition but all are alike. The disciples are without easy recourse to the standard and the legitimate. They trust none but their own and often not even those for the truth may be various revealed, heresy anyplace. Still the committed find in their revolutionary ecosystem an inefficient but real haven that makes a struggle possible. Secrecy, a vision beyond reason, a flawed cause, an unresponsive constituency, the great power of legitimacy, and the continuity at the center assure that the struggle if protected by the underground also is likely to be inadequate and always vulnerable to coercion. The central command at the core of the galaxy is responsible for everything, must risk or lose, might in risk lose, must act when acting is not possible, must always surmount enormous operational difficulties. Consequently the pragmatic and sensible are rarely found underground.

There is never really any hope but the faith—not a distant patron, not a visible haven, never a second chance, only the dream. The dream must be made real: Cromwell made a regular army when mere conviction lost in battle but most rebels must make do underground.

So none possessed of the need to impose a new future by an armed struggle are exactly alike. Again those zealots on the edge of the Beirut international airport dispatching the yellow Mercedes may have patron, less need of secrecy, some of the assets of the legitimacy but are still resident in an active underground. An IRA volunteer would understand their perceptions if not their aspirations. Analytical dissection is an intellectual exercise not an exact science. The Holy Jihad fit because the faith and the arena conditions impose similar imperatives on their armed struggle and their perceptions.

Simply operating in secrecy in alien country while engaged in an unconventional war does not give entry into the world of the revolutionary underground. One can emerge from the dark night of the soul only if there has been revelation—visitor's passes are not issued. Going covert is not the same as acting because of revelation. Going underground is for the faithful a process—and undesirable. It is not a career option, not a special operation, not an opportunity to wage war without restraint but the only course remaining. The faithful want power not to be romantic, not to wear berets, not to learn tradecraft or even at times to pursue operational success. Most want to be redeemed. They are never professional but often required by reality to learn their trade—more or less. This is not true with a British SAS team operating in bandit country where professionalism is often frustrated not by the skills of the opposition but by their unconventional, unexpected, and unprofessional actions. It was mostly not true with the Contras in Nicaragua paid in dollars to pursue an irregular war that many supported but few saw as required by history or a compelling faith. Many could have been paid not to pursue the war. This is obviously not true for the driver of the yellow Mercedes or those who die on hunger strike in Irish prisons. It is not true for many operating in unconventional arenas. They feel they have no choice, the dream has chosen them. All are proud to be volunteers and all else follows: craft, campaign, and sacrifice.

What makes an underground unconventional is not what those involved do and how they do it but why they must so act. What they believe compels them to act against the odds. Reason requires that they create a perceptual ecosystem that makes persistence possible despite all—despite the assets of the center, despite the reluctance of the avowed constituency, despite the illogic of the ideal. For the faithful such a course is a clearly defined responsibility with assured re-

The Design and Determinants of the Dragonworld

turns. To serve some learn the trade as best they can. And the more congenial the tactics the more difficult the challenge shaped for the orthodox who employ military tactics and systems against aspirations and make-do hidden underground.

Underground time runs different, place and data are different, reality is read through a glass darkly. What matters is what is seen to matter not what can be counted. And so none can readily tell the course of history but only that every Dragonworld sooner rather than later engenders an armed struggle. The ecosystem makes the struggle possible but at great cost. For what the Dragonworld gives—persistence—it takes away—competence—and so maintains a precarious balance between hope and despair, triumph in the heart and frustration in the field. Inside the underground, where such a balance is a matter of life and death, the involved exist in quiet desperation, exalted and engaged in a struggle for the future, a future all know but none can imagine. There hidden from view if often in plain sight, they live in the long dark night of the struggle and there are tried and usually found wanting. Persistence is usually insufficient. Incompetence leads to defeat. Idealism and aspirations seldom divert history. In the process those underground are shaped and shape a world beyond reason and the everyday and from this Dragonworld from time to time emerge as triumph over history and the odds, become in turn real.

5

The Dynamics of the Armed Struggle

All undergrounds have a communality but not the one that attracts the analytical attention of the center. The conventional are apt to focus on counter-techniques: each unconventional war, protracted or not but always limited, appears to unfold operationally in similar ways. The center, all centers, see a universality in techniques and tactics. The ambush, the hijack, the use of extortion, or the cell system engender primary interest. Violence, visible as well as lethal, is of first concern. Thus, the sharp edge of the struggle is the focus. Interest is generally not in the core of the galaxy but on the ragged and lethal edge of the secret army: the war of the flea or the subversion of the guerrilla. Consequently armed struggles are a nexus of unconventional tactics and techniques in the service of illegitimate aspirations. The ideas driving the violence also attract concern, for the content of the ideology—communism or nationalism—is assumed vital rather than the mere existence of a faith. So for those threatened an armed struggle is a matter of a mix of unconventional violence shaped by special ideas.

Each classical armed struggle, indeed, arises from a core of those driven by a dream. The dream generates energy and urgency. Each armed struggle is the product of the transformation of the energy contained in a denied aspiration into the available means to change the future, a means inevitable shaped by the tangible power of the enemy and maintained by the faith of the committed. The faith is crucial not the techniques. Without the faith there would be no armed struggle. As far as the dynamics of an armed struggle are concerned, the content of the faith is irrelevant only the attraction. For each galaxy, the faith differs and so too many of the tangibles but not the process, not the dynamics, not in Lebanon and not elsewhere. The People's Will of Russia, the *Brigate rosse* of Italy, the Irgun and Fatah, all were struggling with limited means to make a dream real. Irregular tactics were a by-product of the agenda. The content of the tactical nexus was shaped by incapacity not free well, often by reality not the received theory. The armed struggle was a complex of whatever means were available. Elaborations of the dream must offer strategic imperatives, tactical authorization but what was most crucial was the energy granted to the galaxy. Inappropriate strategic assumptions and tactical blunders might even be made good if the driving energy of the faith compensated. Thus the orthodox focused on the nexus of irregular means and the content of the ideology while those underground concentrated on the meaning of events. Neither were especially concerned with the dynamics of the struggle or the sheltering ecosystem.

Each dream converts a core of the most faithful, those most apt to be engaged in the total pursuit of power and so supply the magnetic center of the swirling galaxy of support. And since those most faithful seldom have the necessary assets to wage effective war, in fact, often lack even the means to begin much less protract the struggle, most underground campaigns fail and fail early on providing another triumphant for the center, for history as written. And the state has enormous staying power. States need neither justice nor a dream, need have no decency or mass support. What is needed is coercive power and most states have ample. Democratic states or authoritarian states and even crazy states are rarely really at risk if for different reasons. History is not easily changed and even less so by those whose great asset is found in their perception not in their power. Power, military power, the power of consensus, the power of the center is usually ample to frustrate any underground.

In the beginning every underground appears feeble and futile. Every galaxy begins small with two or three, with those converted by a dream. Converts come within their orbit—seldom are convinced at first exposure but seldom remain uncertain for long. Day by day, a month, perhaps two, and the new volunteers are as one with the originals. They add to the magnetism of the dream. And each dream is both imagined and actual, reflects reality and seeks to interpret it. And reality is the arena, partly tangible, a matter of mountains and swamps and cities, economic systems and private interests, and partly the ideas there prevalent. Each arena is different. Lebanon was and is and will be quite unlike Italy or Central Africa or Ireland. Each arena offers special conditions, a special history, a menu of ideas, indigenous and imported, a litany of grievances and aspirations. Lebanon was especially rich in all of these—dozens of exclusive dreams, different history for everyone, universal ideals, and tribal totems. In contrast to France or the United States or Mexico, the Lebanese were over burdened with those eager to play zero-sum games, those driven by a dream that could be achieved only at the expensive of others, and those without dreams but eager to exploit for more parochial purpose the opportunities of an arena without a compelling and effective center. Most arenas have but a few dreams and the most acceptable aspirations are usually in the hands of the center. Tyranny has staying power and democracy consent and those outside the circle nothing tangible to offer.

At the center generals and economists, ambassadors, analysts, those eager to impose order are apt to imagine the arena as an ordinance map with an addenda of statistical charts, most matters measured in miles, percentage points, pie charts—ethnic origins, oil production, Christians or Moslem, mountains or swamps. Most revolutionaries, the career bandits, the idealists, and the warlord seeking capacity focus on enthusiasm and energy. These are all difficult to quantify and often difficult to find. The ensuing struggle is inevitably asymmetrical.

Elements of the Struggle

Commitment

A map of the underground displays on the dispersion of the faithful, the galaxy. Here the writ of the dream runs. Those within the secret universe see through the lenses of the faith. The policeman sees a mob, the rebel the risen people. The government notes a new factory, more employment, more profit, development, progress; but the gunman finds in these exploitation. The people may vote for the system but the people have no right to do wrong. The faith does not create what one sees: the factory is a factory, the vote took place, the soldier's map reflects the terrain. The faith explains what those tangibles mean. The rebel eye is cold in operational matters and yet always glazed by the ideal. Perceptions are shaped and imposed by the dream. Thus prison can become opportunity not punishment or a flawed operation not a disaster but engine of new martyrs. Always what matters is what matters to the faithful.

Ideology

The faith is organized as ideology, the central beliefs of those who will not tolerate the direction of history, the persistence of imperialism, or the pride of the Great Satan. Mostly, such an ideology arises from a long and complex intellectual history, the texts and interpretations of many, not always local but always addressing crucial local questions. What is pivotal to the ideology is that it be reductive, applicable, available, and satisfying. What each dream does is answer the three vital questions: What is wrong? What is wanted? What must be done? In Ireland what is wrong is the British presence, what is wanted is a Irish Republic, and what is to be done is resort to an armed struggle. The last answer is always the same—often the last choice but the inevitable one—and as for the other questions, the dream varies, Allah is the answer or a free state wanted, imperialism is the enemy and freedom the aim. It is all very simple and very powerful when pared down.

The simple must be able to grasp all the essentials even and especially when reduced to slogans. To work the faith must be pared to the basics that can be sewn on banners or shouted in unison, "Islam is the Answer," "Peace, Land and Bread," "Liberty, Equality and Fraternity." For many *Venceremos* is sufficient: all else will follow. The advocates and analysts may elaborate and extend such basics; but at the center of each armed struggle is a simple truth: We believe and therefore we act—and act Only Thus as the Irgun slogan proclaimed.

Sometimes the village truth may be subsumed by more elegant ideals, sometimes the village truth may be sufficient; but most contemporary galaxies engaged in an armed struggle have a faith and an explanation that shapes the arena and the movement. All ideologies are drawn from experience and the tenor of the times. For much of the time, most undergrounds are dominated by middle class idealists, often out of universities and none far from learning. Undergrounds need organizational men but not to organize revolutionary war but rather to explain the dream. Those good with words and denied recognition often find a congenial arena underground. Those merely denied, those who suffer injustice, those with tangible systemic resentments need not so much arms as explanation, rationale for action. The dream is made tangible with words and words may thus be given far greater weight than deserved and so lead the leadership astray. So many leaders learn war as an aside, an elective, and this like all else underground is a penalty of the ecosystem.

Much of the past, much tradition, much that is popular is neither intellectually elegant and so fashionable nor appealing to those about to go underground. Ideas, new ideas, revelation all are apt to matter more than operational matters, more than the reality of the arena. Those underground seek change not to deploy the old ways. The underground leaders often feel that they have left the pieties, the provinces, the parochial when they read philosophy or sat in seminars. A few do focus on the operational struggle as did Grivas in Cyprus but most prefer to concentrate on the word, on past grievance, on future prospects, on the ideal. All of this means that the dream disappears underground in a flurry of paper and explanation, is often shaped by theory, and moves forward as a constant discussion: unconventional war by conclave and dialogue. Often this dialogue is mistaken by the involved and the innocent as a matter of analysis. The heavy detailed analysis is not shaping an emerging and effective strategy and seldom determines operational matters. The analysis is a ritual required by the faith. Peasants may sit about reading the Red Book and intellectual gunmen deconstructing the state between operations but what they are actually about is burnishing the dream. Academics, analysts, the specialists in counter-insurgency are apt to seek revelation in the content—insight into Mao through the Red Book—but what matters is the energy of the dream not the content.

An enormous investment is made over the course of any struggle, by far the greatest dispersion of energy, in refining and reinforcing the faith. The faith supplies energy and demands it back. The pragmatic inevitably think that such time might more effectively be spent on arms classes or bomb making or even operational details. Few understand the necessity for the time invested in ideological disputation and recitation. This is a case of capital gains, reinvestment. The great asset is the faith not competence. Faith may not move mountains, may not impose operational capacity, but does empower the dedicated. Energy is the one great gift of the dream, the one asset, the foundation of the armed struggle. And this is true whether the movement is a relatively simple nationalist one like the IRA or an intensely ideological complex like those of the Euroterrorists in Italy and Germany. Ideology filters energy into the armed struggle, empowers the leadership, invigorates the faithful, is transformed into an armed struggle.

Analysis does not win wars but the rather that faith makes them possible and so must be cherished and encouraged. Without the faith, packaged as ideology, there is no will and no way into the future. Anyone can bomb but without effective; for to be effective one must bomb for the faith, for appropriate reasons, bomb out of conviction and assurance. The huge truck bomb in Oklahoma City that killed 168 people in 1995 was not a deed shaped by faith but the end product of marginal drifters driven by demons, malice, and ignorance, mass murder under false flags. So too the acts of American presidential assassins who seek personal vindication through violence, so too the murders done for hire or out of habit.

The dream gives meaning to murder, makes possible reasoned atrocity, encourages at times an armed struggle, often dirty but at times merely irregular: but always shaped to the ultimate end, the acquisition of power. Such power is to be deployed to change the future not to ease personal anguish, not to turn a profit or even to ameliorate specific injustice. The dream may seem to others nightmare. The faith may be foolish to the orthodox. The murder of innocents, the bomb in the cafe, the random killing may be treated as criminal, often is criminal. Yet, for the involved all is justified, explained, again and again, made legitimate and necessary. All know what is wrong, what is wanted, and what must be done. So the dream is all. And ideology is merely the means to make the dream available to the galaxy.

The analysis of the dream, the disputation and discussion, the tolling of the litany of grievance, the parsing of the Red Book or the words of the leader, the endless, endless talk is vital to every armed struggle. Without these rituals, the dream would die. Without the dream, murder would simply be murder as it was in Oklahoma City, as it is in murder for hire or drive-by shootings. The dream gives conviction, explanation, legitimacy, urgency, and mission—it does not offer lessons on bomb making or extortion. So ideology is not, as it often seems, endless and sterile analysis of reality, long, long sessions over a text or the Koran's principles, but is rather the fueling of the entire organization for war. Analysis is crucial, ideology establishes both legitimacy and prospects.

What the gunman assumes allows a campaign, encourages a campaign. Gunmen seldom can go on parade. They must fight wars without banners but always with conviction. Many relish the risks—the prospect of coercion is not threat but opportunity. Underground values are never those of the market place or most regular armies. The underground is dangerous, unpleasant, undesirable, but somehow exhilarating. So indicated Abraham Stern, a classic revolutionary, Zionist, martyr, poet, gunman, at the end alone of all his kind, trapped and killed in a small room in Tel Aviv:

In days that are red with carnage and blood
In night that are black with despair
We are the men without names, without kin,
Who forever face terror and death.

And so at the center of conviction, at the center of the galaxy of the committed, is the faith, burnished by analysis and repetition and so adequate to define the arena, to encourage persistence, to reassure, to effect events.

Arena

Any battleground of the armed struggle is in theory peopled mostly by those congenial to the faith—the Irish or the workers, the Christians or the peasants, the nation. The faithful believe their faith should be shared, that those loyal to the center are misguided and so the center inherently fragile, a castle erected on injustice and illusion. Most of the people in the arena are befuddled by the visible power of the state, befuddled by innocence and ignorance and habit; but all are the constituency of the faithful. Thus the IRA assumes that Irish Protestants are Irish because they should be no matter that all of them, all, rich or poor, insist on being British not Irish. The republicans persist, assume also all the Irish, Catholic and Protestants, unionists and nationalists, rich and poor, North and South, in time will be dedicated to the ideals of the republicans—despite two centuries of evidence to the contrary.

Everywhere the few believe they act for the many. Thus guerrillas in Angola or Zaire acted for all Africans not simply those tribal groups that composed the movement. The student terrorists of *Brigate rosse* killed for all the proletariat even when denied by many, in time by most, denied in an Italy not always enthusiastic about either the government or the system.

Recruits

The underground is apt to assume the arena is the same as their constituency: they struggle for all the Irish, each Basques, the whole of Islam, or every Tamil. This avowed constituency may be diluted, misguided but must still supply the recruits. And the recruits represent not the ideal but those attracted to radical change. The IRA struggles for all the Irish but is composed of the unemployed, working-class Northern Catholics and *Brigate rosse* killed for the proletariat but was largely composed of middle-class university students without work experience. The reality of the recruits, however, does not transform the analysis of the faith although it may shape the course of the struggle: in a post-industrial world peasant movements or even Belfast working-class gunmen face serious obstacles. Thus the personnel attracted to a movement, almost always to a movement involved in action, varies from the proclaimed constituency. The convinced and converted seem not to notice that the people are not peasants, only students or Catholic unemployed or Shona. Some national resistance movements reflect a whole people, involve various classes and categories and talents; but many do not.

The Palestinians of Fatah began with a few university graduates but after the traumatic impact of the Israeli victory of 1967, the guerrillas become the national hope—and a fashionable Arab cause. Thus the movement evolved into the Palestine Liberation Orgaization, incorporated, more or less, various guerrilla groups, huge support organizations, established all the agencies of a counter-state even as the armed struggle was coerced and confined so that by 1970 no options were left but terror. By then the Palestine Liberation Organization represented a nation denied, a movement of all sorts, all factions and all classes. Most armed struggles find that their recruits are more special: the middle class or the Chinese peasants, those of one tribe or only some of the Kurds or Lebanese Christians. Each struggle offers a profile: recruits come from certain classes and categories, all of one religion or tribe, none from the other. Most are young and many with tangible grievances of class and caste. Every struggle is at first composed of those who volunteer in the service of a dream.

In recruitment certainly at the beginning the volunteer is not recruited except by the power of the dream and the reality of the grievances. The primary motive is always idealism and so the great attraction is to the few elders who cannot stand injustice and to the many young people who cannot imagine its persistence. The more effective the campaign, the larger the movement, the more intense the struggle, the more conventional the volunteer. And, in fact, the time often arrives where recruits are drafted, support demanded, even in desperation potential recruits kidnapped from a village.

Growth often means dilution of purpose so that the larger the movement the more time must be devoted to instilling the dream at the expense of operational gain. The more orthodox the movement and the more traditional the agenda the more likely volunteers will have conventional motives as well, defend the village or secure vengeance. No gunmen, of course, is entirely unmindful of self or family, gain or loss. As the struggle grows, the underground becomes more conventional. One of the first indications is that the gunmen have salaries, conscription appears and uniforms. The volunteer becomes a member of an irregular army with rules and titles. Mao began with fleeing gunmen and returned with generals and Red Route Armies dragging artillery as well as Marxist categories and commissars. Once there were only the faithful few who would obey orders because ideology insisted as much as because the direction was logical, but after a generation there were only those soldiers who obeyed direction. Orders were orders and given to be obeyed not debated, not weighted as congenial to a textbook theory. The Chinese communists had marched into the underground, their Dragonworld, reinforced the faith, persisted on will, accumulated tangible assets and emerged bit by bit as Red Route Armies deploying armor and divisions. The commissars had uniforms and salaries and career prospects.

Leaders

Those in charge of an armed struggle vary enormously except in their commitment to a dream—and that shapes each in a special way. Unless there is as second generation arising from a protracted campaign, as has been the case in Ireland, the first in conviction are usually at the core of central command at the end. Cyprus, Algeria, Cuba, even China saw those at the beginning there at the triumph.

In Cyprus Colonel George Grivas invented his organization EOKA and his struggle from 1955 to 1960. In the Palestine Mandate Menachem Begin was brought in by a small High Command in 1944 to give coherence and direction to their revolt. Fidel Castro and Guevara and the others of the first boat dominated the Cuban struggle and Mao and the old communists were there when the new communist state was proclaimed. Arafat is in Jericho and Mandela in power. Those who lead share not only primacy and the faith but also generate a seriousness of purpose, a total commitment, great courage. They are often found not so much during operation but rather at the center of the galaxy. The few are determined to persist even at enormous moral cost. Leaders have no time to use a gun for their mission is to see that guns are used. All are apt to be tunnel blind to other options. Once the struggle begins there was and is no other way. And so to stop without total victory, even to negotiate for victory, proves very difficult. Dreams may inspire institutions but are not the same at all. The leadership is shaped to underground needs just as is their secret army and an underground seeks the ideal. So the gunmen is not easily made regular not because violence has become vocation but because dreams are elusive. The leaders cannot easily adapt to an orthodox agenda or even orthodox priorities. Practicality must be ignored by those with no assets.

The leaders early on often appear not just impractical but also marginal to real events: Mario Curcio of *Brigate rosse* talking into the night while a student or Ho Chi Minh running an elevator, eking out a living in Paris with menial work. James Connolly, the Irish Socialist martyr of the 1916 Easter Rising, once could be found passing out smudged leaflets on a street corner in Troy, New York. His Irish colleagues were cranks, school teachers, and bad poets—dreamers. There was Lenin sitting along at his cafe table in Switzerland, Trotsky and Garibaldi in New York, all the exiles in one safe haven or another with few friends transforming dreams into analysis or proclamations. Each zealous but with no prospects. All spent their early years upstairs in back rooms filled with texts and leaflets and hope while tomorrow was planned. Today there are still back rooms filled with conspirators, proclamations, and unpaid bills.

To make tomorrow real takes a singleness of purpose not readily found in the everyday world where prophets receive short shrift. This singleness of purpose is what all at the center of the galaxy of belief share. After that there is little in any leadership profile to fix a type. If all undergrounds are different so too the leaders. Most are middle class because most active in politics are middle class: the Irish Republican movement being an exception. Those most apt to be idealists—students, those unhappy with the present system, the young, those with aspirations beyond expectations—are most likely to be attracted to a faith with compelling answers. Many will share a common course, agitation, prison, conspiracy. Each galaxy has but a few at the center.

Always in the beginning the core is small. Often those with some leisure and exposure to ideas prove resistant to conventional careers. If the post-Mao terrorists in Italy were often university students or professors as were the Germans and the Iranian Mojahedin, the Basques and the Palestinian militants more nearly represented their inchoate nation. Peasant movements are likely to be led by lawyers without clients and organizational men without organizations or recognition, disgruntled school teachers, frustrated clerks, idealistic students. Only in the outback, at the margin, or as a result of special circumstances do those without managerial skills, exposure to ideas, and the habits of the middle class emerge.

Sendero Luminoso, the Shinning Path, in Peru was dominated by Abimael Guzman, the Fourth Prophet of Marx. Before the underground, he was a philosophy professor with his dissertation from a provincial university. Later there would be others at the center, a national organization but until his capture *Sendero Luminoso* was the creation of Abimael Guzman. His word was the sacred text. He was the Fourth Prophet of the Faith, the organizational key, the last word and the first. When he was captured, the movement was shattered—not destroyed for the form was sound, the agenda not unreasonable, the volunteers true. Yet, Guzman was a typical leader, singular of purpose, bold and cunning, sly and elusive and yet charismatic. His arrest was catastrophic, as is always the case with avowedly ideological movements actually driven by charisma, but not fatal. *Sendero Luminoso* persisted even without the Fourth Prophet.

Each leader is different, some conventionally charismatic, others dominant by sacrifice and example, some devoted to action and others to analysis. Begin was a lawyer like Castro and George Habash of the Popular Front for the Liberation of Palestine was a doctor like Che

Guevara. Grivas was an army officer. Most of the Euroterrorist never held gainful employment beyond university stipends and payrolls. The Latin American guerrillas attracted as many priests and nuns at the top as peasants. And any sampling of the leaders would also reveal a variety of personality types—taciturn and warm, egomaniacs and consensus men—and mostly men but never a type. Begin kind in person, rabid in oratory, unyielding in the revolt. Habash aflood with ideas and political analysis and Grivas interested in neither absolutely convinced in his capacity, experience, and prospects. Che made revolutionary war an adventure—outward bound in Bolivia. Islam has generated orators and conspirators, the cunning and the convincing, each different. A few have found a vocation in revolutionary leadership like Arafat, for what choice did he have but to persist. A few leaders seized the opportunity with enthusiasm like Grivas, but most— Arafat or Begin or Curcio—took charge because there was no one else. Each who led was dedicated to a dream and so blind to other options: the armed struggle was the answer to what was to be done.

Volunteers

Those who followed the leader may simply be more of the same, merely a little later underground, more Belfast unemployed, more university students, more Arab middle-class exiles. The faith may reach out to involve those with grievances, those within the proposed constituency—peasants and those in the ghetto, the alienated, the factions and tribes. In Iran the constituency of the radical Mojahedin was "the people" and the enemy the Shah and his system, but the organization came out of the universities not out of the people unlike the followers of Ayatollah Ruhollah Khomeini—the men of the bazaar and the mullahs of the mosques. The Mojahedin were special in Iran, Western, educated, and not unmindful of Islam. Those without Western ambitions, without certificates, the ambitious and denied Muslims, whose enemy was the same as the Mojahedin's, had a goal far more compelling and yet more parochial but far more attractive to the many who from habit and experience easily grasped what was wrong and what was desired. One Iranian underground was led by traditionalists and the other by revolutionaries and neither was an exact fit for the Iranian constituency.

Any armed struggle is almost inevitably unrepresentative of the constituency and so too the leadership and often so too the volunteers, who may be little different from the leaders in class and caste, in residence, education and ambition—and identical in aspiration. The volunteer arrives underground as representative for all but every time is all but isolated, alone in pursuing conviction to the logical end, different from the others left behind in that conviction. Those more traditional movements acquire more traditional recruits, those who follow tribal mores seek to defend their own, are not unmindful of gain. For the archetype armed struggles all the volunteers are driven to aspire to sacrifice and risk, seek to serve. And any movement that offers change and opportunity to serve presents those without qualifications or position a mission and an attractive role. The attraction may bring all sorts.

Most volunteers are simple idealists but a few are criminals out for profit and a few driven beyond reason by the logic of revolution. Some come for adventure, lack of alternatives, the urge of the moment. And some find an arena for private ambitions. Kozo Okomoto with two others members of the Japanese Red Army volunteered for a mission not in Japan, not for Japan but for the Popular Front for the Liberation of Palestine. In the name of revolution the three murdered pilgrims in Lod Airport in Israel. Okomoto explained that he would become one of the stars of Orion. "I believe that some of those we slaughtered have become stars in out sky. The revolution will go on and there will be many more stars." He went on to prison, to mutilated himself, to serve time, to repent, and ultimately to be released: not ever a credit to revolutionary rationality but example of the appeal of the dream, of service, of sacrifice in an imperfect world. Mostly the underground discards the crazed and keeps the idealists.

The central core is inclined to suspect those too enthusiastic about the craft of rebellion. What is wanted is sound volunteers, sound in ideals. In a few cases the result was not quite ideal. Andreas Baader, who gave his name to the German Baader-Meinhof Group, appeared as much a delinquent as revolutionary. The Irish National Liberation Army, a tiny violent splinter, fell into the hands of those attracted to killing, corrupted by power and ultimately destined for criminal careers and internecine feuds. Some volunteers see chaos as potentially advantageous: keep the returns of the bank raid, use the gun to trade in drugs, extort money from the neighborhood. Even the tried and true are at times hardly paragons, may be crude, disobey their parents, drink to excess, or lack sense and sensibility. Some make good gunman although few last if they simply enjoy the power of killing for that endangers all. Aspirations for power and glory are not the motives the leadership seeks but rather sacrifice and service, discipline and the faith. Most volunteers are, indeed, touched in some way by the faith, are not mad, not bad but armed and dangerous.

The key volunteers are always those most attuned to the ideal. This is readily denied by those who see only the slaughter arising from such commitment. The state label all criminals for each is engaged in criminal action: murder, extortion, arson, theft. The center does not recognize the legitimacy of the dream and so seeks to treat all rebellion as criminal, guerrillas as bandits. Murder is simply murder. The targets and victims are one in their scorn of revolutionary ethics and ideals. Norman Tebbits who was injured and his wife paralyzed by the IRA bomb attack on the Grand Hotel in Brighton in 1984 meant to kill Prime Minister Margaret Thatcher saw no redeeming virtue in the IRA volunteers who had been shot down in an operation in Gibraltar. They were not engaged in revolution but like those who set the bomb involved in pointless crime. They were

> ...common thieves, blackmailers and bloody psychopaths. I don't believe the terrorists who died in Gibraltar were Irish patriots in love with Ireland. I believe they were sick, violent criminals, in love only with cruelty, filled with the lust for violence and corrosive hatred for what is decent and good. They sought their gratification, not in the unification of Ireland, and the abolition of an increasingly irrelevant border, but in the enjoyment of the power to kill, maim, terrorize and bereave.[1]

His assessment is denied by many anti-IRA Irish nationalists. The volunteers were, indeed, killers but not mad or psychopathic but idealists descended from Irish history. The volunteers might be ruthless and wrong but they were within a tradition when they deployed physical force against the British. They might perform ghastly acts, criminal acts, acts without prospect of effect, but the shrewd in Ireland—and even in Britain—recognized the motive was a dream. The dream might have become nightmare but the IRA volunteer would not be dissuaded by harsh words. Tebbits, however, realized the necessity of denying the dream and thereby transforming idealists into sick, violent criminals. For him the faith should not count, did not excuse, was a false flag over atrocity. Only victory or history can give such faith general validity.

The Dynamics of the Armed Struggle 141

The volunteer rarely enlists for gain but often for glory, rarely is a criminal or a madman—neither make useful participants in an armed struggle. Each feels the dream within and each so moved feels impelled to act, to volunteer. Not all who volunteer are paragons or without unsavory aims. Those recruited to the underground are mostly like their own, like their contemporaries in the schools or law offices or hill farms except they have been touched by the dream. And this makes them very different even as they look the same, speak with the old accent, go through the outward and visible routine of the everyday. Each, leader and follower, perceives the world anew, radiates possibility and responsibility. All is enhanced with a new understanding. To a greater or lesser degree every volunteer in every armed struggle is, therefore, alike. Yet, each is different for personal reasons, for parochial reasons, different because the arena is different.

The difference between those who lead and those who follow is often a matter of timing: the most faithful were there first. The others volunteered. A protracted struggle may produce new generations of leaders—the IRA has a long history and so many generations. The leaders are more apt to be judged at first on their faith than on their talents, on their commitment to the ideal, on their enterprise in pursuing the prospects of an armed struggle. During that struggle those with operational skills may emerge, may assume leadership positions as in Italy and Spain. Those involved must organize their aspirations, find an appropriate structure, encompass the mission, allot roles. Romance, the smell of cordite, the exhilaration of life on the dangerous edge has little currency in such an endeavor. There are texts for such undertakings. Some texts are fashionable, read and applied intact. Others must be skewed to fit local conditions. Some groups underground simply evolved from the form of the founders, seminar or party or parish, no text consulted. The form is not a formality, for structure can inhibit or encourage and has at times assured persistence or incapacity.

Organization

Underground structures are all quite different despite the requirements of all struggles, but to last each must be appropriate for existing conditions. The movement must be capable of waging an armed struggle in fact not in theory. A charismatic leader, a great idea, keen analysis, money in the bank, and arms in dumps are wasted assets without a movement capable of molding an armed struggle to purpose. What generated ideas or riots or votes no longer will be adequate—nor at times will a cherished text. Some have had few guides. Begin could look only to the Irish and Indian experience and find nothing that fit his organization while others have a library of proposed structures as did the Sandinistas in Nicaragua. Some go by the book, some make do, some do this and claim it is that. What works works.

The organization chart may, indeed, follow form; but control may actually rest with charisma not the central command. The leader may deny a cult of personality but direct the campaign without consultation. The involved may insist on their craft and cunning but operations are often haphazard, succeed because of compensating errors and surprise—do not fit the organization book, the ideological text. Much can be subsumed in interpretation, doing one thing entitled another, but in the end if there is to be continuity those underground must adjust theory to fact and the fact is that little can be conventional. The forms of the unorthodox are, however, limited, limited mostly by the enemy and by the arena. A conspiracy of gunmen can be called a secret army but those in charge had best shape their armed struggle to reality or risk quick ruin at the hands a conventional army.

Those underground do not want to be unconventional. They yearn for convention, orthodoxy. The revolutionary wants above all else to be taken as legitimate, to be legitimate. Thus a universal ideology has appeal. Thus the armed struggle that matches the text is sought. Everything is, if possible, made conventional. The leaders bestow grand titles on gunmen, appear as presidents of liberation movements, command brigades and liberated zones. The volunteers assume that they are soldiers or volunteers, fedayeen or guerrillas, never bandits or just men with revolvers. All want power and thus legitimacy and hence may pay too little heed to the requirements of an effective organization. The organization that appears legitimate has great appeal. The very first hurdle for the movement is the allure of the inappropriate form, sometimes the ideological fashion of the moment or a hurried copy of the successful elsewhere, too often a child of haste not necessity, chosen from the popular models of the day: a liberation front, a *foco*, a secret army, or a vanguard party.

What generally has worked best is a comfortable form that fits not models but habit and history. Then grand titles can be given and

The Dynamics of the Armed Struggle 143

ideological rationalizations found. Since habits and history are everywhere different, so too are effective underground structures. This is even the case when the trends and fashions of the moment are formally accepted. Theory can supply a patina over necessity without eroding the advantages of the parochial: the secret army of ranks and titles may be no more than a traditional rural clan of night riders. Some armed struggles and many of the more traditional unconventional campaigns make only perfunctory use of the ideas of the time—they ride at night and have no command titles; but often even then warlords have pretensions and militia defenders award military titles and scavenge for uniforms.

During the "Emergency" in Kenya the Mau Mau were reduced to hunted bands hiding in the wilds but were led by generals and field marshals often dressed in a muddle of British-issue uniforms displaying various medals, varied regimental buttons and badges, berets and slouch hats decorated with plumes as well as tribal symbols. Yet the Mau Mau was never military, had never been shaped for a guerrilla campaign, and so for much of the "armed struggle" consisted of small bands of Kikuyu hiding from the security forces. These bands were led by field marshals.

In the attacks on Rhodesia the Zimbabwe African nationalists moved across the Zambezi River wearing battle camouflage with brand new boots and carrying assault rifles, looking like text-book guerrillas and so highly visible, easy to hunt, vulnerable and doomed. Those involved mistook the visible form for effective practice. Later the guerrillas wore T-shirts and flip flops, hid their weapons, looked like everyone else, and in time took power. The Mau Mau had neither capacity nor legitimacy only a form imposed by necessity that allowed nothing but persistence. In Zimbabwe the nationalists had legitimacy and capacity that once adjusted not to orthodoxy but the arena achieved, if at great cost, power.

What is needed by the underground is a valid dream. Such dreams require structure adjusted to arena and capacity. Such a structure to be effective may be very parochial, neither fashionable nor orthodox, not military in form nor in appearance, not what Mao deployed or what worked in Algeria. The organization of the dream is often as an elusive a goal as is the appropriate mix of unconventional tactics that imposes a strategic victory for the rules. There is no valid text for effective organization.

Grivas prepared to launch a campaign for the union of Cyprus with Greece—*Enosis*—in 1955 with a small core that he would dominate by age and experience. He was professional officer with experience in the resistance to the German occupation and then against the communists during the Greek civil war. He had directed a special underground group that he used, more rather than less, as model for his EOKA operations in Cyprus. His was a special kind of underground experience: conspiratorial, political, and yet directed with the assumptions and confidence of the regular military. For his Cypriot campaign, he had the support of a few friends and the enormous and often undirected enthusiasm of many, particularly the young Greeks, on the island. He also had as an opponent a British Empire determined to maintain the island for strategic reasons, coupled with a Turkish minority on the island adamantly opposed to *Enosis*. Stubborn, arrogant, Grivas nearly fifty-seven deploying volunteers a generation younger, had no intention of sharing control of the armed struggle. He would draw on the young Greek Orthodox nationalists of the island. He would be in charge. And his EOKA was unintentionally a replica of a Greek family acting as a partisan band, a tiny army with one father at the core and a scattering of older sons as lieutenants who were a generation younger and absolutely loyal to him. He thus directed personally small groups of very young idealistic Greeks eager to sacrifice, eager for orders. And Grivas gave every order. Isolated in hides and digs, his voluminous correspondence and occasional appearance kept the EOKA family operating despite the island limitations, the queasiness in Athens, the huge British garrison, and rising Turkish opposition. There were urban terrorists for awe and a few rural guerrillas for show and no prospect of escalation given the size of the island and the isolation of the struggle.

On paper his EOKA organization appeared little more than the fantasy of an aging martinet: the good colonel controlled everything, every maneuver while hiding in a basement. Yet the organization EOKA represented Greek Cyprus. Grivas was absolutely opposed to a Marxist-Leninist underground theory of cells and masses and classes; instead a partisan campaign was created that could generate noise, fury, outrage, indignation, and horror rather than casualties. The British could hardly be defeated militarily by a few badly armed volunteers but a visible armed struggle could lead to justice, to Cyprus as Greek, to history being adjusted. His armed struggle was, all things considered, slight, a series of violent and bloody incidents, a mix of urban terror and mountain ambushes. The students rioted and abroad spokesmen urged British concession while on the island Archbishop Makarios symbolized overt support until his exile.

And the final result was not as anticipated either by the British or the Greeks and certainly not by EOKA. Grivas won independence not union, an end accepted by Makarios and most of the island Greeks, tolerated by the Turks, and all but welcomed by Athens. The British kept their bases and something of their pride. Many Greeks assured Grivas that he had won—which he knew was not the case: but he had not lost. What the campaign had revealed was that he had against reason and practice found an effective form for the initial strategy. Then the arena changed while Grivas was isolated underground and Makarios in exile, because the government in Athens was willing to deny unity, deny Cyprus, and so to deny Grivas victory. Makarios accepted the inevitable if not Grivas. His EOKA had indicated that there is no proper form for the dream, for an armed struggle. Some forms fit and some do not. There is no key to the key component of revolutionary structure only indicators that forms congenial to the arena, to the capacity and mission and infused with the power of the dream have more prospects but no assurances.

When later with EOKA-B Grivas tried again, time had eroded his prestige and virility. He was a tool of the Athens regime he assumed would—this time—sacrifice for *Enosis*. He chose to ignore that he was in command but not in control. His lieutenants had aged, producing other families, other loyalties. So he could structure only a conspiracy, penetrated and manipulated by Greek intelligence, attractive only to the callow and the old romantics. And such an organization had no effective strategy, ran on nostalgia and the agenda of the Greek colonels. And Grivas died in his hideout, a husk that Athens had used to set Cyprus on a disaster course. The new EOKA-B had reflected not Cypriot reality but a cabal of the marginal lacking not only appropriate form but also a sense of the possible. The dream had died. Grivas had become a myth and his new organization an illusion, not a revolutionary form but a branch office of the Greek central intelligence agency.

On another island another local solution, imposed not selected, worked and again not as intended. What had really happened in Cuba served as organizational text for a generation. Except no potential revolutionary could discover what had really happened in Cuba. After Castro came down out of Oriente Province into power in Havana, Latin American revolutionaries sought to dissect his formula for success. A few Cubans had landed and fled to the hills so they too would land on the target shore and begin the revolution. After several expeditions had been shot up on Caribbean beaches, there was an agonizing reappraisal. What was needed was a carefully prepared rural insurrection spreading out from the wilds to gobble up the center, a means not alien to Mao and seemingly close the Cuban model. So rural guerrilla armies were formed, often by city students, who found peasants difficult to lead into insurgency and Indians unwilling to be led at all. For a decade the Cuban rural experience led throughout Latin America to rural rising, brutal repression, defeat, and despair. No organizational adjustment or tactical novelty worked in the countryside.

Other theorists noted that Castro and Che from their perspective had underestimated the part the urban cadres played and so the text was rewritten, new mini-manuals circulated, and the Latin American urban guerrilla appeared, blossomed, and was slaughtered in dirty wars. Those who despaired of ever finding the proper structure for an effective irregular war, who would not wait and could not start, rationalized a return to the few, a *foco*, a detonation point for a latent revolutionary explosion based on little more than hope and the denial of previous experience. Every such *foco* failed every time, even before Che Guevara's futile death in Bolivia. The workers and peasants seemingly, no matter how miserable or exploited, would not enlist in any form of revolutionary movement so that in one country after another the middle class children died in detention, in the bush, or on the streets, unable to emulate Castro in form or function. The Cuban experience turned out to be very complex, very Cuban, not for export.

The Latin Americans from the first recognized that one of the obstacles to rebel success is inappropriate structure—a form that fits neither the revolutionary strategy nor the nature of the involved. Unconventional crusades are not a matter of a charismatic leader flourishing or a party gone underground. And each unconventional crusade that proves effective, even if not victorious, appears different, some more different than others. Sometimes national characteristics offer organizers special benefits—German efficiency, Japanese consensus, or American individualism—or produce useful variants of a societal model—the Irish church or the Italian family—but with adjustments so special that imitation is difficult unless one becomes Irish or Italian. Many revolutionaries have ignored the special, assumed a universal form for revolution, and sought to organize their form and their volunteers into the model as perceived. The underground is isolated and lonely. So unity with others, with ideological elegance, and with a certain legitimacy of form have been alluring.

Revolutionary fashions may change but tend to share several factors. Usually the organizational model is constructed on a perfect past example, all the rough stitching, thin spots, and patches airbrushed away by the idealists. Lenin came to power in a tumultuous coup not by tracing his own dictum: he grasped power when the opportunity arose without reflection on proper means or appropriate structure. A great many Leninists subsequently preferred the learned learning, the written word, the Collected Works, rather than the actual 1917 events in Russia. And most revolutionaries who enter an armed struggle with the movement organized on an ideal alien model tend to fail. Such idealists are, however, not simple minded ideologues. They want to win. They are going to take their dream into harm's way. They believe that the ideal organization will produce the desired results. It is tried and true, not only the trend of the moment but also the best way to do battle as did the Irish in 1921 or the Irgun in 1946.

Such universal trends for the revolutionary party, the liberation front or a *foco*, the forms of partisan insurrection, all offer the leaders real returns as well as psychic benefits. The Malayan Chinese communists with a scattering of Malays in the 1950s opted for an armed struggle, inspired in large part by their reading of Chinese events—read haltingly, at a distance and without reference to Malayan reality. They created a party as army. More than anything else the centralized control of the leadership, so crucial in a communist party, assured ideological and operational control from the center but at great cost. Initiative was lost, flexibility was lost, and the benefits that might have accrued by a structure reflecting the nature of the Chinese diaspora family was lost and, of course, Malay recruits were lost as well. From the early days of surprise and deep concern, the British center coped, initiating new tactics and techniques, fashioned a political alternative to imperial control, isolated insurrection, and rewarded support. The communist rebels, forced back deeper into the jungle, increasingly unable to appeal to the unstated ambitions and expectations of potential recruits, simply persisted. Neither in victory nor despair did the leadership reflect on their constituency in form or agenda. They preferred the comfortable, the ideal party form that proved inadequate for the task. The pure and politicized died or lost heart, dribbled away, returned home, and were not replaced. The central commanders commanded fewer and fewer and ultimately, almost as an afterthought, the British announced the end of the Malayan Emergency.[2]

In various parts of the world, those who refuse, even subconsciously, to organize in response to tribal factors, the national character and local practice, the existing social forms and values, mostly find their crusade in difficulty. In Italy, whatever the politically correct form, the commissars and columns, *Brigate rosse* ran as an Italian family or a collective of families and so persisted despite all, despite blunder and brutality, despite the isolation from most support, despite the coercion of the state. Italian families work rather better than states or parties or artificial organizational imports. There was a previous Italian example of the power of the provincial. In Italy in 1944–1945 partisan bands unwittingly structured to parallel an extended family, the essential Italian unit of functioning trust and obedience, proved brave, ruthless, dedicated, persistent, brutal, and bold in combat. So, too, did *Brigate rosse*. Imports and theory can be dangerous, lethal. In Italy the communist text was displayed but the faithful family was the key to survival and capacity, a fact largely hidden from all including the volunteers bemused by the endless seminars, the constantly shifting arcane language of ideology, by the babble of analysis. These enhanced the dream but did not shape the form of the organization. Elsewhere ideobabble may be taken as insight and reality. Instead, almost always, the local is a better model.

Revolutionaries, some with the most elegant contemporary ideologies, some with simpler aims, have in the Radfan mountains of South Arabia, the Eritrean highlands of Ethiopia, and the wilds of Afghanistan successfully relied on bandit structures and historical grievances and ambitions. The IRA in Belfast is part lay order and part street gang but in the country a family of night riders. The German Euro-terrorists ran seminars in dormitory apartments: the university as the world. To each his or her own as long as the form fits the function.

In Iran the Shi'ites organized their revolution as if the country were a mosque: a congregation of the devout who on the orders of the mullahs, instructed by the ayatollahs, would sacrifice for the faith. The other Iranian rebels, the Mojahedin, were first-generation university students, torn between the appeal of Marx and the national virtues of Islam, who represented none but themselves, not the country, the bazaar, nor the masses. The Mojahedin found their underground form unable to cope once the mullahs were in power. As conspirators and urban guerrillas, they could persist for a time but recruit no more to their faith when the country was in the hands of Islam, the true faith. Cell structures and Western forms allowed the armed struggle to be turned on the new government but could not maintain that struggle for the Mojahedin form no longer fit the necessary function. It is, in fact, unlikely that any form would have been effective given the conditions of the times, the correlation of forces. Mostly nothing will propel the armed struggle to triumph against the odds, but always a congenial structure shaped to a viable strategy will allow persistence. As for the Mojahedin, they were engaged in a asymmetrical contest with another dream and one shaped in opposition and in power to great effect.

All of those committed to such a dream are united during an armed struggle not only in the cause but also in a movement that allots responsibility and missions. When the faithful are few and the galaxy tight—as was the case with the Stern Group in the Palestine Mandate or the Baader-Meinhof in Germany—everyone does everything. The command center is everything and is often isolated from the constituency. At the beginning, when the beginning is small, most undergrounds appear very similar, a few men in a room, a seminar over coffee, a dozen refugees in the wilds of Oriente Province or exiled in Dar es Salaam. The movement may grow, may like that of the Palestinians move from the cafe to become a counter-state in control of not only conventional assets but also often territory. Liberated zones, safe havens, exile bases can add conventionality to a movement just as can recognized structures within the movement—front groups, legal political parties, authorized publications—but someone must always be in charge.

Command and Control

At the center the shape of the organization usually depends not upon a chart or theory, not on form as announced, but the requirements of the mission and the tangible arena. If theory does not work, then neither does the command center. Every organization has a command center, usually a tiny group, sometimes one man as with Grivas, sometimes a committee. Grivas is example of the meld of factors, the individual as organization, as charismatic leader, as strategist, as the controlling factor even while hidden in a basement. Under pressure from the center without haven or the benefits of any sort of legitimacy, most underground organizations reveal, certainly at first, quite similar solutions to persistence and discover quite similar problems in shifting strategies or even tactics. Command and control like all underground factors is more about persisting than succeeding.

Nearly always there is a central military command with those responsible for operations, intelligence, arms and finance, and propaganda. There may be a more general council or commanders without guns, but the agenda of any center is always the war. Beyond the center with a command staff and a secret army is found the peripheral often unstructured component that maintains the struggle. These may be commanded but are often merely employed. In theory the movement is compartmentalized but in practice those in one cell know others unless the movement has cloned independent centers of resistance (which may happen when coercion is heavy) or has spread beyond easy communication (which often happens in rural insurrections). Much is make-do, old contacts used, old houses reused, old friends coopted, the everyday in service of the dream.

Command and control underground arises from possession of the faith that legitimizes strategy and tactics, orders and response. Whatever the theory may be, control underground is not at all like that experienced within orthodox institutions. No one underground wants to give an order that will not be obeyed, wants to command those who are not fully dedicated. Volunteers must be volunteers. Movements that kill their own, forcefully draft recruits—some of those engaged in Mozambique's civil wars kidnapped whole villages—or maintain their campaign by coercion usually have a short shelf life except in chaos zones. There is in chaos no prospect of legitimacy, no center, no authority in prospect. Elsewhere the underground is shaped by faith and those with none are unwelcome.

To command and control there must be a consensus constantly reinforced by an agreed ideology or by the impact of the charisma of the leadership. The greater the weight of personal charisma the more likely the innocent are to find the movement ideologically committed: in reinforcing the dream with personal appearance and personal example the commander stresses ideological conformity in a way that shapes the movement to his example. In a movement with a historical commitment in the midst of a protracted campaign, the control center can be changed but usually is replaced by those little different in profile or commitment. Thus the chart of the command center may fit that of an underground militia or a national liberation front, a secret army or a communist party but always there is a tiny core, the faith at work, a military command with missions and roles, and finally beyond them is the galaxy, large or small but always in some part representative of the avowed constituency.

While the armed struggle may benefit from the various missions of the movement—politics or publicity—may, indeed, require such action to persist, the core is always the secret army, the volunteers with guns. Commissars must carry guns not just red books, for if the party is to be dominate the leadership must be armed and involved.[3] At times the whole strategy becomes the operational deed when this is all that is available—propaganda is achieved through operations, the Tsar assassinated or a symbol like the World Trade Center bombed. At other times the military although at the core becomes merely a factor if the vital one: thus the IRA after declaring a ceasefire at the end of August 1994 for seventeen months allowed the political leadership of Sinn Fein to control the strategy, the pace of events, the focus of the movement while the secret army stood down and waited. The London bombs of February 1996 indicated that peace negotiations had always contained the option of renewed war and control was back under the direction of the Army Council. Then a second ceasefire in July 1997 allowed Sinn Fein to play the lead role, a transformation accepted sullenly by many volunteers dedicated to physical force and dubious of British intentions.

The decisions concerning such matters are always tightly held within central command. The IRA runs by consensus. The *Brigate rosse* ran by continuing caucus, a pragmatic consensus reached by endless ideological analysis and the influence of a few key operators. Some movements, especially in the Arab Middle East, are shaped by the charisma of the leader—often disguised from all by his seemingly highly ideological explanation of the dream. Most ideological disputes reveal an alternative leader who in turn and in time emerges as the central figure of a schismatic group. Always those most committed are near the center, the still point of the turning world. Always the driving purpose of all is to persist. The limits of physical force may encourage parallel movements and alternative strategies—civil disobedience or overt politics; but in any armed struggle it is force, the secret army, the gunmen, that must be organized and deployed and this is the prime purpose of central command. And such a center is dominated by those with the greatest sense of purpose or with the greatest allure. Undergrounds, even those with the most extensive assets or liberated zones, are run from the top by the few.

The underground commanders, limited in control and hampered by the flaws of their craft, the cost of secrecy, the power of orthodox, must do the best they can. They must deploy what does exist to advantage. And always, always, this means first: protect the dream. Heretics must be sought before operations can be mounted. Volunteers must be instructed on the catechism before they can be armed. Risks cannot be taken, no matter how alluring, if the faith may be damaged. The dream is crucial for it supplies the necessary energy for an armed struggle.

Thus command central far from being free to act, to deploy murder and atrocity, to break all the rules, to control events is cabined and confined. Thriller writers, film producers, the romantic and some anti-insurgency experts and most of those threatened by an armed struggle assume that the unconventional has enormous capacity instead of limited prospects. The underground deployment is an exercise not in resource management, not the shaping of strategy or tactics but rather one of desperation. The unconventional is all that is left and even this is dangerous: there is so little that every investment may be the last. The underground is an exercise of minimal risk, the strategy imposed by scarcity. Those involved have learned the trade on the job. They know, if nothing more, the dangers and fear the price of failure. They must conserve what exists for in persisting the dream is kept alive. Everything possible must be husbanded for tomorrow. Even during escalation an armed struggle always lacks necessities and so deployment is piecemeal, determined by the day, dependent on the parochial.

Maintenance

It is largely the faithful who supply not only the impetus for the armed struggle but also the means to wage the campaign. Those under attack and those within the underground are apt to neglect the enormous cost of maintaining a struggle; for it is neither a romantic exercise nor often coherently organized—nor often visible to most. Everything is difficult underground so that even a telephone call made or a letter sent, a revolver repaired can be a task requiring risk, cunning, and time. Operational flexibility is limited because what seems easy to a professional, to an academic, or analyst is underground an enormous chore. To eat breakfast is a problem and a risk much less planning to fire a shot. Everything is difficult, uncertain, time consuming, and a risk. And the risk imposes an intensity and meaning into the everyday that can be addictive. Although for most who maintain the active service units, their lives appear normal, above board, conventional, the usual round of school and office and football, that is illusion. Every gunman, each volunteer, any messenger or driver lives on the dangerous edge.

Almost all of the faithful in prison or out, in an organization or not, are engaged sooner or later, more or less, in aiding the gunmen: a word said, a newspaper bought, a word not said, a car lent, a door unlocked. An armed struggle can be maintained only by the accumulation of services, some obscure and small and some vital and complex: secret surgery, the acquisition of complex electronic equipment, a nod and a wink, the use of computer time, the purchase of fertilizer, or a note passed on.

After a long protracted struggle, much of this activity is almost natural: the map of the faithful always shifting but known to many, a constant resource, safe houses, safe men and women, skills to co-opt, money to be collected, messages delivered, information received and no tangible cost. All this takes place within the haven of the ecosystem. The galaxy supplies not simply funds and volunteers, a net of supply and distribution, but also medical service, import-export facilities, ideas and prospects, banking and brokerage, contacts abroad, technological and linguistic skills—everything required for a postmodern campaign, even one dominated by those largely without middle class skills, without certificates but with cunning and dedication and the assurance that their own will be faithful to the end. To that end the central command deploys the tangible assets of the underground.

The two compelling factors beyond paucity of assets are the distribution of volunteers and the nature of arena. The center can only command those who exist and deploy them within the existing arena. Thus a movement may have to remain entirely underground, the entire central galaxy hidden. If there are wilds as haven or even liberated areas and no-go zones, the movement armed and visible can exist as irregulars. Journalists can visit, films can be made. If there is no haven, if the underground is deep and closed, then there is no access, no visibility but graffiti on the walls and the fear in the market. Thus in 1999 the rebels in Egypt seeking an Islamic republic must be hidden amid the general population while the irregular armies of Kosovo can be seen on television. Some areas are bandit zones where the guerrillas contest the center, often at times at night or in the rainy seasons, arenas liberated only to be lost with the daylight or the sun. There visuals may be arranged but most armed struggles are covert, require craft for the faith to survive.

Tradecraft

Those who command inevitably find theory no help. They must learn on the job for survival does not come from a text nor seminars. What is needed is an exposure to those already experienced, a chance to learn, audacity, cunning, pragmatism, a sense of the immediate arena, and luck. Even with all else one must have luck to learn, luck to survive, luck to persist and when taken, all will be lost and the next lot forced to learn on the job.

It is easier to persist, gain experience, learn enough of the trade to survive if the arena is congenial. Thus tradecraft is most easily deployed where support is firm regardless of illusions about the constituency. In Italy the *Brigate rosse* could be found in certain cities, rarely in the countryside, and almost never in the south or on the islands. Where pursuit is difficult, the guerrillas can only be deployed out in the wilds, for a different volunteer is needed to survive in the cities. Always the underground must operate underground in secret, operate without experience or often without a useful operational theory. No matter the deployment, there is inevitably much hidden from the state, perhaps the entire movement, certainly much of the secret army, and always the world of the faithful. In this hidden world—the real underground—the movement must learn to operate, learn sufficient craft to survive and do so despite the compelling desire not to need craft, not to be forced to be unconventional.

The faithful are sent into harm's way because they are the most faithful, so far fortunate, and available. And they are sent over and over until lost and new volunteers are sent instead. And so the unconventional has enormous cost—it is impossible even to discuss such costs within a revolutionary ecosystem with ease or dispatch.

Communications

Secrecy assures confusion, limits the lessons of experience, and more than all else clogs communications, not just a matter of tradecraft but a crucial dynamic within the underground. To command and control, to recruit and train, to deploy and persist all the involved must be involved, given orders, encouraged, appraised of reality, reassured, deployed. Yet without secrecy there could be no campaign. And if such a campaign is protracted the habits of secrecy, the necessity for secure communications become second nature, a craft absorbed, natural. Nothing revealing is ever said on the telephone—never give a name, a date, a place, and never explain, never in fact use the telephone at all if possible. Those who are feckless are soon lost. Someone always calls home, makes a note in the address book, takes the wrong plane, drinks too much.

What is true about communications underground is true about all aspects of tradecraft. Those more cunning and more cautious may survive to provide example—no one has time to run a course on tradecraft. Those who know how to do are so engaged. The best that can be done is exposure to those who have survived and are still deployed, still active, never professional for none wants to stay underground, but rather survivors. A dedicated revolutionary must survive to serve—risk is inherent but not to be sought and if at all possible to be evaded. Sloth, exhaustion, the flaws inherent in some cultures make underground communications difficult; but even with the most admirable of habits and experience, the state still has most of the assets, can interdict, often monitor, and at times corrupt communications. Everything is difficult underground and everything is learned at great cost. The prisons are filled with those who failed their final exam, often failed their first test.

The state seeks to discover what is being done and said and ordered, taps phones, follows suspects, reads the mails, plants sensors, buys informers, probes and analyzes. Underground communications must be downgraded. Inevitably the messages move more slowly, more rarely, and more crudely: but communication is maintained. Tradecraft is learned, orders are sent, encouragement dispatched, prisoners contacted, and the Dragonworld kept informed. Underground communication is always faulty, confused and apt to error and maintained at great cost. To a degree those movements relying on consensus like the IRA have an advantage over those shaped by charisma or even by an ideological imperative where decisions must be constantly reinforced. With consensus the faithful know what to believe and most of the time what to do without exhortation or explanation: both will be needed if there is to be a new direction, new leadership, change—and so change is unwelcome underground.

In time those underground hardly recognize how complex a secret life is, how much the clandestine world charges for cover. Operational commanders may have to wait for an order to be delivered by three men, a bike, and a small boy. Messages to far places take a long time. Agents sent abroad cannot report back to headquarters when they find the police waiting for them at the airport or their local telephone contact number a digit shy. Operations take far longer than the conventional imagine. Many are aborted—the word not passed, the gunman left without a gun, the target early or the target late, the car with a bomb but no detonator. Everything must be put in place again and again. Perhaps equally important the nature of the covert world is that narrowed communications also narrow the essential personal contact that eases the workings of the underground. Thus fearful of display or discovery, the responsible send short notes, make cryptic telephone calls, rely on a nod or common sense and so lose personal touch.

Even with consensus contact is crucial to the faith as well as to operations. Discussion erodes difference, raises morale, encourages all. Every man on the run enhances each safe house as living presence of the struggle and each hour spent over cold tea debating old grievances and new prospects is capital invested for the future. Thus the difficulty of communicating hampers operations, hampers the oversight functions of command central, hampers the reinforcement of the faith that contact assures. Friendship need not, often does not, flow from social contact but rather renewed dedication. The faithful often do not find each other attractive or interesting, only sound and trustworthy, fellow crusaders, all different and all the same.

While the general tendency is to focus down on more personal and primitive means of communication that are more difficult to monitor some movements have access to skills or resources that allow recourse to more advanced technology. Thus Fatah employed fax contacts when the fax was novel. The use of computers, portable telephones and buzzers adjusted for security, piggy-backing on the communications of others, adapting, adjusting, even stealing from the state are standard. The rural guerrilla movement Revolutionary Armed Forces of Columbia, FARC, had a website operated out of Mexico City and still sends E-mail. Now many movements do. There are always ways to communicate with the underground, within the underground and even the most limited movement can learn.

The IRA in Belfast—a working class movement with little access to technological assets—has managed from time to time to monitor the communications of the security forces. At one stage, the police discovered in a hidden roof-space an IRA command post filled with radios, unscrambling equipment, sophisticated monitors, military-style transmitters, position-fixing device, telephone taps routed through the British Telecom network from sophisticated leads—an entire complex of monitoring equipment as effective as the state had to deploy. The IRA was way ahead of the security forces and had been so for years. And thus IRA communications could be secure as well as the intentions and capacity of the security forces monitored. The state's assumed advantage was for once reversed.

Mostly the IRA and the other undergrounds rely not on hi-tech but on low-tech means, rely on observation, routine, and the newspapers. to penetrate the communications—or the intentions of the state. This is often ample. No hi-tech skills were needed to find Prime Minister Thatcher's schedule or Lord Mountbatten's vacation plans. And often the communications of the state are as vulnerable as those of the IRA difficult. Some movements, the Euroterrorists or Latin American urban revolutionaries, have middle class talents unlike the IRA; but what none have is open communications easily achieved. All must scramble to communicate, to exchange views, to give commands, to acquire intelligence.

Intelligence

At times targets are chosen for ideological reasons, for symbolic value or for vengeance sake, but most operations occur because they are possible. What makes them possible is intelligence, data that comes to the attention of command central often rather than data that is sought and found. The militants and volunteers note prospects and pass along the word—or act themselves against unlocked gates, routine patrols, unguarded armories, against a vast concrete structure at the Beirut airport patrolled but not guarded.

Reliance on the natural, the comfortable, the everyday pays ready dividends and nowhere is this more true that with intelligence. Underground intelligence like much else is the product of the committed, the universe. Always the faithful rely on the faithful. Each of those who support the struggle is apt to pass along observations or prospects. Data may filter up without real system but often with great effect. Intelligence officers at command central from time to time direct specific operations, especially against particular targets; but in a very real sense most volunteers and most of the faithful are part-time intelligence agents, seeking capacities and vulnerabilities and options and simply passing along the results to those in control. Secrecy may hamper the passage, lack of tradecraft may limit the material. Some useful data is lost in the shuffle, passed to the wrong man, forgotten, misinterpreted. In the end much gets through to the center.

Then it is generally used, often promptly, or discarded, lost, forgotten—only rarely as in the case of Mountbatten is a watching brief maintained. Underground the present consumes all. A secret army keeps scanty files so that not only is almost all intelligence closely held—those in the know knew and when arrested all that was lost—but also there is no archive, no resources to tap. Yet much less is needed for most operations than imagined: the appeal of a target and then access and escape. And patently what is needed has often been found even if the involved did not entirely escape. After the Mountbatten boat was bombed, there were two arrests made the same day. One conviction resulted: a poor return for the loss of a royal. After the World Trade Center—a symbolic target with a parking garage for the truck bomb—the authorities rolled up the scattered conspirators, all but one caught, most without craft or cunning but with the deed done. It is easy both to discount the clumsy and often ineffectual system of collecting bits and pieces as it is to overestimate the capacity of the underground to know all. Operationally all was not needed to bomb Thatcher, to bring down this airliner or that, to kill Sadat or kidnap Moro. Often only a little did well enough.

In the underground intelligence sought and intelligence received intermingle with command central hardly aware of the difference. Both kinds of data arise from the galaxy—so that the more faithful to serve the cause the more data is apt to be available. If the movement is very grand, then intelligence will grown more orthodox, open branch offices, keep records, rely on full-time directors. If the movement is small, every operational volunteer is also an intelligence agent. No matter how small a movement, how limited the underground, most operate within arenas filled with targets. There are always ample vulnerabilities for the enterprising.

What the underground has to hand is not so much diligence and enterprise as enormous amounts of time and absolute dedication. The soldier may go on leave but not the gunman. The soldier may depend on the system but the gunman is the system. And spectacular operations succeed because no one thought that anyone would blow up the World Trade Center or that Moslems would murder an Egyptian president who prayed so often that his forehead revealed the bump of piety. None in London imagined that the bumbling Irish would implant a bomb with a long-running elegant electric timer into the wall of a hotel months before the prime minister was to visit. In America two marginal men detonated a bomb that killed 168 people in Oklahoma City: for them intelligence was a matter of a few manuals and the name of the building. Nothing else was necessary, not skill, not a dream, not practice, nothing except the fantasies of conspiracy.

The first time is often easy, very easy with the intelligence available in an open society. When matters are made more difficult then the underground must look elsewhere for vulnerabilities and opportunities—rely on their own, on luck, on public sources. Like orthodox intelligence forces the underground would like technological quick fixes but mostly depends on means familiar for centuries. And the old ways are often not necessarily the best but are often sure, work, will do, and this is what matters. Any intelligence officer would like to find penetration agents, moles in proper places, insight into the heart of the state. The rebel may or may not have monitoring devices, use E-mail or read the president's mail but is regularly beneficiary of those converted to the dream, secret sharers. There has often been a spy in the castle, a policeman with a rebel heart, a clerk with access to the needed files.

The innocent and some professionals are apt to exaggerate both the efficiency of the underground intelligence net and the assumed presence of hordes of hidden agents. Most data is simply acquired, easily available for any to discover. The facts are there for those who will invest time and take a risk. The shoeshine boy outside the presidential palace sooner or later will find out something useful. Often someone finds a vulnerability without looking—the door unlocked or the general on the steps of his club each Friday. Much of this drifts into command center. And much is lost along the way. Intelligence underground is like all else: difficult, the solutions temporary, assets hard to find, the triumphs often a matter of luck and patience and a few keen men or women.

The one aspect of intelligence that absorbs an enormous amount of energy and is only rarely structured by command center is that of the informer—the apostate within the core of the galaxy. Each convert is assumed pure and yet capable of betrayal. Thus any underground, every ecosystem is permeated by a free-flowing paranoia: a heretic lurks, has done so, could do so again. Flawed operations are parsed for betrayal, the cat walked back. A failure to comply with the routines of ideology, too great an interest in the personal, oddities of action or nothing much at all, each can serve to engender suspicion. Each ecosystem is filled with free-floating, subliminal suspicion, some are paranoid, all are vulnerable to apostasy. Everyone is always suspicious and so each is always engaged in counter-intelligence. Some movements have a form for this, some create one when an informer is found or suspected but all invest time and trouble in keeping the faith. Thus a state has counter-espionage to prevent effective intelligence penetration while the underground has a dreadful fear of the corruption of the dream. The revealed truth can be betrayed—is certain to be betrayed, certainly by those seemingly most faithful. Everyone is faithful and so underground everyone is suspect. Such paranoia, shaped to habits of the mind as much as to formal structure, safeguard the galaxy's core, reinforce the dream, and involve all in a world closed to others that makes possible a campaign to redeem history.

Campaigns

The pursuit of the future with an armed struggle—the campaign itself—is shaped by time and place, by a special history, and by the particulars of the moment. Each is different but all seem to share certain characteristics or differ in special ways. Various analysts have assigned categories by motivation or ideology, by the arena, rural or urban, by class structure or by patronage. Each has a history, a beginning, a span of struggle, an end: often this can be a matter of hours and sometimes decades.

As a campaign evolves so too does the movement, recruitment may grow or contract, there may be a shift from city to country, arms may become scarce or available, times change, the center changes, allies come and go—if, and always if, the struggle can be protracted. Most campaigns are swift and futile. Castro was captured during his first attempt. The second attempt left a dozen stranded volunteers in Oriente Province without prospects but a will to persist. The campaign incorporated other rural centers of resistance and then urban action and national support culminated in a rush to Havana when the unpopular and increasingly ineffectual center collapsed. The exact formula was never clear. No campaign is quite like any other—all rural communist insurgencies are different from Arab urban conspiracies but both share much. Differences in a campaign are easy, each a singular case-study bounded in time and place as the Latin Americans discovered with Cuba. Similarities are more elusive.

All campaigns even those narrowed to personal terror have a strategy declared if not always employed by command center. Most assume that the will should in time erode actual power—that history is on the side of revolutionary justice. All are willing to begin with too few assets and all—communists, nationalists, tribes or clans, Tamils or Moslems—must persist until the will triumphs. A few command centers have found that will does at times overpower the center—this was especially true for some of the armed struggles against overt imperialism but not all. The British opted out when challenged—the armed struggle in Palestine or Cyprus, South Arabia and even Kenya was not unduly protracted. The Portuguese on the other hand held on because in Lisbon the will to do so was not challenged for a generation. In Algeria the French won the military campaign in the Battle of Algiers in 1957 but withdrew in 1962—the will to pursue coercion as a policy did not evoke a national consensus or offer great gain. In Indochina the communist-nationalists took nearly a decade to erode French will. Laos and Cambodia became arenas for others struggles complicated by external intervention. Then the second war of liberation in the South consumed another generation at enormous cost. The era of the twilight empires was special. Mostly even protracted noncolonial campaigns fail. Those campaigns that have not in Eritrea or Cuba or China are special. All campaigns seem to fail in the same way—the center snuffs out command center or drives the faithful into exile—but the few succeed in special ways. To find a special way is to shape a strategy and most strategies cannot be greatly adjusted. The campaign is simply continued.

Most campaigns, even the most extended, last but one revolutionary generation. Such an irregular conflict to all the involved may seem protracted and this is especially true to those on the narrow edge underground but an end does come. A few rural insurgencies simply persist as a "bandit" problem beyond reach of the center. In Latin America there have been repeated failure but no end of rebels seek a shining path into the future. The wars of Asia go on and on: the governments still cannot quite pacify northern Burma or East Timor or all the ethnic rebels of the Philippines or the guerrillas at the margins of Cambodia. And the most protracted of all struggles has been that of the militant Irish nationalists. The republican movement has pursued a united Ireland for two centuries deploying physical force that rarely can exact concession and never triumph. In an effort to escalate, the campaign has been extended to deploy against any perceived vulnerability: the Fenians invaded Canada more than once without effect and the Provisional IRA has attacked the British Army of the Rhine, sought targets in Gibraltar, sent letter bombs to Washington, exploited any vulnerability.

While few armed struggles may have as extensive a history as that of the IRA, the Basques have persisted for over a generation and the Kurds for as long. In Columbia the Marxist guerrillas still soldiered on after a campaign of over thirty years, able to escalate every time the government in Bogota is distracted. Their campaign is a form of institutionalized insurrection without real hope of victory but with wages to be won and profits to be made and pride in place. In contrast most of the Latin American *focos* were snuffed as isolated infections without constituency or too deep underground in the outback to attract support or notice. In fact a campaign may not even be necessary if the center is too vulnerable as is often the case in Africa. No one bothers with the outback but goes for the palace as did Castro in his first Cuban foray. Then the strategy is simple: take the center. Mobutu lost slowly in Zaire because the guerrillas gradually realized his vulnerabilities and their luck but they still had to walk to the capital to replace him at the center. Their long march ended shortly with a little help from their friends and Mobutu's enemies: no strategy necessary for such a campaign. A year later a replay was frustrated by foreign intervention.

Strategy does not so much determine the course of the campaign as reveal the aspirations of the involved. Ideology often has most of the

The Dynamics of the Armed Struggle 163

answers but the arena is often immune to the faith. The agenda of the faith, however, determines the original deployment. Whatever shape the campaign intends, whatever form it actually takes, the center is apt to react in similar ways, close down toleration, drive the conspiracy and the gunmen from the streets, push the guerrillas into the wilds, and deny the underground all legitimacy. At times, at most times this is sufficient. When it is not, the underground persists if rarely to victory. And at very special times the center does not hold but the rebel campaign does not lead to triumph. There can be a collapse of all order—a Liberia prey to wandering warlords or Sierra Leone with the palace surrounded by armed bands without ambitions beyond loot and excitement. In Lebanon by the time the Marines arrived, there were many campaigns and many centers and much chaos. There was a need for defenders and opportunities for the ambitious.

For there to be a real campaign, an armed struggle, there must be order, coherence, structure underground, coherent opposition by the nation-state and so a lethal dialogue, always the same and always different. This is the campaign. This is rarely the campaign as planned, for strategic intentions are usually lost in the confusion and alarm of reality. If not, the campaign may actually unfold as intended but not with the result assumed: Che did just as planned and was killed in the process, the yellow Mercedes went off as planned and the Marines withdrew but Islam the next year, the next decade, still did not provide all the answers. The Great Satan proved more difficult than imagined but remained a convenient enemy. So those still determined deployed symbolic operations, bombed the World Trade Center, bombed American barracks is Saudi Arabia, assured vengeance if not power. And that diffuse, protracted campaign pursued by a variety of competing groups and individuals without central direction but compelling purpose continues, fundamental, lethal, arising from historical legitimacy and a present agenda.

Once underway an armed struggle changes the arena. Although eager to escalate, most movements must focus instead on persisting and many on coping with declining capacity. Thus steady state campaigns tend to occur on the margins of society or as institutionalized forms of protest. For a generation the correlation of forces in Northern Ireland has encouraged stalemate: the IRA can persist by adjusting to rising security competence but cannot escalate to the point that London would feel compelled to respond with serious concessions. Much the same has been the case with the Basques in Spain. Concession may, indeed, come as in Eritrea where the underground slowly constructed a military force that proved capable winning a conventional war against the center at Addis Ababa, putting the rebels in control of the whole country not just their own Eritrea. And negotiotions may open as in Spain and Ireland and South Africa.

Some central governments simply cannot afford to destroy the opposition and must suffer lost of control at the edges. In Zaire those edges proved lethal when encouraged by Mobutu's enemies and the rot at the center. Some limited wars are a perpetual stalemate despite shifts in fortune, gains, and losses as in the case of Angola. Some undergrounds as in Serbia and Croatia turn from defenders into regulars and some regulars as in central Africa become wandering guerrillas without banners. There are all sorts of campaigns. Mao's may be the classic. The FLN in Algeria or the Viet Cong in Vietnam are in every text. Grivas and Begin are apt to be forgotten and those seeking to defend the center parse instead the example of the Euroterrorists and the failed *focos*. There are ample variants for all, most beyond texts. Who recalls the Bretons, pays any attention to the Corsicans, much less remembers the violent anagrams of liberation, ZAPU or COREMO, MIR or FALN?

Underground campaigns are largely kept underground, the secret army stays secret—and so there is an element of mystery and romance to go with the ferocity and visible brutality. It is difficult to explain to the threatened or the media, or those eager to exaggerate the provocation of the underground, just how desperate, haphazard, inelegant, and clumsy are most campaigns. Most times, most underground stumble to success because of the incompetence of the state. Mao faced a corrupt government weakened by warlords. In Cuba Castro had a brutal, greedy, and inefficient target at the palace in Havana. The anti-imperialist liberation movements were opposed to empires past their due date, imperial centers lacking the will to persist. Even the persistent who have learned their trade underground like the Vietnamese tend to fail—as did the Viet Cong with conventional escalation of the Tet offensive—as much as succeed—as did the Vietnamese communists when the United States withdrew and changed the balance of the arena. Most underground must persist for years and years at great cost, learning much less than imagined, hoping for a shift in the correlation of forces, just like the Vietnamese.

One of the great contemporary classics has been the struggle for Palestine, a multifarious campaign that evolved through every tactical and strategic option but despair. The fedayeen tried guerrilla-revolution and terror, built real armies and encouraged children to throw stones. Gunmen were deployed and diplomats, publicists, hijackers, lawyers and killers, and elegant poets. And when with the Jericho-Gaza agreement arising from the endgame seemed the final act, a new fundamentalist Islamic generation emerged to persist at least in part with physical force. Terrorism returned with martyr-bombers and a campaign focused not on victory but denying accommodation, shifting the nature of the arena so that Zion would still not be secure, never be safe.

In tactical matters undergrounds can be spectacular and so are often judged effective. Suicide bombs in Tel Aviv and Jerusalem are spectacular, technically awesome and tactically effective. With all such spectaculars, there is awe and amazement but too often the effect is achieved not with care and preparation but because the unconventional always surprises the orthodox. Who would have imagined Palestinians as walking bombs? The IRA mortar-bombed Downing Street. Moro's kidnapping shocked a nation. The World Trade Center bomb horrified as did each jet airliner destroyed by a bomb. Yet, only minimum tradecraft is needed and often that is not deployed. The IRA did not hit their target at Downing Street. The *Brigate rosse* snatched Moro despite dubious planning and operational error. The Palestinians of Hamas still kill themselves with premature explosives, are swept up by Israeli skill—and in the end operate without tangible victories beyond the symbolic persistence that requires limited skills and great dedication. To acquire the skills of the underground requires longevity and luck and natural talent. Most underground never have any luck, have limited talent and so a short time active.

Friends and Allies

Underground the flaws of craft are often ignored in a desperate search for tangible assets to pursue the struggle. Those underground want what the state has—a recognized banners, power, guns and technological assets, money and friends. Obviously any campaign can run more smoothly with aid and comfort and so soon each underground seeks out patrons and advocates, even allies. The first sweep is through the far reaches of the galaxy and out into the diaspora: overseas Chinese, Jews for Israel, Irish-Americans, Armenians or Moslems or Tamils. These ought in theory and often do in practice become one with the galaxy. Others may be attracted because of a real or imagined ideological identity: communist states, free African nations, Moslem governments, or enemies of empires. Thus for the underground there is a search not only for one's own friends and those who agree with the premise of the struggle but also the enemies of the enemy. The IRA found as anti-imperialists they could become recipient of Colonel el-Gaddafi's regime. Nearly always the center has enemies and the underground friends. The more isolated the underground, EOKA on Cyprus or the Irgun in the Palestine Mandate, the more difficult it is to transform friends into comfort. And the easier such comfort, the more open patronage, the guerrillas fitted out by this state or given haven by that government, the more difficult underground independence. Patrons if possible like control, a return for their investment. And there is a cost to exile havens. In exile there is tendency for the underground in part and often in strategy to become more overt, open offices, publish newspapers, have bank accounts, and finally become a government-in-exile. All this risks transforming the faithful into conventional forms and so at worst a client of the orthodox and at best at the expense of pursuing the armed struggle as the major axis of the action. Generally those without assets are willing to take the risk. Since most undergrounds have proven losers, rarely do patrons overly invest, seek no more than to meddle. Still, very little interference can cause serious problems underground and an exile focus can seriously hamper an armed struggle. With and usually without friends, the underground must cope against formidable odds and so any help is apt to be welcome. Accepting too much aid and comfort is a risk gladly taken. All is a risk underground anyway, even friends.

The Campaign as Armed Struggle

With an entrenched enemy in control of history and no assets but the dream, any campaign at all is a triumph. Many all but sleepwalk into a struggle bit by bit until so deeply underground, so committed, they cannot return to the everyday. A few have a text for revolution. None have much experience underground and those, like Grivas, who do often have the wrong kind, draw the wrong lessons. Each case is different and so each campaign. Each that persists has had a beginning and most will come to a bad end.

Mostly campaigns begin with the deployment of what is available as central command seeks to exploit surprise and encourage the faint of heart. Yet there are all sorts of beginning just as there are all sorts of campaigns. Grivas in Cyprus had EOKA detonate a few bombs in 1955 to attract attention, to enhance a bargaining position—ready to continue, eager to continue, he yet began small. Begin in Palestine had to begin small because the Irgun had neither men nor equipment—in fact the Stern Group could only opt for assassination in 1944 as the most cost efficient option, really the only option. In Angola the rebels tried to start everywhere at once while in Malaya the campaign began not with a bang but with gradual escalation—the British hardly recognized a threat until there was severe danger. Mao began by retreating into the countryside, for the communists had failed with conventional conspiratorial strategies. Once underway, urban or rural, the classics or obscure tend to be similar. There are texts on what to do, the conventions of the unconventional. While there is no simple campaign profile, some begin and fail, begin and persist, begin and sputter on with varying degrees of intensity, some escalate and some few win simply by persisting, the course of the campaign is quite recognizable. What each underground finds special is the necessity to impose the will over the assets of the center. And this like the tactics and techniques is everywhere the same, the central dynamic not just of the campaign but of the entire armed struggle.

Asymmetry

Almost all enemy centers possess similar assets, if not similar liabilities. The center is in possession of history, legitimacy, the assets of the state or the system, the responsibilities of power, and the avowed respect of the international community. As an enemy the center has the tangibles and often an equally compelling dream and the support of the people. No enemy is either static or simply like all others—a fatal underground error where the faithful are apt to accept their analysis as reality. It is the central power that imposes on the underground and so the campaign an asymmetry. In the armed struggle the conventional army is everywhere visible and the rebel nowhere to be found—sometimes it is not even certain that there are any rebels beyond a few riotous boys, which is why the threatened can pretend if need be there is no threat at all. Few can balance the reality of the unseen and feared and the conventional statistics of the state. It is relatively easy to weigh and match the assets of the orthodox: NATO divisions, the Israeli airforce, or even the strength of the Afghanistan irregulars armed with artillery and tanks. There are so many men of such a quality, so many tanks with varying capacities. In a war, a real war, much is neutral, the weather or the terrain, deception and surprise can be factored into scenarios of battle and even morale and national will. Past records, present quality, and special considerations can be included to arrive at a relatively sound basis for comparison of conventional forces. Some analysts have even given such factors numbers, quantified war to several decimal points. None of this is quite as clear about the underground, although estimates can be made as to size, armament, capacity, and morale. The key factor, however, is the power of the dream, the energy supplied to the galaxy that could make the state's tangibles ineffectual.

Dragonwars are complicated. One side has all the tanks and the other no planes. And no one can count on next year being at all like last. Even tangible assets may shift rapidly, upset the assumed military balance. In Indochina Vo Nguyen Giap had no artillery pieces until March 1954, when he acquired over two hundred. Dug in around Dien Bien Phu, his new asset was unknown to the French until too late. The IRA suddenly, secretly, and unexpectedly received Libyan arms that made possible not only an intensified armed struggle but also in time a more politicized strategy opening the way to accommodation. It is usually the underground that is suddenly at risk by a shift in the arena. The previously secure ecosystem is put at risk by new technologies, devised by the security forces or purchased abroad by the government, new tactics, or new ground rules imposed by the center. Gunmen may be killed by soldiers with night-sights or caught by infrared devices. Nets of surveillance may sweep up the active or prevent movement. The army may attack safe havens, attack exile bases. Grievances may be addressed and supporters offered sinecures, informers paid, and dividends offered. In turn the underground must seek some unexpected way to respond. In most unconventional wars, no matter the shifts, now the rebels have rocket launchers or the regular army nightsights, the struggle remains asymmetrical across a broad arc of perceived and real assets.

The Dynamics of the Armed Struggle   169

In matters of technology, the orthodox enemy always seem to have more advanced, more elegant, more effective things: jet-delivered smart bombs opposed to rifles or simply the same kind of rifle for every soldier. The deployment of hi-tech weaponry, whether enhancing mobility or intelligence or battlefield communication or like napalm or B-52 strikes the capacity to kill and maim, may prove of less than crucial value. There is seldom a quick technological fix to the problem of law enforcement or an armed struggle. The new and the elegant work but limits, does not destroy. A single fix not matter how complex and expensive rarely can be found. And often there is not even an appropriate technology at all or an effective means of escalation: using airstrikes into safe haves did not end the Kurdish insurrection in eastern Turkey any more than the Cambodian invasion ended Vietnamese infiltration.

Despite all its technological assets, the United States found it difficult to project power as far as Teheran during the hostage crisis or protect the Marines at the Beirut airport—or even revenge their death by deploying air and sea strikes. Despite the night-scopes and listening devices and computer banks, the British could not pacify the IRA in Northern Ireland. As long as the arena allows persistence, the underground can usually find a way to wage limited war. A match may be a guerrilla's best friend and a main-battle tank of little use in pursuing guerrillas into a swamp. Word can be passed by mouth and murder planned over tea. An ice pick in the dark is an effective assassin's tool, easily purchased, easily used, and easily discarded. The state can change the rules, adjust the arena, resort to counter-terror, kill everyone who might be suspected. And in brutal and efficient dictatorships everyone can be reached, anyone can be destroyed, and no underground can persist, few can even be imagined much less established. Much of the world, however, is not controlled by those both brutal and efficient, much of the West is open and the security forces restrained by law and inclination and so eager for means that are efficient but not brutal, effective but not illegal.

Orthodox anti-insurgency technology simply makes the task of the unconventional more difficult for the rebel—but rarely impossible. It is simply one more—and one highly visible—indication of the asymmetry at work during an armed struggle that always puts the underground at a disadvantage. This is usually apparent in the actual size of the forces deployed. Given the realities, the underground is often narrowly limited in the numbers that can be used without risking an overload. Thus the British have maintained a large force, tens of thousands of men and women in Ulster, supported by an array of hi-tech anti-insurgency devices. There are regular army regiments on renewed tours, a huge paramilitary police, reserves in uniform and intelligence agents in considerable numbers, and two out of three civilians supporting the security forces—not to mention vigilante loyalist paramilitaries. All this is in one way or another deployed against an IRA of four hundred active service volunteers, most leading much of the time conventional lives, and their advocates. After a generation the IRA still cannot be defeated, the dream killed off even by other Irish nationalists, or Northern Ireland made secure. Many in London, as in the IRA, hope that a non-military strategy can achieve an accommodation since anti-insurgency had not been able to impose one congenial to London.

At rare times during Dragonwars the insurgents have the numbers but the capacity always, always rests with the center where the professionals and skilled are retained, trained, and directed in a coherent campaign. More does not assure triumph, for if there is not will at the center then all the things, the men in nifty uniforms, the money in the bank and the materials, the helicopters and hospitals and electronic gear, will not matter. The British had one hundred thousand troops in the Palestine Mandate to oppose a handful of Zionist, urban guerrillas. And the British evacuated without imposing order or even delaying the establishment of Israel. The Americans evacuated Vietnam and all the residue weapon systems and trained regiments could not keep the communists from winning. The will does count.

Many at the center accept this, see the key not in force but in persuasion. Such analysts want to limit the purely military focus. General Sir Gerald Templer, High Commissioner and Director of Operations in Malay during the emergency noted, "The answer lies not in pouring more soldiers into the jungle but rests in the hearts and minds of the Malayan people."[4] What the British did was deploy sufficient troops to isolate the guerrillas militarily, isolate the rebellion as Chinese, and meanwhile began to concede control to the Malayan people. The army won because the ecosystem contracted, become isolated and irrelevant to a changing battle arena. All of this may not have required pouring troops into the jungle but did require men, material, and money and political initiatives elsewhere far beyond the capacities of the rebels. In Northern Ireland, along with the men and the money, the investment in the future, the myriad of proffered political options and economic rewards, London has always stressed the British will: no concessions to violence, no accommodation that erodes pride defined as responsibility to the Protestant, unionist Ulster, no erosion of resolve. The British have managed an affordable level of violence in exchange for very considerable psychological returns: vindication of their will and way. And London assumes the establishment's will is resolute. The IRA does not believe London—the enemy is ever perfidious—and so persists, waits out the peace process.

Sometimes the center simply does not need too much or very much at all to respond effectively: the best response is inevitably the least response. There is in any case no tried and true formula. Each arena indicates different priorities, special agenda, varied investments in coercion and revolution. The conventional army may be very much smaller, as was twice the case in Congo-Zaire, or very much larger as was the case with the British in Palestine and Cyprus and Northern Ireland. Usually the closer the two force levels the more conventional the war: Tet was almost a set piece battle with the Viet Cong deploying light infantry in classical patterns. North of the Caucasus, the Russian siege of Grozny saw an ineffectual regular army deployed against an effective irregular force: in time the simple power of the regulars drove the Chechens from their capital to resist successfully in less formal formation. This resistance, not easily met by the Russian army, achieved a stalemate: to win went beyond Russian will—not Russian assets however frayed—and the reverse was true with the Chechens prevented from conventional victory only by a paucity of military assets.

For the conventional what works works—the hearts and minds of the people may be beyond reach and the fate of the gunmen solely dependent upon deploying state power. In Israel the answer to Islam can hardly be co-option and may be beyond accommodation so that power displayed, power deployed is but an indication of the will of Zion. The problem for the orthodox is to deploy against what cannot be seen. The unconventional at least can judge something of the struggle's impact if their opponent deploys more. What is nearly always true is that whether or not one side is much more numerous, the conventional can be seen and counted while the underground cannot. No one knows how many are in a secret army, often not even the commanders of the underground. Numbers do not matter greatly underground, only operational capacity, and that is inspired by commitment and opportunity. One gunman can kill a head of state or a boy with a match burn a palace.

In armed struggle, then, rebel numbers slip and slide: the proper question is not how many exist but what can the existing do? How does one factor in those who only wait or hide or collect and transport, the part-timers, the yet unused? A revolutionary ecosystem often contains but a few gunmen, only a small number on active service. Even when the unconventional have underground regiments with ranks and roles, there are still always the others above ground supplying aid and comfort and hidden away as agents and assassins. They are a force although not easy to count.

The Enemy

For the underground whatever the count, the enemy remains both awesome and vulnerable, both in control and at risk. The enemy is seldom really known but easily imagined: the Great Satan of Hezbollah or the Imperialists State of the Multinationals of *Brigate rosse* in Italy. In the everyday world of war and peace, allies and enemies, each citizen can determined who is who, what is what: they are them, the French or Israelis or the communists. The more compelling the dream, the more truly revolutionary the struggle, the more apt the enemy is to be symbolic as well as actual. The irregular tribesman finds the enemy in a police patrol and the defenders of the village those in the next village with differing ancestors. The gunman in a Belfast lane shoots not at a British soldier, a lad from Liverpool, but at the British imperialist system still in place in Ireland after 800 years. By your enemy shall you be known, so that the more universal the dream the more grand the enemy. And the greater the enemy, the more universal, the more difficult is any adjustment short of absolute victory—Liberty or Death is not simply a slogan devised by Patrick Henry as American rebel but an interpretation of the eighteenth century world when the Rights of Man confronted hereditary despotism. Enemies are necessary to the faith. Central command and the whole galaxy require both God and the Devil. This often makes an accommodation at an end game difficult but does make persistence in the armed struggle easier. And revolutionaries thus perceive a world that others do not see, a simpler, more compelling, more demanding world of harsh and unforgiving forces. It is difficult to leave this world behind and emerge to the everyday, to negotiate at the end not justice but specifics, to deal not with the great tides of history but the numbers on municipal boards. Thus the Great Satan is an asset to militant Islam, the British Empire to the IRA just as the Empire of the Multinationalists was to *Brigate rosse*: monster enemies imagined, enemies needed.

What need not be imagined but managed is the capacity of the monster: the tangible assets, the technological advantages, the power of the state. It is apparent underground that the dream alone, the secret army alone may not assure swift victory—and the faithful even in a protracted struggle feel a sense of urgency. So to persist and certainly to escalate, the enemy must be countered with what is available, the unconventional assets and reliance on the incandescence of the dream. The faithful must persist beyond reason until the long tides of history wash away injustice.

At times such perseverance is neither unreasonable nor futile. After all, the old empires are gone. The guerrillas in China became route armies and the hunted in the hills in Cuba and Nicaragua government ministers. Some time the end game is simply driving up to the palace in Saigon or more likely signing bits of paper under the aegis of benevolent powers. What gives the underground pause is not winning all but winning something: negotiating away the dream for reality. For much of this century Irish republicans have disturbed the island peace because reality has not matched aspiration: each new essay in seeking an accommodation has failed to convinced sufficient of the militants to bar any future struggle. For others there has always been hope that sufficient power will be ceded to allow the dream to be achieved by more conventional means: concessions on the installment plan or acceptance that the maximal demands of the ideal were not necessary. Conciliation results not in compromise or even an agonizing reappraisal but a shift in the asymmetrical assumptions of an unconventional conflict. The faithful do not discard the dream but the old enemy. What mattered before must matter less because what now matters is no longer what once mattered. Thus, for there to be an accommodation, an end game must involve a shift in underground perceptions—an enemy transformed—just as the old enemy must conceded legitimacy to the underground in order to negotiate. As always, the exchange is asymmetrical but in this case establishes an equality of agenda, each takes what is given to replace what is conceded.

Conclusion

What matters to those within a Dragonworld is a perception determined by the dream. Time is different and so too the people, the nation, the intangibles. Some see a hired crowd bussed to the square and others a nation awake, some see a deluded and demented puppet, and others a martyr driving a Mercedes truck. The world underground is rich in symbols and meaning and concepts invisible to the common place. Castro could from Oriente see into the hearts of all Cubans, a virtue possessed by most in the grip of a dream. Giap knew that once the people rose during Tet in 1968 American resistance would collapse: ideology insisted, Mao's example insisted, reality as read insisted, and so too the dream. And they did not. The Viet Cong were slaughtered, the work of years lost, the campaign crippled. And so the underground must be refashioned, reinforced, must persist until reality met the dream. And yet at the end of the day, Giap had been right—a military battle victory at Tet only indicated in America in 1968 that battles would be necessary in a war supposedly won and so support eroded and the campaign to protect Saigon and confounded Hanoi could not be maintained. There was no longer the will to do so. The Americans withdrew in 1972 and the Saigon regime collapsed ahead of schedule three years later in 1975.

Just as persistence requires faith and a special perception of reality so too does victory require different spectacles, a different agenda. Underground optimism is vital. The gun is not to be questioned for this has been the last resort and the key to the Dragonworld door. The past is shaped as means and the future promised as certain. In the meantime without real evidence, without adequate tangible assets perception is everything: the Irish Protestants are Irish, Allah will explain all, the nation is mobilized, and the people committed.

Those underground always assume their perceptions are accurate, their reality real, the tribe not tribal but class, the Imperialist State of the Multinationals as real as real, the system evil, and the faithful certain of victory. Life for the dreamer is not simply enhanced with new meaning but all is changed, utterly changed. If this were not the case, then history would be prologue, the system go unchallenged, grievances seen as inevitable, and none find solace amid dragons. The dream inspires but also transforms reality to warrant inspiration.

Everyone underground is driven to some degree by such inspirations. The more classical revolutionaries are apt to have more elaborate ideals, a more complex ideology and so are more driven to change history. Others driven underground or to the margins, driven to defend the village or the faith, may have profound convictions but often have as mission persistence rather than revolution. All underground if engaged in shaping the future through an armed struggle must also conform to certain factors that impose system on every campaign. A dream energizes an armed struggle that inevitably inspires the center to impose arena conditions that shape every campaign. Secrecy is as costly to the IRA as to the Tamils of Sri Lanka—or any organization. All undergrounds driven by the necessity to find means to act against power, seek by necessity not choice what is available, what can be organized and deployed, what will allow protraction and with luck escalation. Such irregular means, the sudden ambush, the bomb tossed into a cafe, the general murdered, all come as a surprise, all for the orthodox are the end product of vast cunning, great skill, fine intelligence, and ruthless brutality. Within the underground each is the result of desperation and often successful because of luck, compensating errors and the audacity of a few. None in command want to be unconventional. Most do not want to be underground at all but in power. Few thus learn their trade.

The faithful and those who would defy the center, defend the margins, must rely on secrecy and the dedication supplied by faith: and mostly this is not enough. The tradecraft of thrillers and films, the professional assassin and the bold gunmen rarely exist in real life. Few armed struggles last long enough to allow the skilled to emerge. Few gunmen or bombers last long enough to serve as exemplar and tutor for others. Few operational planners reveal natural talents and few intelligence officers can operate effectively without records, without research, without corporate memory, or access to the advantages of technology. Most underground campaigns end badly if for good cause.

In order to cope those underground rely on stealth and what can be brought to bear, on the faithful and luck, on the haven of their perceptual world. And because most vulnerable societies are filled with targets, because the Dragonworld proves sanctuary, because the dream is compelling, because the only available means is unexpected, an armed struggle can at times persist, impress, and so display one more special case of the general dynamics of the armed struggle. And so they do and so they have: thousands of armed struggles each arising from an underground, each patterned and confined by the conventional arena, and each driven by similar dynamics.

Notes

1. *Irish Times*, March 14, 1988.
2. The Malay insurgency became a text-book case with British advocates inclined to point out that victory came largely because of the effective deployment of military force and a centralized command, somewhat neglecting the transformation in the arena and the errors of the rebels.
3. Orthodox communist parties have almost never initiated armed struggles not only because success was unlikely but also because a covert, armed struggle places a priority on guns not ideas, operations not analysis, an uncongenial agenda for even the most subversive communist.
4. Gavin Bullock, "Military Doctrine and Counterinsurgency: A British Perspective," *Parameters, US Army War College Quarterly*, Vol. XXVI, No. 2, Summer 1996, p. 16

6

Responses to the Armed Struggle

There are all sorts of responses to an armed struggle, popular and political, analytical and academic. The reaction of the police or the public, the peasants or the premier vary enormously. Some chose not even to recognize that there is an armed struggle. Others deny its validity: criminals, the work of outside agitators. A few refuse to note change, step over the bodies, go about their business while others make it their business to exaggerate the threat, shape a career in response to the challenge. Sometimes the outbreak of violence is assumed to be traditional disorder—the revolting Irish again, bandits from the outback again; and sometimes it is seen as unique, a terminal dispute with the future under siege by Godless Communism or the Great Satan. So each resort to an armed struggle engenders many responses: the police get new powers and weapons—or do not, the professors have another case study or the peasant a champion. From the first shot the response is often to explain as well as to act. For in an armed struggle what matters is what is assumed matters.

There are rings of response moving out from the central core of conflict, not ones easy to place on a chart nor as clear in real life as in analysis but indicative of degrees of involvement. Thus from the first are those with guns, those responsible for order, those who have gone underground or command the security forces. Almost from the first, there are as well the explainers: the police have an explanation, the guerrilla comes with placard and theory as well as assault rifle, the first body bag has meaning for the observers, for the reporters, for the people. And everyone, people and participant have not only a perspective, a response but are avid to discover those of others. What is going on? Who is who and what matters most? Those to whom it most matters are those most vulnerable, those involved—and like everyone else, they have definitions and explanations and responses, some immediate, some instinctive but all deployed forthwith.

I

Obviously those who assume the struggle matters the most are those at risk, none more so than the targets of the underground. Despite the enormous diversity of armed struggles from classic terror to bandit forays, the response of the center tends to be similar—and the options of response even more so. Unconventional threats are met in conventional ways. Often there is no clear beginning, riots become covert agitation and then the most militant are underground preparing to use guns. The armed struggle emerges not in stages but imperceptibly. The conventional at times may have difficulty recognizing the unconventional. On the other hand, an armed struggle may begin with a bang—with bombs and murder. The bandits can unexpectedly acquire new banners and new patrons. The students can become terrorists and their bodies soon found in the gutter.

The most ambitious underground may snatch the attention of the media with atrocities but still not unduly alarm the system. What can be done in any case? The system will not fail if the king is killed, the hostages taken, the bandits converted to Islam. Some threats are truly traditional, some minor, some require a response or the semblance of a response to quiet anxiety. Only some armed struggles are from the first taken as serious. Those regimes with limited legitimacy, limited assets, and long experience with rebels are apt to recognize the intensity of the threat. At times the regime seeks a threat from the fringes to legitimize repression or the system. The great empires were always alert for unruly natives and outside agitators. In time those responsible to respond even to "bandits." Authority must say "No" to terror or come to the aid of the starving poor or the frighten investors. Some in power must do something about the chaos seen nightly on television or rumored in the market place. Any response is a mix of perceptions, self-interest, and experience arising in a special place at a special time.

If there is a tangible beginning, instead of an imperceptible slide into violence, the center is often surprised, appalled at such pointless and illegitimate challenge to authority. Given the imposition of empire and in many places the tradition of native resistance, most colonial regimes were still surprised even when trouble had been predicted: the imperialist felt that not only was the system legitimate but also it was beneficial for the colonized. Any insurrection seemed to defy reason as well as interest. Of course, many governments, even the most regal, were experienced in matters of threat: the British had been subject to the revolting Irish for centuries, the Tsars had secret police, any Latin American state exists with peasant truculence, nearly every African government can expect a coup if not dissent away from the center. Yet every time, each rising or insurrection or revealed conspiracy, the government, the establishment, is apt to be surprised that anyone dare, that any alternative vision exists.

Understandably the less legitimate the center the more those in power feel any challenge must be considered a serious threat. In fact some very legitimate monarchies like the Russian assumed that potential anarchy and revolution was the natural order: royal power must be protected by constant diligence. The Marxist successor were for different ideological reasons just as suspicious of potential dissent if more effective in their diligence. Those governments democratically elected often find a resort to arms outrageous, an affront. They have assumed suffrage engenders people's governments open to change through accommodation. Armed factions and dissidents are perceived as an illicit surprise and so difficult to co-opt, to isolate, or even at times to destroy. Thus the Germans—the new Germans—found their few terrorists intolerable: illicit, unrepresentative, zealous millenarians disturbing the nature order of the country. The Americans too find militia dissidents outside expectation or even rationality: paranoia as a political cult. Those most authoritative and those most democratic share only surprise at insurrection—the Tsar surprise that they dare and the President surprise that they desire to dare. Both share, as well, the necessity to respond to the threat.

At both ends of the spectrum of authority, the rebel is perceived as a danger. And even when expected and especially when not, the reaction of the establishment, the government, the responsible is surprise, outrage, and indignation. The immediate reaction is always to deny the dream, the legitimacy of the challenge. The challenge is a merely police matter or the result of alien forces. The gunmen are terrorists, the guerrillas bandits, all are mad or criminal or manipulated by outside agitators or ideas. There is no legitimacy in their struggle. And often there is great weight to such assumptions. Few rebels have polled their avowed constituency. All have made the complex simple. And some are less driven by a dream than by special interest. All are, of course, by definition, criminals—patriots only in victory. All are motivated by a dream alien to the center and embarked on a campaign that to the sensible appears less than prudent: a mad adventure. And all are apt to pursue a flawed dream, one that takes no account of the Arabs in the Promised Land or the unionists in Northern Ireland or the form that is to replace a Raspberry Reich. The reaction to such threats is then shaped to need, to the tangibles, and to the desirables, shaped by the perception of the involved.

For the generals and governors, the media, the romantic, the television audience, the analysts, the scholars, and the constable at the corner, the crucial factor is conflict. The violence is visible. Murder is real, can often be seen on the evening television, is not a matter of conjecture. The gun battles are undeniable. Yet what is happening—not just why or how—is a matter of interpretation.

The dream is always denied. No matter the cause—Zionism, Irish nationalism, an answer in Islam, the dream is flawed and so the outrage at the center justified if irrelevant. Even nationalism as advocated is seldom as pure as the underground assumes. And as for those who speak for the wretched, few are wretched; and in power the wretched have usually seen first to their own control and so their own interests, climb into limousines to attend the show trials of the old regime. Revolutionaries causes close up look suspect. What if all the peoples in Europe sought states? Do all deserved states—the Walloons and Corsicans and Scots, the Bretons, the Lapps and the Occitans, the Catalans? And what if land is twice promised: France and Normandy, Wales and Britain, as was Palestine to the Jews and Arabs? And what to do with the minorities, the dissenters, the Protestant unionist in Ireland or the Russians in the Baltic? Romantic nationalists would create a Europe of quarreling clans—did so in Bosnia. One aspirant's dream seems another's nightmare and so the power of the center, the present, responsibility is apt to deny all proposed dreams.

The people seldom vote for dreams but candidates and policies and parties. Yet dreams persist—and realization is not without prospects: the world changes. For generations, many European national aspirations were mere romance, denied reality and at times would have denied the rights of others. And so such aspirations were denied. Denial is not forever. Slovenia appeared and Slovakia. Latvia is free and Moldavia. The Irish republicans persist and so too the Basques. Legitimacy and the practical are in perception and those underground assume they can see what others cannot, the reality of an alternative future, a virtual reality more compelling than the present.

The armed struggle is waged about the future. Those who have their own future guaranteed are in the capital, and some in the countryside want no future only tomorrow to be like yesterday. Order and stability have advocates. Most people want no trouble and dreams are troublesome—and most important to the responsible are irresponsible and so criminal. Because the convinced sacrifice and die for a dream, only makes it less rather than more appealing to the orthodox. They advocate continuity and coherence. The great asset of the orthodox is not only are they in place with banners flying, regime recognized, assets to tally but also that the persistence of the present is enormous. The system can be strained but still exist. Society under threat manages to move along, the children go to school, there are films and concerts, and the mails work. Somehow the village can cope with transitory armies, irregular war, and fire fights in the fields. Not always, sometimes the last bare bones of civic order snap—the villagers become refugees, the rebels slaughter their way through the provinces, carpet bombing drives the peasant into the cities. Mostly the center holds, the society persists, the underground fails to escalate, fails to persist.

Some spectacular violence is transitory, out of context, and soon out of mind: the wave of assassinations and bombing by revolutionary anarchists was a threat to life, to the individual king or president but hardly to the system. The West was awake but only to the spectacular not to any threat—such a threat could be found in a naval arms race, the frailties of the Hapsburgs and Romanovs, or the survival of the balance of power. In 1890 or 1910, those most concerned about revolutionary bombers were those responsible for policing. The great and powerful went about managing society—all that shifted was the potential cost some might pay to anarchist assassins. The Great War came because of great interests not because an assassin killed Francis Ferdinand. After that war the potential terrorists extorted their way into power in Germany and Italy, operated on the margins of Europe, could hardly be found even in the colonies. And after the second great war, the same was true: terror in the core of the West hardly mattered. Britain has far more pressing priorities than IRA bombs. The empire was defended from rebel aspirations on the cheap until this small investment did not pay. The French could leave an African empire after the turmoil of Algeria and return on hidden currents of influence, co-option, mutual interests, and the discrete use of the army. One kind of empire replaced another and the new empire was at the end of the century as difficult to dismantle for dissenters as it had at the beginning.

After the Palestinian spectacular at Munich in 1972 terrorism became highly visible as incident and as strategy. Terrorism was not limited to the margins of the liberation struggles of the anti-colonial Third World, but in Italy and Germany arrived at the center of the post-industrial world. *Brigate rosse* or the IRA did not greatly threaten world order but did compound the problems of governance and the assumptions of consensus. Many violent threats are not terminal even when they evolve into triumph. A free Zaire was still a Western colony after Mobutu came to power and may be after he has gone. The French did not leave Africa when they left Algeria. Vietnam did not prove after 1975 the fatal domino assumed. Castro stayed on Cuba until he was irrelevant and the Euroterrorists faded away, an incident. All such unconventional threats are assessed, often unconsciously, sometimes around a table, and given priorities. What matters to the threatened is what they assume matters. Not the killing but the context. Assassination has implications because of the context, the values of the victims, the intention of the killer, the resonances of the deed—otherwise one person is simply killing another.

What matters to any response is what the involved assume matters. And *that* is often adjusted at the end of the threat—usually by the victor: this is why the revolution triumphed and this is why the center held. Che assumed he knew what mattered and failed in the Congo and died in Bolivia. The Pentagon drew congenial lessons from the Vietnam debacle and the British pretended the empire was translated into Commonwealth out of magnanimity not need—only Nasser and Suez revealed the illusion of imperial capacity. Those involved perceive events from their own special perspective.

Those who wage or advocate armed struggles describe the killing in terms of strategies, stages in a people's war, or the inevitable triumph of a risen people. The theorists remained committed to theory, and those activists who adopt theory regardless of application are apt not to survive. Those who cope with or without theory may in a few cases not only survive but also win. Then come the revolutionary memoirs where all the pain and terror, the corruption and betrayals disappear. The analysis of real generals arising from conventional assumptions and exposure to an armed struggle are often even more divorced from reality. Most of the traditional imagine the underground a flip image of the orthodox, an interpretation not uncongenial to old gunmen seeking respectability and legitimacy. Both project what matters as what they perceive has happened.

In retrospect the underground triumph is embellished, shaped as conventional and inevitable, or the defeat is parsed and dissected as effective example of counter-insurgency tactics. If history were so easily read, then the Cuban experience would have transformed Latin America and Italy would not have suffered a decade of terror and confusion. Still Castro did win and the Italian Republic survived. In these cases, in all cases, the concerned understandably focus on reality, the numbers of death and incidents, the dates and events. Yet, what the involved say mattered may not have mattered at all. What mattered was their assumptions about the dynamics of the asymmetrical conflict—and such conflicts can rarely be effectively analyzed quantitatively—even and especial by social scientists or scientific socialists. What is necessary is an insight into perceptions of reality. An uncertainty principle means that perceptions are not only the subject but also involved in the analysis: shape the actor and the critic imposes meaning on an forever elusive reality.

II

Those involved in an armed struggle edit not only their experience but also in retrospect their initial reaction to events. The perceptions of the times are thus adjusted once the results are in. This product has for them, for the future, an enormous importance. If what has happened is known, then the future can be adjusted. Thus even a history of an armed struggle is not a matter of putting down the plain tale of events, rewarding skill or memorializing sacrifice, but in mapping the way ahead. The nature of an armed struggle is never really stable: perceptions shift and so do responses. What happens is what should have happened. What was done was what should have been done. What matters at any time is what matters most.

The British fought a generation of colonial wars that, whatever the operational returns, ended with the closure of the British Empire—except for the Irish issue. The lessons, mostly tactical and operational, learned by the army were at each new opportunity deployed as doctrine with great confidence. And yet the empire has gone and the Irish are still revolting even amid a peace process. Somehow the proper lessons escaped the specialists, the planners, the system. The British still like to assume the withdrawal had been planned, adjusted to the winds of change—Nasser remains a villain but not because he exposed the limits of British power. That power is still somehow assumed real—the Falklands war won and the nation still tutor and mentor to the Irish where duty must be done and can be done at an acceptable cost. And so the British establishment have written the past to the present's need—as have the Irish and all those who were deployed against the empire. Reduced to the margins the neo-Marxist in British provincial universities still shape imperial events to class analysis: they too know what they know. And any rebel is no less eager to shape history, knowing what he knows.

Those who spend the campaign underground do so in an intense cauldron, bonded to their own, stretched to the limit, each day a risk and a triumph, and so emerge with the armed struggle hard-wired into their psyche. What has been learned shapes all else, exile or triumph, the new regime or a determination to continue. What has been learned varies, evolves from the special perspective of the underground. Castro and Che though their *foco* not only crucial, which it was, but also singular—and so discounted history, the times, even other Cuban arenas of the struggle, for all these lay outside their day-to-day experience. Thus a generation of Latin Americans looked to Castro instead of Cuba and tended to fail.

In all cases the response to an armed struggle, the immediate response, the analytical response, the interpretation and application is largely shaped by perception even and often at the expense of objective reality. The reality is what one imagines it to be and sometimes this is the case and sometimes not. The British army, tactically always more experienced and astute, won the war in Malaya, which is true but the changes in the arena outside military control transformed the appeal of the communist dream, a fatal change that lay outside military capacity or some subsequent analysis. The *foco* appear to Che and Castro to be the center of the Cuban struggle and so in time many guerrilla analysis would assume no matter what happened elsewhere in Cuba the *foco* was crucial—and they were not wrong if not fully right but most important were read as offering a Cuban way into the future. Much analysis focused on armed struggles does, indeed, seek ways and means, not dreams and ideals but examples, tangible strictures for future wars.

It is the tangible, especially the chronology of violence, that absorbs most interest simply because intangibles are invisible, malleable, assumed. The underground is largely imaginary, site of qualities—terror and sacrifice and risk—that are not subject to quantification. The orthodox often miss the curious nature of the underground medium that imposes a dynamic unlike that of the visible world. To explain motivation and intention within the hidden ecosystem is assumed open to conventional analysis. The ecosystem is assumed to be no more than a cover for conventional dynamics. Those in the grip of a dream are merely conspirators, some reasoned, some not, but all acting to the same imperatives as soldiers and generals, suffering the same pain, restrained by the same considerations.

Those underground often appear irrational to the conventional. Why would one rebel against legitimacy? Most important, why would anyone do so when prospects so poor? The odds are so high on triumph, the pretensions of the gunmen so hollow, the center so much more responsible that only zealots could be imagined engaged in an armed struggle. Hidden zealots, however, are still vulnerable to conventional security tactics, the conventions of anti-insurgency that assumed a military campaign necessary, a symmetrical struggle with the state in command of the high ground. Those not in the grip of recently revealed truth cannot imagine the imperatives it imposes or the perceptions it establishes. Thus what goes on underground remains secret, leaves evidence only in the bodies left in the gutter or the truculence of the people. The visible world focuses on what is visible, those bodies, the dead, the violence, not the dream and not the dynamics imposed by it. The ecosystem is secret and is a haven but it also represents the armed struggle by shaping it: an inward and invisible determinant and the source of great psychic energy.

The gunmen kill, of course, but the body count only matters as indicator of the faith, as indicator to the tides of history. The rebel dream is the great asset not the tangibles. The energy supplied by revealed truth is hard to weigh, hard to imagine. The state may measure and balance but always seeks tangibles to weigh. And the scholars and analysts deal in tangibles as well but the tangibles underground are different. For those within the ecosystem time spent over the faith instead of the weapon is not time lost but energy generated. Time so spent cannot be understood by the apparent demands of the campaign and so is easily be factored into the considerations of the conventional. The asymmetry of interpretation is between the will and the power just as the conflict is between the orthodox and the irregular, the visible and the covert.

Why things underground happen is less crucial that what things mean, who is winning and who losing. Yet what is most tangible about an armed struggle is the violence. Analytical description begins with fascination of the techniques of the campaign and then the tactical impact of an armed struggle. The spectacular nature of the unconventional—the surprise and the horror—hides the lack of underground tangibles. The gunman is assumed both awesome and inept, motivated by error and engaged for gain. The violence is judged both as to intensity and to legitimacy, judged as to effect and as challenge. The violence is what matters. These low-intensity campaigns of violence, for generations have fascinated observers: the use of the unexpected by the unknown and unseen to achieve power over the strong. A vast literature has arisen focused on the irregular organized as revolution.

The spread of the armed struggle in various forms by the end of the century displays a huge spectrum of action tactics from assassination to sit-ins and strategies employed by rural communists or urban nationalists, by terrorists and by religious zealots. Some unconventional struggles have been various and others not. At times, as in Vietnam, almost all means were deployed and at others the underground is left with only propaganda of the deed or persistence in exile. With the collapse of empire, irregulars gained new prominence, tribes contended for national flags, and village defenders evolved into armies. Without the bipolar labels of the Cold War, the new chaos has still engendered the same responses. And the focus has always been on the use of the gun by the secret and illicit. Ideology is especially important in anti-insurgency if it reveals how the gun might be used. Thus the strategies of village defenders, mutinous officers, mad assassins often fall far outside great concern, not easily categorized. Revolutionary terror, however, has for two centuries been a constant, not always a threat and never easy to categorize but still there, a concern that may matter, has mattered. It is a matter that induces a response and the response is apt to be military even when the state recognizes that the gunman does not simply act irregularly as does the village defender or the bandit. The gunman even after two centuries is alien to the military mind, the analytical mind, to the everyday, is the stuff of thrillers and films.

Revolutionary violence has thus maintained an exotic charm for the everyday, those in universities or before their television set. An armed struggle is a muddle of romance and horror and involvement: the more terrible, the more sensational; and the more sensational, the more riveting. There are thrillers and films and plays about the struggle. Some exploit slaughter and some are art works. And those real struggles that take place for the camera attract the greatest interest—the Palestinians choreographed operations for prime time. Television imposed the necessity to respond, the opportunity to respond. It was possible to sit in Gaza and watch on television an IRA funeral dissolve into violence, tumult, and rioting not to the advantage of the security forces and see example and so ancestor for the *Intifada*. So it is not simply the everyday viewer that responds to the televised event but the involved: and these can program their own deed. The fedayeen used televised terror to create a nation but not a homeland. They could not transform media prominence into political power but they could act—an armed struggle of a few desperate terrorists watched by hundreds of millions. Israel was no weaker nor the West more sympathetic by the time the Palestinian struggle dribbled away into negotiations seeking to achieve something before all was lost to the Islamic militants of the next generation, but the response had been enormous.

In the glory days of terror, the Palestinians assumed they mattered because they dominated the evening news but in fact this did not matter at all to Israel. As always the armed struggle was asymmetrical but in this case the perceived military successes of the underground were illusionary while the subsidiary impact of the campaign became primary. Spectacular atrocity went far in the creation of a people, not a new and revolutionary people as Franz Fanon suggested, but a Palestinian people. So the Palestinians dream became a Middle Eastern reality—a welcome strategic triumph but not the one sought.

After the Palestinians came the Iranians with wider dreams and so the world had televised jihads. The Great Satan and the world of the heathen was punished not simply in Beirut but at the World Trade Center and on CNN. There was a new wave of gunmen and assassins, bombers and hijackers to take up the methods of the unconventional.

The orthodox in all such struggles are by necessity concerned with the tactics and techniques not the dynamics of the struggle, often not even the avowed cause. The academics and many analysis focus on what is congenial—published ideas, the ideology. Everyone looks at what can be looked at that fits existing priorities: policy people study policy and soldiers the techniques of ambush, the intelligence officers dissect cell structures, and the social scientists factors that can be quantified. The police round up the usual suspects and so do the historians. All outside the underground usually can focus only on the visible. Thus analysis is almost always the rough edge of violence and motives and grievances of the engaged. And much analysis is policy oriented: given this, what is to be done? The visible is real and the challenge is real and so too the analysis. Disinterested analysis is apt to come later and be interested not so much in what mattered then but what matters to the analyst, a method, a product, a predilection.

A complication for analysts is that what does not condemn seems to many to condone. A disinterested response is apt to find no favor. In those circles where revolution is advocated, those who do not support the cause are assumed enemies. Many assume that beneath each text is a subtext of advocacy. So for those threatened, disinterested treatment of the gunmen is betrayal of the system and decency. The reaction to an armed struggle even at several removes is still part of the confrontation. And nearly always the local and provincial campaign is assumed to have greater meaning, even the bandits on the border or the religious zealots have implications for global order. Much of the analysis simply describes the past, a history going down one hill and up the next, the chronicles of the campaign. These are not action texts or analytical history but war news. After a time, when the underground emerges or dissolves, when source and documents and interviews are available, history can be written. Even this is no easy task for much of the struggle is undercover and the revolutionary ecosystem not easy to imagine. And those who have been involved are no longer in the grip of revelation and often are too engaged in adjusting the past to present considerations, transforming the irregular into the orthodox.

Analysts and their customers in government, in publishing, at the university want orthodox history, methodologically sound analysis. History is often assumed identical with reality. And no wonder since control of history was the prize sought during the struggle, history as lived, history as written, history that would control the future. Even many social scientists are apt to assume that a historical record is real rather than an art work often made for purposes beyond anesthetics or academic reward. Yet, even and especially, very contemporary history becomes part of the problem, shaped by events, judged by the involved.

For decades the great struggle has been between the great Cold War powers with interests everywhere. Dedicated to world revolution, radical change, and so turmoil, the communists of one flavor or another all but monopolized revolutionary language and explanation often leaving their opponents the anti-theories. These negative theories generated handbooks like the French in Algeria with *guerre revolutionaire* or the Americans' more pragmatic and less effective anti-insurgency manuals. The non-communist texts and guide books, except that of Grivas in Cyprus, tended to be memoir or manual rather than theory. Anti-communists tended to know what they opposed but hardly recognized the power of their own dream: Greek nationalism or an Italian fascist heritage or the strictures of Islam. Western sponsorship of insurrection or resistance, particularly American, has tended to be atheoretical: the shipment of arms, transfers of money, the purchase of peasants. The most visible initial response to an armed struggle, to an irregular challenge is often not based on analysis or even what is perceived to matter but on reflex: violence engenders repression. Those in power concentrate on operations. What most anti-revolutionary theorists advised was to take the techniques of the underground without discarding the tangible assets of the state. For many in the West national liberation was a communist false flag for Soviet interests. And for many unconcerned with Marx or Mao but their own aspirations, scientific socialism offered a means, a text, and assets. For much of the Cold War, the West assumed revolution was a threat and the East a tool: titles and labels took care of the exceptions. And everyone took their cue from the two great blocs. Only in the last decade has there been need to explain what mattered and in the meantime, as always, those challenged responded without hesitation to provocation.

The Soviet responses to such provocation in Angola, Ethiopia, or Afghanistan showed no more theoretical concern that that of the old colonial powers. The new Russia made all the old mistakes in Chechnya without proper assets. The center often forgets nothing and learns nothing but the skills of tradecraft and the bush. Thus much of the enormous literature on insurgency, terror, the armed struggle, revolution, and revolt is advocacy: how to win or how to persist, how to respond. What evolved was the conventions of the unconventional: how to respond to what was perceived. The orthodox analysts often seemed to assume that all undergrounds could be ruined if the proper formula could be found: an integrated command, the hearts and minds of the people, isolation or compelling force. In turn all undergrounds seek a similar formula, an ideologically congenial solution, that will assure ultimate triumph: the nation mobilized, the party as vanguard, a *foco*, a people's war, or the stages of escalation.

III

Those who seek to change the future by recourse to revolutionary violence or to prevent such change are the first to respond to a conflict. They are the academy of the committed immersed in that struggle. Those threatened produced tactical analysis and those who threaten tended to seek ideological authorization—although there are exceptions. Since such analysis is shaped by the medium, by the reality of the struggle, by need as well as by assumption and perception, the result is part of the lethal dialogue.

Most in the contemporary underground until the last decade have been descended from Marx and Lenin and later Mao however faint the lineage. Their ideas translated and adapted were deployed by many who wanted authorization from afar. The determinists in the university or some seedy movement office simply applied formulas without need to generate proof. In so doing they tended to impose the language of determinism on all revolutions. Marx offered legitimacy, a rigorous scientific analysis, and an organizational chart sympathetic to the central core—and all this was valid for those who were not Marxists at all but nationalists who still went to the mosque or had tribal priorities.

A few underground had alternative ideological explanations, usually simple nationalism but occasionally the ideals of fascism or religious particularism, some other faiths. Some anti-imperial liberation movements tried to shape an ideology that would appeal to all—fronts that covered quarrels or tribal interests or religious differences. Some did not. The Egyptian Moslem Brothers were not Marxists or simple Egyptian nationalists or merely religious fanatics but had their own special aspirations if no book of tactics nor strategy of revolution. They sought power for Islam in Egypt with a gun. They sought to

Responses to the Armed Struggle   191

strike first at the symbols of the old regime and then at the charisma of new Egypt of Nasser—kill the powerful was a simple imperative, do so for Islam. There have been other armed struggles launched without operational texts but with appealing ideologies. Each has an explanation of what the struggle means and so too their supporters: they respond to what they perceive.

Most activists have little time to read once the shooting has started. Most gunmen, except the intensely ideological assigned to publicity and propaganda, have a tendency to read one book, often the wrong one. Since most of those who do the shooting are not by nature bookish, they respond to reality rather than to ideas about reality. If the gunmen read, it is often in prison where ideas may do more good than harm, maintain the faith without encouraging blunder. Many university students, however, talk of books while planning murder. This is an aspect of the litany of the faith. The student, the scholar as gunman, offers an explanation of reality that is more shaped by ideological lenses and the isolation of the school or university than the world experienced by the aggrieved, the avowed constituency. Society is imagined as the university grown grand—just more corrupt. The operational erosion of the campaign often does not even concentrate ideological minds: the theories grow more ethereal and so less relevant to the killing business. Those underground that do the real killing no longer have time for theories and books, are on the run but not from reality but toward the dream. What is happening during the campaign is not found in tracts or mosque sermons but the assumptions of the wanted, in the frantic who must keep the faith. That faith imposes a reality others cannot touch.

Those who must oppose the ideas and ideals of the gunmen also have their ideas about evolving reality shaped not by the faith but by loyalty to the present and so opposition to the means of change. The politicians, the generals, the detective at the corner need not be immersed in ideas but must take proper precautions to see what is held, civility, stability, the system, is not lost. So only as the campaign grows protracted does the center focus on ideas and ideology: first comes violence.

The second great school of revolution can be found not within the struggle or the academic camp followers of enlightened gunmen nor their opponents but among social scientists. They, too, seek more general explanations in relative deprivation or communication theory or the problems of political development. For two generations many deployed the strictures of scientific socialism and even those who did not see Marx or Lenin as prophet were inclined to see their value as tools of analysis. And some of the most conservative saw such tools as contaminated—and so were rewarded with the collapse of the Soviet Union and the isolation of neo-Marxists to provincial universities. In fact academic analysis has always been both isolated and academic.

Few of those involved in academic analysis have been personally exposed to the underground although those who have, as in Italy, do no better or no worse than the others who remain in the library. Once the faithful use a gun, tactics overshadow strategy, survival absorbs opportunity for reflections. All those involved in an armed struggle arrive with an ideology set, with assumptions in place, the large questions settled, know what matters. And academics like ideas as well as quantitative models. Those at a greater distance, safer, less involved, analyze away from the reality of atrocity. Yet, one need not lay an egg to tell a rotten one—if much analysis is not first-hand, little history can be so either. Academic analysis is suppose to be academic, perhaps disinterested but need not arise from experienced reality. The scholars of revolution, despite the isolation of the academy, often suggest interesting, if not always applicable, ways to consider the armed struggle—noise in a communication system or a by-product of modernization. Content analysis reveals much about the underground that was not intended to be revealed. A variety of behavioral strategies lead to intriguing and often improbable conclusions. Social and cultural anthropologists, economists, and statisticians have contributed to the literature and at times an understanding of the underground. Threat analysis has become a field. All academic and much analytical work is apt to ignore the enormous complexity of the real world that is immune to quick fixes, the common wisdom, even the most simple suggestion. If texts could make winning generals, all generals would win: maps are not the terrain nor guerrilla guides assurance of victory. Some, much work is academic, academic to the policy maker. Some is addressed to policy makers who will never read the well-meant strictures that assure tenure and research grants and the odium of the rebel. And some analysis is swiftly applied: psychological portraits or the capacity to decode threatening letters as dangerous or benign. At times the anxious will deploy academic theories and fancies either from conviction or desperation. Vietnam was for America a great arena for good ideas and bad ideas and often the arrogant innocence of the learned.

Excluding the complexity of the terrain and the simplicity of most analytical maps, mostly the social scientists, like the communist ideologues, have a mirror-image problem, assume the underground to be not unlike their own. When the Italians took this perspective into the underground, saw the world as a seminar, reality did not conform to the lesson plan. The noise of battle hid the reality of Italy for some years but ultimately the map did not match the terrain at all. Academic models seldom work very well except on a level of little use to the gunmen. Analytical fashions may be more appealing to the orthodox, who often find rationale for pragmatism in the university. There ideas can be hijacked out of books and run up over cherished bureaucratic policies or military programs. Such ideas offer depth, legitimacy, and reason to institutional self-interest and are so deployed against the armed struggle not because they are valid but because they are congenial. The university, the research centers, the conferences and conclaves and professorial network do, still, produce both useful work and also at times revealing work—and also many ephemeral exercises—the common wisdom with statistics, pop history, acultural assumptions, and ahistorical conclusions. Academies reveal the good, the bad, and the ugly. And those involved in an armed struggle, the gunmen and the generals often take their choice.

Academics are not only apt to be too academic but also too trusting. Since they are reasoning and analytical, they assumed too that the rebels have rational agendas and so are amenable to conventional inquiry. Class, cultural, the inherent incapacities of those underground, the contingent and unforeseen, the blur of events are discounted in a search for models and rules and coherence. Undergrounds are assumed fit subject for the tools to hand. And why not? One need not be a mystic to study the phenomena. Still, certain aspects of the underground—or mysticism—are apt to be ignored because existing methodologies are clearly inappropriate. To a considerable degree all prefer the alien to be shaped as congenial and convenient—and the underground is neither.

Seemingly a basic analytical and academic assumption is that the gunman could, if properly advised, increase the efficiency of the movement, accelerate events, deploy theory to advantage. Few in front of a computer understand the nature of revealed truth that energizes those underground. Few understand the enormous amount of energy the underground consumes, the unstated priorities of the galaxy and the individual, the reality of being hunted without hope of tenure only with the prospect that the next moment will be the last. To exist on the run much less function, to perform while exhausted, uncertain, frightened, badly led, and poorly armed leaves no time to reason out priorities, leaves no time in fact to learn the trade much less discuss the options. It is easy for professors to contemplate the construction of nuclear devices, the theft of plutonium, the destruction of liquid natural gas containers or computer sabotage but not as easy to imagine a world where breakfast presents a problem and making a telephone call is dangerous. There is never safe time, rarely any time. And if there can be time, it is spent not even on tradecraft but on the faith, the slogans of the red book or the history of grievance. Underground life is hard, dangerous, exhilarating, and brief.

A life in the underground—imagined—does have its attraction. The faith moves a few—once nationalism or anarchism, more lately that of Islam or even the tribe. A few scholars come and are converted, can be found involved in urban conspiracies or tramping with the peasants in the wilds. They are sucked into the galaxy and leave behind the conventional. The underground offers intensity, meaning, a time on the dangerous edge. It can be addictive as well as dangerous and exhausting. The thriller writers and the film producers grasp this no less than the involved. Regardless of the dream, the doing appears electrifying—almost melodramatic in contrast and color. No more so than when the running is over and it is time for memoirs: for all wars tend to appeal best in retrospect when the blood dries, the pain stops, and tomorrow is like yesterday.

Any armed struggle is sure to engender mixed appraisal, mostly focused on grievances and aspirations rather than the course of the campaign or even the romance of the vocation. In America there were those who sought analytical insight to encourage revolution and so Free Africa and also those who sought analytical insight to frustrated revolution and so defend Western interests in Southeast Asia. Few are untouched when focused on a revolutionary underground, for a law of contamination exists in underground matters. To seek, even to see, is to become involved. And most find what they seek, a congenial movement, excitement at a distance, the wave of the future, a hot topic, or more data for the model or institutional evil. Mostly the observers, analytical or not, remain at a distance, watch television not the real thing, avoid involvement, seek decency and accommodation.

IV

Beyond those focused on explanation are those charged with displaying reality as found: the media. If some theorists by choice and many analysts unintentionally become involved in the underground, the media is from the first part of the campaign often co-opted by intention but also unwittingly by the nature of the trade. News as consumed is spectacular, novel, dramatic. The media does not simply display but intrude if only through the priorities of the medium and the industry.

Television in particular by presenting events assures that not only will some events be crafted to be televised but also that all events will be exaggerated through distribution. What can be seen in color, personal and close up, in the living room is part of the campaign, an aspect of the armed struggle not merely fair comment or disinterested presentation. The messenger makes possible the message and to a degree the reverse. So those underground are apt to choreograph their messages for the networks. There is more than coverage. What is displayed by the media is going to be perceived as important and urgent so than some armed struggles are perceived by the public. And the media in an open society is trapped: free speech, free comment, the news is vital. So events must be covered, terror reported. None in the West, within the system, in control of the press and television, *want* to encourage terrorism and violence only reported what is important.

All atrocities are not equal in a post-modern world. Some horror is photogenic, conveniently located, and explicable to network executives. Those horrors that take place in the outback beyond camera range and general understanding, those small wars in obscure places, those horrors are not documented. This fact is a given. The powerful care only about themselves but everyone loves a spectacle. What appears on television matters. The image counts. Many in command of an armed struggle realize this and so adjust. The West is not interested in far horrors unless very spectacular so horror is imported to the West. What is needed is to attract attention, to punish and to alarm the powerful, to send a message. And so most governments under threat want no coverage—Margaret Thatcher wanted no oxygen of media freedom to reach the IRA. And the rebels want access to such oxygen, to BBC or CNN. Visibility indicates significance—and the perception is what matters.

Propaganda of the deed is a central aspect of most contemporary armed struggle. Those noticed matter and those not noticed sacrifice, if not in vain, to lesser purpose. So those marginalized by geography, the trends of the time, bad luck, or under exposure are desperate to attract the cameras and thus escalate the struggle. Some terrorists to bring their grievances and violence to suitable stages and so shoot passengers in European airport lounges or set bombs in London or New York. As long as the media in open society is driven by novelty and sensation to hold the consumer, then the terrorist or the gunmen can choreograph events that the media will seize, shape, broadcast. In most of the West, the best a threatened government can do is narrow distribution, impose mild limits. If the consumer is not the determinant and the government decides, then an open society is apt to be closed. Even then only the most efficient and authoritarian regimes can impose closure: there are satellite broadcasts, fax and e-mail, and the web sites.

The underground tacticians realize that their deeds can all but guarantee an audience and so have an impact quite beyond the capacities of the few isolated underground. Efforts, especially during the more violent terrorist years, by the Western media to mute this effect missed the point: once the operation is underway and covered nothing else matters, not who wins or losses, not justified horror or general distaste, not criticism or swift justice. All the terrorists can be killed on camera and the operation is still a success if carried live world-wide. Terrorists produce and package, the media distributes.

So many undergrounds, fearing power will never come, have tended to focus on media spectaculars—strategies of terror. All, however, recognize the tactical advantages of manipulating the media. Thus, on the edge of analysis, the media's response is part of the process as much as comment. And almost none of those within the free media are aware just how telling is their "analysis." Nothing is apt to shape perceptions more, impose importance on the obscure and urgency on events than coverage. And this coverage can never be disinterested and is often manipulated. A free media in turn must inform the public or fail in responsibility. The problem is there is no solution. The scholar may explain what is really going on or the committed what should be going on, but the media does not so much show the event as shape them for distribution—the medium is both message and analysis and always apt to act as anticipated, attracted by the novel and the spectacular. Thus the reaction of the media to an armed struggle is a part of that struggle, an integrated factor in the arena. And no tinkering with rules of reportage can alter the basic reality.

The consumer, those who without special interest, follow the armed struggle at a distance, ordinarily through the lens supplied by the media can become actors as well. The concerns of the struggle may seep out to involve the distant and in turn the distant may impose perceptions on the involved. This is especially the case when the media can bring events to a billion watchers: who has not heard of Vietnam or Palestine, more people know of the IRA than of Yeats. And so the violence takes place involving actors, including the media, in a great usually tranquil pool of observers. These in turn imperceptibly may touch the direction of events at the center without intent, without planning and at times without great impact. It is often this "public" that is a target of operations and announcements—world opinion, national opinion, local opinion, the perceptions of those with nothing at stake. A gunman may need no more that public toleration to act—but wants approval. The state may need more but cannot impose enthusiasm only obedience.

V

In an unconventional conflict the response of the involved if apt to follow patterns also generates a singular pattern: the asymmetrical dialogue. This interplay of perceptions seems inevitable, a visible synthesis arising from the clash of oppositions. The dialogue is most obvious in the more classical armed struggles but even in irregular wars, colonial insurrections, tribal violence, as long as there are two asymmetrical centers, then the dialogue emerges as a response to perceived reality. What the involved are doing, what is reality, is a matter of explanation: the response of those knowledgeable or concerned or so assigned to explain. And, of course, those explanations rest on perceptions of reality—a reality often difficult to discern since the armed struggle is in significant part secret, a matter of assumption and expectation. This reality in turn is shaped by the perceptions of those involved: a dialogue with perceived reality undertaken unconsciously. Still, the response of the explainers to all armed struggles differs from any response to any dramatic or violent event only in the degree that the events involve observers. Those responsible and those involved, the soldiers and police and provincial governors, the television producers and the economic advisers at a think tank, must deal directly with the campaign. And, of course, those in the center of the galaxy of the faithful do nothing else. Those in command, if not always control, may find explainers useful to the degree that the results can be used to advantage but their own response is more immediate. Despite the enormous number of campaigns, there are basic similarities in the response of the involved to the gun entering the arena.

There is seldom a time spent on quiet analysis: the campaign comes as a surprise to the center and once begun leaves the underground short of seminar time. Most government policies, most new directions, most unexpected operations, most military shifts or legislative packages seeking to win hearts or minds rest on assumptions about reality. These are mostly in place—have been for a long time at the center and often so too for the underground as the movement moved toward the struggle. Those in power do have time to think on such matters, authorize research, consider options but most results can be easily predicted. Those underground are as easy to predict except in tactical matters, the visible aspect of the unconventional. Understanding is rare, communication through action, meaning assigned. The dialogue of response is matched by a dialogue of violence. What is special about the dialogue of analysis is that inevitably it reflects the asymmetry of the involved, the qualitative against the quantitative, will against power. Thus there is an asymmetrical perception of reality with a basic structure found in nearly all armed struggles and often in much unconventional conflict.

Such an asymmetrical response is not crafted by analysis but seemingly arrives from the nature of the clash between order and aspiration. Those involved immediately know what they know. The result is that most assume what they know is ample to explain what is happening—no further insight is necessary or available or needed. Those underground are hampered by the blinding vision of the dream that reveals all truth. The truth is not so blinding that gunmen walks in harm's way down the center of the road. Only the paranoid and the primitive court harm. The gunman evades danger but not at the expense of conviction. The convinced knows what needs to be known—which is why there is an underground. And so too do those threatened by such an underground know what they know. The threatened are hampered because they are largely ignorant of the underground dynamics. The conventional cannot know the covert only assume what is to the orthodox obvious. The conventional must assume that the underground is not unlike the ordinary world except secret, that irregulars are reasoned if undisciplined. What is truly secret about the underground is not the living gunman but the hidden dynamics of the galaxy. The gunman sees too much and the generals and analysts not enough.

One of the clearest examples of the contrast in assumption where reality is interpreted is the perception of contrasting assets. Both accept that the struggle is asymmetrical: one side has certified legitimacy and the other relies on history, one side has orthodox armies and the other relies on will, one side controls the land and the other the night. Both sides of an armed struggle are likely to overestimate not only their own assets but also those of their enemies. They, the enemy, possess the crucial advantage—not sufficient to determine the struggle but ample to protract it.

The Lure of Dragonwars

While conventional soldiers often express a contempt for their "unmilitary" opponents who are cruel, terrorist criminals, murderous incompetents, guerrillas on the dole, psychopaths, they are attracted to the rebels' perceived freedom from restraint, the quick-fix murder. The regular soldier and often the government feel that an armed struggle offers the rebel special privileges denied the conventional. They need not wear uniforms, we must. They can retreat across borders, we cannot follow. They can go home at night, know the country, find comfort in the underground, we are strangers in our own land or in this alien world. They can torture, we must not. They must be captured, tried, convicted, and imprisoned, we are shot in the back. All feel the advantages of an unconventional war lie with the unconventional.

Soon the center may be defended by state-terror, may wage of war that contaminates legitimacy and, worse, often fails of effect. To pursue the war underground, often necessary, tactics and techniques evolve that erode conventional practice: dirty wars dirty everyone. The arguments of pragmatism, all true, are apt to ignore this. The state pretends all is normal and so erodes both credibility. Many profess not to care. A terrorist shot down is a terrorist dead. A terrorist murdered by the police is also a martyr, a weapon against all police power, against the pretensions of the state, a loss of legitimacy which is the center's great asset. In fact often the more ready a state is to kill beyond the law, without compunction, mercy or apology the more uncertain the legitimacy, the more uneasy those who rule. The cunning despot is apt to kill those who might be rebellious rather than wait on time. Once time has brought an armed struggle, the pressure on all governments, on all the threatened is to respond firmly, reveal competence: crush dissent even by unpalatable means. After all the inevitable result of the asymmetrical dialogue is the assumption that the other is not only vile but also deploying unfairly. The center sees the underground as cunningly reaping the advantages of secrecy and amorality, free of the restraints of responsibility.

Yet, at the same time, the orthodox military always suspects action without discipline, special operations, anything beyond hierarchy and conformity to rule. Independent formations cause trouble because they are independent. Elite units bleed the regulars for diversionary purpose. Often it is the political center, less prone to discipline and more attracted to effect achieved without great investment that support special operations. And special operations are apt to rest on false assumption: that all is possible once secrecy is achieved. And the possible will be cheap and effective.

The orthodox, covert world does, indeed, have charms: no parades and the pride of purpose clothed in urgency, no routine, no shaving or regulation haircuts, no inspections, seemingly little limit on necessary expenses, the immediacy of action, and the admiration of important patrons. Most government employees so involved run only minimal risks even when active abroad. Most know that their pensions are safe, their government supportive, and their acts legitimate. Most important nearly all like the life: the best of both worlds or so it would seem. From the CIA director violating international law "secretly" in the name of need to the SAS sniper in Northern Ireland shooting to kill a "known" IRA gunman, there is the assumption that the underground is without conventions, without regulations. Free will can run and the orthodox piper is never to be paid.

Contemporary history is filled with efforts to deploy the irregular piper. In Washington in 1981, for example, immediately after the return of the American hostages from Iran, the Pentagon in a momentary flurry of concern for enemies operating below the threshold of conventional war decided to set up a special unit to strike back militarily at terrorists anyplace in the world—and leave no traces. There were already other military units focused mainly on intelligence or on swift reaction like Delta Force to terrorist provocation; but the Special Operations Division was to be very special, very covert, a secret even within the Pentagon, very unconventional. The unit had friends in high places, most notably General Edward C. (Shy) Meyer, Chief of Staff, who perceived a need for action within the Defense Department, where the high command continued to dislike elite forces and special operations. Increasingly to Meyer and others there grew a perceived need for action on the irregular challenges: the new Special Operations Division. Colonel James F. Longhofer, an experienced combat soldier practiced in low-intensity warfare, a hard, driving man concerned with results, became commander. There was soon sympathy within the Reagan administration. The director of the CIA William Casey was a special operations buff and the National Security Council would take particular interest in responding to the unusual with Colonel Oliver North, another can-do combat officer. Special operations were romantic, assumed effective, always deniable—and inexpensive. In general the new Reagan administration seemed eager to counter terrorism, a high priority according to leading officials including Secretary of State Alexander Haig, another Vietnam veteran with hard-edged ideas on communist provocation.

Colonel Longhofer had not only friends but also money. Between 1981 and 1983, funding for the unit was $325,000,000 dollars, money that often arrived through a fake firm Business Security International, code named Yellow Fruit, to be spent on various missions assigned but seldom closely monitored from above. In a Pentagon budget such a sum is not great if not inconsequential—in the underground the IRA ran their campaign for twenty-five years on considerably less. So there was support and money.

The unit set up the secret Seaspray airline, bought and equipped other planes and helicopters, invested with the CIA in a freighter, ran some forty missions in two years mainly in Central America but also including Bashir Gemayel's trip to Washington via an Egyptian airport and a stop in Shannon, Ireland—a venture that also involved

Israeli cooperation on the Middle Eastern flight laps. Special Operations Division was operating at worse as a redundant operation but at best to advantage, by extending Defense capacity and pursuing American interests.

Most interesting, almost from the first Longhofer and his deputy Lt. Colonel Dale E. Duncan and the others began to assume the airs and graces of the covert, drew on the assumed privileges of the romanticized Dragonworld that sanctioned operational freedom and personal peculiarities in the name of pragmatism and realism. The orthodox colonel's vision of the underground was no more true than that of the radical students mimicking an armed struggle, war reduced to costume and righteous slogans. The special operations of the CIA often seemed lifted in time and rationale from thrillers and the Pentagon proved no more immune to Boy's Own Adventures than did the National Security Council with Colonel North unleashed to effect events—secretly.

Within the Special Operations J Division, uniforms were discarded, new identities assumed, and secret business trips laid on with first—class accommodations, accompanying wives and even elegant luggage acquired as cover. Hair grew long and clothes trendy. Expenses were run up without accounting. Lt. Colonel Duncan spent sixteen thousand dollars and submitted his voucher without a single receipt. Formalities were not necessary in the authorized underground nor regular hours nor concern with civil liberties nor budgetary restraints. More fancy equipment was acquired. At home Pentagon rivals were set up. Abroad operations, like those of their allies in the CIA, became more elaborate and potentially more embarrassing if revealed. There is always an innocent assumption that what is secret today is forever so, what is done today will require no accounting on the morrow: especially since the purpose is vital to national interests and the means pragmatic.

Operationally the regulars assigned to the covert mistake special operations for secret operations. For those more experienced in the gray world of intelligence "covert" means only that there can be plausible denial at the moment, which almost always arrives, when cover is lost. From Casey to Longhofer, there was an assumption that what was done undercover would stay undercover. And necessity was ample excuse if there was no oversight to penetrate the cover. In an open society cover is hardly ever forever.

In the American case most of the Special Operations Division's activity became public knowledge and brought the commanders to unexpected grief. The Special Operations Division people ended in court—the rules did run. Longhofer was sentenced to two years in prison, reduced to one. Lt. Colonel Duncan, whose Federal court conviction for theft was overturned, was sentenced after his court-martial conviction for filing false claims, theft, and obstruction of justice—under appeal—to ten years at Fort Leavenworth. Longhofer and Duncan saw themselves as scapegoats in a Pentagon power struggle and victims of the system's distaste for special operations. Prosecutors saw them as individuals seizing an opportunity to operate without rules or restraints. Few noted that the Special Operations Division was simply one more, not especially important instance, when the conventional misunderstood the nature and prerogatives of the underground and so responded ineptly within the asymmetrical dialogue.

The Limits in the Dragonworld

Each dragon campaign both appalls and attracts the center—and each dreams differs and so too the nature of restraint. A different morality may run for Shi'ite soldiers of God, for Palestinian hijackers and IRA gunmen. Some dreamers do dreadful things within cultural bounds. Others must kill close up where states can kill at a distance. Some rationalize the unsavory but all operate within bounds—the few advocates of any horror have produced proclamations and theories rather than campaigns. There is no absolute free will. In some cases chits must be filed, car bombs accounted for in duplicate, and expenses authorized by voucher. The gunmen must account not only for the bullets but also the bus fare. The killer can not kill this one or that. There are rules that limit the arena: no IRA operations in Scotland, the Viet Cong did not send assassins to Washington, the Italian terrorists killed in Italy for universal communism. Atrocity is most often found when the coherence of the arena is lost, where both gunmen and maniacs can operate as irregulars, kill the enemies of the village, claim authority and cleanse the province of Serbs or Croats, seize the state and murder all the Tutus, all the Cambodians in cities.

The desperate and incompetent may kill by error, fecklessly and brutally. The desperate may kill the innocent by intent but within cultural and ideological bounds—bounds that for others seem infinite. And the determined and frantic may violate cultural norms with intent, for purpose, slaughter the innocent for a dream: then there are no innocent people. In Algeria the frantic finally began killing anyone for Allah, everyone they could reach, the women and children, those at the mosque and those on the street, anyone would do and so everyone that could be reached died, another victim for Allah: murder in a delirium of rage. The horror, the secrecy, the audacity, and arrogance of any underground appalls the conventional, forces a response in kind. The state feels impelled to act and often act as irregulars, become irregular.

It is in fact the state that is apt to introduce escalation of atrocity within the dialogue of the campaign or in forestalling any such campaign. Murder by mass has been the province of the state where, if there is shame or cunning, secrecy can be imposed rather than sought. The responses of the state, however, are easier analyzed, easier imagined even when dreadful. Neither the killing Gulags nor the special operations team share the gunman's hidden environment. It is because this world is hidden that the conventional respond to imaginary dragons with real lances.

Such are the lures of the assumed underground privilege that sensible politicians and professional operators are persuaded that such operations are possible in an open society. Professional French operators sank the Rainbow Warrior in New Zealand on July 10, 1985, and Paris could not quite cover it up. The security forces in Northern Ireland have repeatedly with good reason been charged with violation of the law in the name of order—suspects murdered, the innocent detained or imprisoned or tortured, innocents corrupted and the unsavory co-opted to authorized violence. Irregular wars are inevitably dirty and the more sophisticated the venue the more likely revelation. In an open society, the Israelis could not evade responsibility for the murder of the wrong man in Norway on July 21, 1973, by the Wrath of God unit seeking Palestinian terrorists. In any case some Israeli special operations were intimidation of the deed. And as the case with many special operations, incompetence comes with secrecy and the irregular: the deed is neither secret nor effective. In October 1997, the Israelis deployed a chemical element against a Hamas leader in Jordan, were caught, forced to make concessions, saw their enemies reap advantage and their friends compounded.

Someone in power always assumes that such deeds can be hidden if need be. And yet all sorts are caught out not just the efficient and ruthless Israelis, but those caught out killing Teheran's enemies or

Responses to the Armed Struggle   205

Tripoli's dissidents or Algerian rebels, those caught pursuing enemies abroad, always caught without cover and to disadvantage. If not discovered at first, they are often discovered later—and if nothing is proved much is suspected. The Spanish security forces were eventually revealed engaged in covert, illicit operations against suspected Basque terrorists in France—unauthorized but approved. The French have for years tinkered in Africa. Some times paratroopers are sent but there have always been secret agents, mercenaries under false flags, advisors and paymasters. Losing in one Congo in 1997, they subverted the other Congo the same year—supposedly the year of French withdrawal but the lure of intervention proved too great for yet another French government. And Islam recognizes no operational boundaries, its most militant advocate little restraint. Under threat Latin American governments have often resorted to state terror.

Most armed struggles generate retaliation in kind. All governments, most armies assume the others, the underground, does the same not in spite but in defiance of the law and does so and for the same reason. In fact the underground has very different priorities and very different restraints. The gunmen must act secretly or be destroyed. They must pay the price secrecy exacts on efficiency—a cost never calculated by the state's special operators. They lack the arms and skills and opportunities to organize a campaign other than the irregular. They would prefer real bombs rather than car bombs, real battles rather than ambushes. They have but a narrow menu of operational possibilities. They note their lacks, every weapon hard won, not enough money, not enough time, no haven and few friends, limited respectability. They yearn to be regular and often give up secrecy too soon, appear as regulars without capacity and so are crushed by the skills of professional orthodoxy.

All the assets of legitimacy and visibility the orthodox take for granted even, often especially, when they go underground. The state supplies the weapons, pays the salaries, organizes intelligence and safe-haven and awards medals. The command center of the underground runs on faith, offers a chance to sacrifice, offers not a vocation but a way into the future. Those underground do the best they can to be legitimate: titles are used and ranks, uniforms are worn when possible, proclamations are printed, peoples' trials held, opponents executed not murdered. There are no terrorists underground only volunteers desperate to be acceptable, eager to explain away atrocities that the state seems eager to emulate. Underground there is no feel of freedom that the special operator finds but of intense commitment, nothing parallel at all, a time curbed by limits and restraints and shortages, not romantic at all and so secret that few understand what is involved. In a sense the cover is perpetual, for even with triumph the underground remains a secret.

The Dragonworld is truly different but not absolutely so. The volunteer does pass through a mirror into a strange and wondrous land but not into a wonderland without rules: a fantasy for the orthodox who mistake secrecy for license, ruthlessness for irresponsibility, terror for amorality. The rebel volunteers have no such illusions about the benefit of free will. The unconventional see more problems than opportunities, fewer advantages. Without uniforms they may be hanged for murder and are often treated as common criminals: murder is murder. Without the power to hold liberated zones, they are forced to flee, constantly on the move, on the run without home or haven. Without courts or police or the leisure to dispense proper justice, they must maim offenders or mete out rough vengeance to suspected informers. Yet such coercion may endanger their fragile legitimacy and risk the alienation of their own. Their view is that the conventional have all the assets: the big battalions, the well-fitted hospitals, leave for the troops, the capacity to make the law, overt legitimacy, the trappings and privileges of power—everything except for the righteous cause.

In sum, during an armed struggle there is an effort to take and deploy the perceived assets of the other: become legitimate, deploy illicit means. Both would like to wear the other's clothing: while for the time being keeping their own. The most patent misperception of both focuses on the apparent necessities of Dragonwar: the distasteful if defensible techniques of violence. The dispute concerns the methods of unconventional war's armed struggle, the irregular means of combat. And almost inevitably the perception of the involved shapes an asymmetrical response that parallels the asymmetrical nature of the conflict, of perceptions and assumptions. Any unconventional war, each armed struggle offers such a dialogue.

VI

The asymmetrical reaction evinced by an armed struggle involves both the qualitative and quantitative. Assumptions matter but more than in most events determine what numbers mean, may even determine the numbers. Those threatened describe what they see and what they assume, what they fear and what they can count. Those underground out of sight, often out of touch, perceived their challenge as a process that need not be examined since the truth is pure. In Ireland the British always see light at the end of the tunnel and distinction gained by persisting in their responsibility of rule. The IRA having transformed physical force all but into an end—the outward and visible form of the Republic—see not light at the end of the tunnel but the necessity to persist as historical legacy. The peace process makes everyone directly involved in the campaign uneasy for the shift in priorities puts crucial roles and missions at risk.

Quantitative factors are deployed only for ideological purpose—although all count, count the votes, the dead, count on the past, count the money to hand, the troops deployed, count on the future. And the Irish experience is not unique. Thus everyone is apt to perceive the course of the unconventional war as proper focus. What is happening. And many, certainly many underground, always know the answer without need of counting. Those at the center count for good purpose but also to weight their assumptions with data. These perceptions form the inevitable dialogue of asymmetry—and medium of response.

An armed struggle nearly always requires a military response: bandits can be left to the police and gunmen jailed as criminals only for a time. Then the armed struggle either recedes into a real bandit problem and the gunmen are all in jail or there is an irregular war, an unconventional struggle made in great part unconventional by the mismatched forces.

Underground there is often simply a man hunched over a basement table finishing a crude device fashioned from flex and brown powder, a cheap clock, an EverReady battery from the store around the corner. His mind is seized on the great cause, a cause so compelling that he may forget and smoke while working, may forget what was picked up from an old text and hasty training. No matter, there the dream lives for the moment, time ticking in a room fouled by the tang of cordite and fear, visible in a device just sufficiently novel to evade the professional disposal team. On the other side, above ground, is the orthodox, salaried, accoutered with rank and long training, who settled in the cockpit of a forty-million dollar airplane has the most elegant smart-bombs hanging in clusters below the sleek wings. He is a professional, a career pilot facing a vast dim-lit, computer-driven digital display that will assure a precision strike in the total dark at full combat speed: a smart bomb delivered on dumb, awe-struck targets.

Yet no one can be sure that the one or the other is winning. They even have difficulty reaching the other with their devices. When they do, the bodies after each bomb can, indeed, be counted by not the will and never the implications of the dream or the staying power of the orthodox. As long as the hand-held bombs can be detonated there is prospect. As long as the center persists in paying the cost of a response these prospects are poor. And until all is clear, the gunman dead, the planes turned over to the revolutionary government no one can be absolutely sure if the military is winning or losing.

What seemingly can be measure is the apparent intensity of the campaign and so the depth of the challenge. Wars generate numbers. Intensity can be given numbers. Yet again when these are toted up there must be someone to give the implications, keep the score, take names. Perceptions of value may indicate what is to be counted and what is to be learned. Those who estimate underground morale from captured weapons may be innocent that weapons are being sold into the process for underground gain. And that even if all weapons are lost all may not be assumed lost by the faithful who may persist without weapons: the Mau Mau in Kenya was reduced to spears and zip-guns and the Bushmen in the Kalahari desert carved SWAPO on the thin wooden bows that had not changed design in millennia. Just as an armed struggle often begins with a few men around a table, the odd revolver, an old rifle and a guerrilla text so too can a campaign last with similar assets as long as the dream has validity. At any rate, all is not simply imaging and hope and blunder for the real is real, the body bags are filled, the countryside unsafe, the prisons have inmates, and the general a list of missions flown and patrols dispatched. An underground threat over time can be charted and graphed by incidents, arrests, weapons found, and soldiers lost.

Counting gives the concerned numbers not meaning. Perceptions matter the most. Terror, especially crafted spectacular terror, inevitably engenders a great wave of popular repulsion unrelated to the few killed. A dozen hostages in Beirut warped the priorities of America. Israel chose to go to war over one shooting in a foreign capital and risked peace to kill in a friendly Arab country. A crucial aspect of the intensity of an armed struggle is the perception of the threatened: the underground always assumes the best and is tasked to impose the worse. The center can weight intensity by need and experience and exposure—can mix data and desire. Some can always shape the challenge they find congenial just as some underground can always imagine since they feel intensely that the armed struggle is effective. Numbers may not matter.

Even numbers may be misinterpreted: the totals hanged or interned or shot rioting may be added to the government's assets by error. Escalation is a sign of rebel success. Reinforcements rushed to Ulster or Vietnam may indicate not strength but weakness. French reinforcements rushed to Algeria produced comforting military statistics: great numbers of FLN killed, captured, exiled, the battle of Algiers won, the countryside tranquil; but in the end, more soldiers, more military triumphs added nothing to French assets or extended the future of French Algeria. Successful operations in Algeria were battles won on the edge of the real war: the will of Paris eroded—and that will could not be counted out in decimals nor could the intentions of De Gaulle, the final response of the French people, or the reaction of the elite. In time the French decided they would not persist. The will had gone, paid insufficient dividends, cost too much. And so the center would cede history to the Algerians and do so not because of tangible losses but perceived advantage. French Algeria was not worth the price. Now, so far, their ultimate successors the generals of Algeria have paid the price to hold the center against the Islamic rebels, are being tested by a ruined faith. The bombs in the streets can be counted, the bodies in the gutter, the village atrocities but not the will of the involved now frantic, slaughtering their way toward a fading future.

Numbers can, do indicate the tide of events. The ideological terrorists of Europe ran out of recruits and so the incident level went down as the faith waned. Matched against reality the dreams of the Italian and German terrorists failed. And the other ideological terrorists, a few French or Spanish, rallied no one else, cadres without party or prospect. Nowhere did ideology work as the text indicated. The Italian state had no heart—there was no Imperialist State of the Multinationals. The Germans radicals ended up murdering Jews—Israelis, Zionists—ideological "valid" targets. Everywhere, sooner or later, the proletariat ideal of the militant left became a nightmare not a dream. In fact the incident total, the target lists, the outward evidence was increasing indication of the implosion of the world of the Euroterrorists. Numbers could be shown to matter.

The Irish problem has persisted at various levels of intensity, often moribund but at times rising to insurrection. What persist is the republican dream: one Ireland, Free and Gaelic. What does not always persist is the capacity to deploy physical force. The British in turn have become inured to the revolting Irish and so are prepared to pay the costs when due: an army dispatched, corruption of justice and official brutality, sterling wasted. The more incidents the higher tangible costs but somehow to the British so too the great psychological returns of duty done even if unpleasantly. In any case, much of the time, the Irish can be isolated in perception, made to disappear, ignored except after a special atrocity or particular disaster. Britain has suffered with the Irish long enough so that there is little intensity to the challenge.

Always the involved see what they choose, what is comforting, what fits neatly perceived needs. How intense the struggle depends on how intense the perception of the involved; but intensity cannot be shaped without recourse to the real, to the bodies in the ditch, the inability of guerrillas to move in the daylight, the number of soldiers or prison cells or incidents. The shrewd can read the implications of incident numbers or lost weapons, the shift in a voting return or the sales of a newspaper; but this is most easily done away from the clash of the conflict. Even then, the disinterested are apt to become interested in the war and adjust the quantitative to expectation. Numbers can lie to the gullible at the center and always to the optimistic underground. There is always light at the end of the tunnel. There are always outside agitators, encouraging reports from the commanders, the reassuring list of targets bombed and bodies counted. The rebels count, add and subtract in their own special ledgers of progress, even more optimistic than their opponents since they have less to be optimistic about.

Only at a distance and in retrospect can numbers be given appropriate meaning, but there is meaning enough even for the dim in a protracted campaign to estimate intensity. Experience and reality is adjusted to assumption. The kill count can be toted up each week, this many dead, this many this week; but the total is apt to be arranged as conclusion not number. And in most cases the counting is fudged, intangibles factored into the end result, elusive data made hard, statistics that average out to advantage. Yet reality can be imposed even on such assumptions.

The Palestinian fedayeen could not operate within the Israeli-occupied territories in 1969. This was neither desirable nor anticipated—nor for a time possible to accept. The theory and practice of guerrilla-revolution, the justice of the cause, the expectations of the faithful were all being denied. In time the leadership accepted the reality: all operations failed. Why and what was to be done was not a matter of counting. What could be counted by all was that no one came back across the Jordan: that could not be counted away. And this for some meant recourse to terror as strategy rather than guerrilla tactic: spectacular hijacks and the murder of innocents was within capacity if outside the Israeli arena, assured fame, persistence, and mission if promising little but visibility. After 1972 the IRA could no longer operate as a militia in no-go zones and after 1975 not hope for quick British concession, but for several years the Army Council still hoped to reverse the direction of events. The secret army had to regroup into cells, could no longer bomb and ambush with impunity but could find soft targets, take the war to England, persist. The losses and obstacles forced the IRA to adjust, cut risks, cut operations, cut size, accept there would be a long war. And the implication of those events never really penetrated the core of the Irish republican galaxy where the truth remained revealed. The long war went on and on, year after year, a lethal stalemate with the British will adamant and the IRA capacity no greater. Sacrifice was sufficient for some and persistence offered a congenial roles but others sought other means, a new direction so that tomorrow offered hope not assured disappointment. No Palestinian or Irish republican wanted to read a future different from the dream. And many would not and so placed their faith in the long war, in time, in will, in history, not numbers and reason. If they had paid attention to numbers and reason, there would have been no armed struggle in the first place. Today might not be as imagined yesterday but tomorrow is another day.

In matters not easily numbered, the nature of tomorrow, the meaning of yesterday, perception becomes all as the analysts on both sides of the barricades measure pain and morale, prospects and intentions. None can do so with rigor or often even accuracy, so none can precisely keep track even of the war's direction. Intensity, in fact, may decline without relevance to outcome: winning in the countryside, winning in Algiers or at Tet does not assure victory anymore than does the body count. How firm is the morale of the generals? How dedicated the soldiers walking point? How sure the man in the palace? How long can they tolerate these losses? How long will Israel live in a state of siege before recognizing there can be no peace without justice. How long before the British establishment accepts Ireland is not a responsibility but a liability? How to ponder the heart of darkness? How to balance the asymmetrical? Who is really winning and what is there to win? Each challenge demands response, response to the little questions, response to the overarching means, responses to opportunities and dangers and data. All armed struggles are a mismatch, a confrontation of the desperate and frantic with the pragmatic and responsible. And so what is happening, what is the very nature of the threat, is open to perception.

Is there a war at all—in Kosovo or Corsica or the Punjab? With another IRA ceasefire in 1998 is there still a war in Ireland? There is no shooting, nothing to count, but what matters is what those at the heart of the galaxy imagine reality to be. And with the Gaza-Jericho agreement what of the war for Palestine? There were certainly still bombs and still those who trust only in armed force to defeat Zionism or to crush the Palestine dream, but are these residue or harbinger? Are the fundamentalists of Hamas open to co-option or apt to become irrelevant tomorrow? Until then how can those eager to die be dissuaded by threat of force? How can twice-promised lands be divided? How can a divided society be united? How can history be rewritten to assure an Arab Nation or be remade to assure Irish vengeance and British pride? How can anyone find advantage in Afghanistan, Sierra Leone, or the South Sudan while the arena is a free-fire zone and the population fleeing? And what of the graduates of the wars of Islam wandering in search of a target, volunteers in a self-declared jihad? They are committed and armed but how dangerous? The future is not easier to read, ever. Who will win and who can win? And who would have bet on the Chechens? Most important, the answers change with the questions: to know what troubles Ireland is to know the answer to the Irish Troubles, to know what Allah wants is to know what must be done.

The big questions estimate the direction of history and are likely to be answered only by history. What did Grivas win in Cyprus? What is the heritage of the Mau Mau? Is Zionism merely to be an incident of Jewish history? Even the little unconventional questions of an armed struggle are just as open to misperception or to rationalization. What did Mountbatten's death accomplish, for the IRA, for Ireland, for the British? What effect did Sadat's death have on the Egyptian fundamentalists? What impact will any or all of those who seek the answer in Allah and the gun have on the Middle East, on the West, on the Great Satan?

What do the violent events, counted or not, of an armed struggle mean? This is the crucial question asked and answered by those who will drive a car bomb to paradise or persist as a rural guerrilla *foco*. No matter what the answer is to the involved, an answer is required. In this asymmetrical struggle how intense is the action—and how can direction be estimated? How fast are we going where? Some armed struggles cannot impose questions for the center answers with effective coercion. There is no question that the rebel leaders are dead in the prison courtyard and their peasants sent back to the provinces. The war is over. Real wars give real answers—D-Day is D-Day, Verdun defended or taken, the defeated driven from the field at Culloden or Gettysburg. An armed struggle remains a Dragonwar where what matters matters. The distant, the analysts in research institutes, the observers are too drawn into the meaning of meaning. Some data is hard but much is not and much shaped to supposition Death matters, data matters; but if the underground dream persists, then those who judge and estimate matter most—and so too their dreams and perceptions and assumptions. In sum, the threat of a Dragonwar evokes in large part a perceptual reaction that is not wholly dependent upon data, upon right reason, even upon advantage but rather on the involved, those with guns, those at risk, those responsible, the frantic, the desperate, and the dedicated who each sees through a dark glass of assumption and desire—sees clearly for present purpose.

* * *

Once an armed struggle exists, once there is a challenge, a response, a special psychological climate exists, different for each struggle, similar only in that a perceptual medium inevitably is there and adjusts. Some of those in power can tolerate little provocation and some rebels, whatever ideology insists, have little support. To be effective, even to be tolerated, an armed struggle must fit effectively into a psychological context: this is one of the most unconventional aspects of such wars. Mere grievance does not authorize violence. An underground must assume that the cause arises from the avowed constituency—the people, the Irish, the faithful, the wretched—which is seldom the case. A reluctant constituency need not matter as long as an armed struggle is tolerated. Those who would act in the name of all—usually without asking permission, need only to be left alone to wage a Dragonwar. There must be a climate of consensus that creates a psychological climate of toleration. Unless the underground can achieve such unstated legitimacy, can read the psyche of the constituency, there can be no violence. No matter how compelling the committed are to change, how vile the grievances or weak the center, there must be at least toleration for violence deployed. Without it, the struggle will whither away.

In a real war the attitude of the constituency rarely matters—the people's morale is important but the weight of force can compensate even for lack of popular support. Thus real armies focus largely on other armies not on the enemy nation despite the investment in propaganda and the occasional strategies of intimidation: bombing meant to terrorize. Dragonwars, just like the guerrilla text indicate, are people's wars; but the people can be intimidated, can be reluctant enthusiasts, can be ignored but in the end cannot be so opposed to the use of force that they deny the cause, inform on the volunteers, enlist with the enemy. In fact at times the center may lack big battalions, lack legitimacy, conviction, text book assets, even lack the will to persist, but still find the underground incapable of transmuting the climate to permit an armed struggle. IRA guerrillas on the streets of Northern Ireland in 1965 would have been a joke and five years later were a reality: the climate had changed and the IRA on the way to a real armed struggle. The psychology of the masses, the people, the nation, the family down the street matters. Any struggle takes place with overlapping, often contradictory, rarely complementary psychological assumptions by the involved.

West Germany in 1970 had emerged from a confuse and dubious historical past. The Germans were unsure of the nature of the legacy of history—Bach, Beethoven and Bismark, the Prussians and the triumphs of science,—and always Hitler. Thus many had been unwilling to contemplate the nature of that inheritance, especially those who had lived within the Nazi era. Most Germans in the West hoped that the new democratic Germany was real and permanent, that traditional order, discipline, creativity, productivity, and enterprise in the service of democracy was the way into the future and out of the dark past.

Hope was not surety and so the rise of the terrorists, a generation who had not known Hitler and recognized hypocrisy, facile assumptions and agreed evasions. The new zealots held all Germany to blame. And Germany with unrequited guilt and traditional assumptions seemingly offered an especially rich medium for a few killers. The terrorists represented no one but rather represented what all Germans feared: the past come to call. It was a challenge that awaked not only conscience but also habits of response thought aberration until the time of terror.

The ensuing chaos appalled the government, the Germans, for the Germans had not lost their commitment to order and discipline as vital not mere tools for governance or productivity. A few reckless and frantic ideologues found not an effective medium, popular support but national outrage and fear. They converted no one and even with support from afar did not last the course—only indicated that if Germany was not ripe for revolution, the past was still beyond reach of the present.

In Italy the present was always aware of the past, a history of tribulation, betrayal, and failure: power at the center was to be evaded not as in Germany cherished. The new terrorists of Rome and Milan and Turin, had believed that the power at the center should be used not evaded. Italian society should be reformed not exploited, that the conclusions of seminars were identical with reality as perceived—as they perceived it. They were truly a new generation if they, like the Germans, had roots and ancestors. Those in *Brigate rosse* and the other radical undergrounds believed that change was possible and good and imperative: the future could be better than the past. Even the fascists gunmen believed that the past could be reimposed to make a better future.

Mostly Italians had felt that if tomorrow were as yesterday no more could be asked. The Italians could easily tolerate disorder and did tolerate violence against a state the vast majority supported. While historically Italian states engendered no great loyalty and the truculent resistance to the center was part of the Italian psychological climate, the aspirations of the terrorists, Right or Left, reasoned or emotional, were a danger to the everyday. And it was the family, the neighborhood, the small successes that mattered, not a dream. Dreams in Italy were always dangerous. In the end the terrorists ecosystems imploded when the dream proved unrealistic. The classroom model that had left bodies in the gutter, fear and anguish and the distinguished dead, failed in real life. Italy had always been about real life.

Italian regimes were not really prepared for such dissent but could respond traditionally with delay, routine, and patience. This too would pass once the gunmen realized nothing was going to change. The German establishment could hardly imagine such an inefficient response to provocative terror: such chaos could not be tolerated. In both cases the armed struggle flourished not because of grievances, real or perceived, not because of popular enthusiasm, largely imagined, not because of underground competence, absent or erratic even in the most violent years of lead, but because the psychological climate was rich with potential.

No one in Britain but the Irish could imagine purpose in the gun. The Spanish had their Basques but most in the West had neither a nationality problem nor a medium to be exploited by ideological revolutionaries. Almost no one in France became involved in terror, and those who did generated a police problem not a psychological one. When bombs and murder came to France, early and late, it would come from Algeria—a nationality problem. When terrorists operated in Sweden or Switzerland, they were transient. And no matter how rich the medium may appear, harsh and efficient coercion in a closed society will prevent an armed struggle, so across the Iron Curtain there were no armed struggles: the East would imploded. Then there would be opportunity for old dreams and new nightmares.

What many revolutionaries seek in an armed struggle is to avoid overt defeat while they win, sooner or later, a psychological victory. Such a triumph may be won in the hearts and minds of the people but not exactly the ways the texts indicate. The people may find the rebel hateful but inevitable. And the enemy is people, too, and may find excellent reasons at some point to end the battle. Thus conventional generals in an unconventional war find that it is crucial to know the impact of their maneuvers or the perceptions of the concerned. Who will suffer the most? Who will persist and how long? Who reads history with assurance? Their military victory, like that of the Americans at Tet in Vietnam or the French at Algiers may end up as disaster politically because of the perceptions of crucial observers. The British always chose to assume they moved out of colonies or mandates willingly leaving tutored governments that would in time send representatives to the Commonwealth meetings to play their part, meet the Queen, show the new flag, sport the old boy ties. Those who did not play, Begin or Nasser, remained British villains—the victims at fault. No less differently that the rebel imagines history or weights the will, so too does the state shape reality to be congenial.

In each case what is going on and thus what should be done arises when the armed struggle is perceived and shaped by the psychological assumptions of the responsible. The underground only rarely shifts basic assumptions, so perception continues and reinforces the existing psychological set. Those under threat, more varied, more complex, more fluid may more readily shift. In fact for the weak to win they must shift. Thus in the asymmetrical confrontation the underground is always more conservative. The underground must conserve the few assets, the will and the faithful, the dream, can rarely afford to be flexible, to think twice, to think again, to risk much, even to risk change tactics. As always the psychological response to an armed struggle is asymmetrical.

The underground, more so than the target at the center, is apt to recognize the importance of psychology—having so little else open to use—but not their own inflexibility. The revealed truth is always assumed an unchallenged asset, a rock beyond erosion even when belief shifts, doctrine is adjusted, and the gospel is rewritten to need. The core commitment to the dream remains static and so too the psychological perceptions of the involved. These perceptions of the armed struggle tend to determined if not reality then the significance of reality. Winning is what the central command and the galaxy says is winning. Death at the hands of the Israeli commandos or the German GSC-9 to the transnational terrorists may be only an incidental aspect of an operation perceived as successful once the television cameras have switched on. Munich was a success in 1972 despite the losses. The martyr who drives the truck bomb wins a ticket to paradise and his colleagues not only a tactical success with the 241 dead Marines but also a strategic one with the Western withdrawal. That withdrawal, that victory was a result of psychological shifts in Washington and America not in Islamic assumptions. The outcome of an unconventional war depends on revolutionaries using their armed struggle as a lever to transform target perceptions and hence speed the transfer of power from the center.

The power at the center is usually real enough. Surrounded by tens of thousands of well-armed troops, whipped from place to place by helicopter fleets, supported by artillery, reinforced at will, opposed by rag-tag terrorists, it is always hard for the conventional commander to recognize that visible military strength, even properly deployed, is but one aspect of unconventional conflict. And unconventional war is to be won by perceptions rather than the weight of metal. There are other powers, other factors that matter in a unconventional war that can often be ignored in real war. Military power, state power, even the power of the people can be brought to play on the dream and the dreamers. Force can impose the psychological assumptions of the powerful on the weak—and power does come from guns.

The operations of an underground in a Dragonwar is apt to produce an impact that is psychological not really military at all. The terrorist of Germany could kill only a few at a time, could threaten many but harm almost no one, but they engendered an enormous response from the orthodox. The IRA could not bomb the British out of Ireland only hope that in time the will of London establishment would erode. An end game of negotiations is a test of wills for the faithful not a process of accommodation. Those who accommodate always deny the dream and to survive must offer nearly all, offer a road into the future toward the ideal. Consequently an end game is enormously difficult for those so long content with a single reading of reality.

The IRA or the Irgun, the rebels or the imperialists always are shaped by psychological considerations not fully understood then or later. All respond to the armed struggle—the participants and the observers, the disinterested, the committed, the analysts, and the gunmen—in special ways adjusting and explaining reality. Those perceptions are shaped at the time and in the future by the general and particular psychological climate. This may change even if the perceptions of response largely stay the same.

The very shrewd may deploy against those perceptions in order to induce psychological change. The gunmen may resort to terror not to win over great power but to win over their own or to appall. The British among other reasons left the Palestine Mandate because they were appalled at the lack of gratitude of Arab or Jew, appalled at the irrational ambitions of both, appalled at the lost of British life, appalled ultimately that none could recognize good will or British decency. So the powerful may be so sicken as to withdraw. The powerful, on the other hand, may be convinced that concession is strength or that the contest is no longer worth the cost or the contest has been won or lost. Britain left part of Ireland in 1921 for a variety of reasons that shifted with perception over time—left in 1921 because it cost too much to stay—and Britain has stayed in part of Ireland since 1921 for a variety of reasons that have shifted over time but mainly because of psychological returns paid for with an acceptable level of violence—psychological returns discussed as moral imperatives or historical responsibilities. Much that is crucial in the response to an armed struggle is thus difficult to quantify. Most often the more conventional, the orthodox, the generals and bureaucrats are apt to discount or deny the significance of what cannot easily be counted. The struggle for the hearts and minds of the people is often counted out in leaflets distributed, speeches made, strategic hamlets opened, or votes in an election. The great joy of tangibles is simply that they are tangible, real. The real matters, often matters the most if actual assets can be brought to bear: the hearts and minds do not matter if the rebellious heart can be quelled by force. Force displayed, force tangible, force in the gunman's weapon or the state's constabulary has enormous attraction.

The conventional assume that what is needed is more the same, greater freedom of action, less restraint and more assets, a traditional, orthodox military solution. Yet the military is not without talent or insight and can recognize that unconventional challenges may require unconventional responses: mostly those responses are adjusted for existing priorities so that the irregular is pursued by the book. Strategic hamlets are built, peopled, and listed—tangibles, the perceptions of the peasants so involved are less easy to quantify. More appealing is the attraction not of ideas and attitudes but the techniques of the unconventional. Thus special operations inevitable, often reluctantly, are given a role. Inevitably the techniques of a dirty war are integrated into the already unconventional campaign. The center wants all the assets, those that can be counted and those that count in the field. The invisible opponent unchanging and impervious to reality persists. The dream empowers, supplies energy, and what more is needed is the tangibles—those assets held by the center. So from the very beginning the underground seeks what is necessary; arms and money, recognition, liberated zones, general admiration, and more bullets.

Each side is inclined to assume that they have more assets than the opponent, the people or the power, but long for the other's anyway, no matter how dangerous this may be. For the conventional to resort to the irregular may mean the erosion of the legitimate assets of a normal, recognized army; and for the unconventional to seek to act as a "real" army in a Dragonwar may waste their real if not easily touched assets. Both accept once the struggle persists that unusual means will be needed. And the center, the state, the orthodox have the graver problem since what they want is hidden beyond techniques or tradecraft. To respond to the Dragonworld requires either great power, effective brutality, or patience and cunning.

VII

The practicalities of responding to the unexpected challenge of a Dragonwar—are formidable even for those intimate with the irregular. Yet, more often than not, the first response is coercion. The dynamics of an armed struggle, the implications of perceived reality, the interpretation of events, the psychological factors, all fade for those who must respond to provocation. Those with real power want to use real power—first. They focus on the evolving campaign, on operations, on force and only later on means to deploy a whole spectrum of capacities. All those underground are apt to pursue their ideal in similar ways, restricted by scarcity, covert, illicit and opposed by the state but focused on first force.

For both the target and the gunmen, whatever their strategy, whatever ideology offered, whatever expectation, the armed struggle when protracted always proves far more complex and far more trying that imagined. Again, even protracted struggles are the exception: most are not complicated at all but only foolish and audacious and swiftly ended. Only a few struggles do persist, escalate into a real armed struggle much less an irregular war. Some few do triumph, even some that arrived at the end with fewer assets than at the beginning. There are always a few to persist despite failure. The IRA arise from centuries of frustration. Those who find the answer to all in Islam are one more wave of the faithful. A century will not erode the national virus or coercion assure the death of a dream. Most dreams that inspired an armed struggle like the struggle fail, some at length and some at once. Carried into exile they are ossified and lost over time: no recruits, no prospects, no conviction. Kept in the secret hearts of the sullen faithful they die at home as well as abroad. The center usually holds, deploys coercion effectively, moves on.

Those who seek to impose order have greater options and may have more friends. The spectrum of response is thus enormous and the ensuing campaigns various. This is why the differences often seem for greater than the similarities of an armed struggle. Yet, the great differences are simply matters of degree: the level of deployment. The factors that determine that level are so diverse that contrast rather than comparison is more analytically appealing. Each center is so different, each arena so various, each campaign so special that general phenomena tends to become clouded. Yet, on the ground what is different is mainly the deployment of force, a matter of numbers and tactics.

Certainly, those engaged in an armed struggle are apt to focus on the immediate and the concrete regardless of the recognized importance of perceptions and the psychological climate. The gun concentrates minds and shapes all practical responses. An armed struggle may generated perceptions, may shift values and estimates, may change minds; but first an unconventional war is apt to demand an orthodox response: the restoration of order. The police round up the usual suspects, the army appears on the street, the courts work overtime, the informers are paid, and the phones tapped. On the other side of the dialogue, the underground simply focuses on persistence, first evasion and then escalation, always focused on the operations of the moment not on the complex strategic disputations of the analysts.

Unless the gunmen can simply be ignored, the center responds. And the more powerful the center, the more ambitious, the more likely any challenge, no matter how distant or where directed, will seem to require a response. Authority seldom is willing to tolerate subversion unless tradition or habit so indicates and always each case is special but not too special. Thus the Italians could tolerate more disorder than the Germans and the British are inured to the revolting Irish. All, however, whatever the psychological medium, the perceptions of the involved, the historical legacy, deployed force and in similar ways—differences in degrees and in tactical mixes.

This coercive response does not occur in isolation. Often despite the isolation and local nature of the conflict, the action of the arena is apt to expand. Others are concerned. Certainly, everyone involved wants allies and approval—even if the government insists the problem is internal without other implications than those of criminal subversion. Very soon any underground will have sought allies and at times been sought out by patrons. Operations may seep over borders. Neighbors may take sides. And if the campaign persists, the government is apt to seek reassurance, the good will of those abroad and in some cases overt aid and comfort. Thus an obscure and special armed struggle may give rise to extensive international concern and often overt or covert intervention. This was especially true during the Cold War when all events anywhere were imprinted with ideological meaning. The new postmodern order is medium for concern for many who support or oppose a whole spectrum of causes, creeds, and concerns. This has always been the case: Napoleon had revolutionary friends and Roosevelt and Taft and Wilson in Washington found that they had revolutionary enemies south of the border. The Ayatollahs have intense interest in matters far from Iran, outside the Shi'ite community, have enemies and friends in the shadow of the Great Satan. Libya is interested in the Moros and Mali, in the Irish struggle and in the Sudan. Angola intervened in Zaire and then the Congo, an Angola still engaged in an internal end game but not so involved that an opportunity to adjust the region could be allowed to pass. Each power center has local, regional, and general interests that disorder is apt to focus.

Such interest and concerns create a second ring of active response: democracy is defended by the dispatch of arms and advisors. Revolution is encourage with access to training camps and weapons, Allah needs allies or the faithful a helping hand. And in time the distant may become intimately engaged in encouraging or denying an armed struggle: Semtex to the IRA and the word of the prophet distributed in Brooklyn. Even the distant who want only accommodation and resolution may find good will entangled in violence: United Nations troops killed in the Somalia or regional peacekeepers dispatched to Liberia or Bosnia. Armed struggles may stay isolated in the outback but often do not, do not for all sorts of reasons often external to the grievances and agendas of those most concerned. Thus the operational response to a Dragonwar is not necessarily a matter of the native arena or even the surrounding area. Vital interests tend to be what is vital today to those in the capital. The French tinker in Central Africa and the Congo. The West reluctantly becomes involved in Bosnia. The Turks have friends in central Asia and enemies in Damascus. An armed struggle has resonances far beyond the killing zone. Even without a Cold War distant agendas may be altered by local violence. Those directly threatened may respond locally, may seek aid and comfort, may find the armed struggle spread or contracting. The distant then shape that struggle to their own agenda—respond from beyond the initial battle arena.

Overt intervention outside formal arenas is projected because the distant assume their vital interests are involved or at times simple opportunity is offered to make the world safe for democracy or indicate that France is still a great power. Reasons differ: the Soviet Union intervened in Hungary and Germany to protect vital interests and in Africa or Latin America for ideological purpose. Some power centers may claim the right to intervene as the United States has often done in the Caribbean or with the Monroe Doctrine deny others such a right. Others may simply and seemingly without right reason become involved in the wars of others: Libya aiding the IRA or the IRA aiding the African National Congress.

The most amorphous challenge and so the most difficult to fashion a response is a transnational threat against the system or the West or imperialism. Those who bombed the World Trade Center, the free-floating Euroterrorists of the 1970s, the revolutionary anarchists at the turn of the century, all could strike anywhere within the system, none had effective sanctuary in the West, friends but no bases, and all presented problems of response. How to strike at the core of the conspiracy? When a patron becomes too visible reaction is possible: the United States bombed Libya and in Lebanon shelled suspects. Symbolic terror deployed against stability is not easily countered by discrete acts. Many unconventional challenges, armed struggles and irregular wars, can more easily produce the responses of the orthodox to subversion and provocation encoded in anti-insurgency tactical doctrine. The center, the state, the regime, the alliance, assumes that what is needed is the usual: deploying more troops, tapping the telephone system, buying information, building fences, running agents, setting up the panoply of security, centralizing command, seeking the good will of the international community. The basic strategy of most target-states is to maintain order by coercion and consensus, seek and destroy their opponents, keep order and prevail. The variation is how to seek and how many to destroy: this has mostly shaped the policy.

And this policy coupled with the intensity of the challenge produces the campaigns of the armed struggles, each different. One patent difference is the degree of external involvement. And while only a few campaigns can evolve in absolute isolation, even those struggles that persist only rarely are shaped by intervention from a distance. Action at a distance, opposing an armed struggle abroad as response, always presents a set of problems. Constant exposure to threat may establish a corporate memory, a text for response as is the case with most empires under threat. During the erosion of the old empires, both the British and French military produced coherent anti-insurgency strategies. These were largely an integration of tactics with political initiatives. The fluidity of military deployment was most effective as a coherent part of a general response. These texts were often read but not always integrated into doctrine or used on the ground noted the role of the people—the guerrilla sea—and the professional weaknesses of their opponents, mice against elephants, the war of the flea. And so noted, the troops could be deployed effectively—as long as the government had long as well as short views, pursued pacification with more than elite units and fitful attention. In any case, the military, not unreasonably, focused on tactics no matter the role assigned to politics or psychology. Strategy was for the capital. And often the capital had other priorities.

All armed struggles are, as struggles, local and all intervention from a distance is apt to alienated those locals. In turn the underground operating at a distance does so not out of strength and the aggressive pursuit of perceived interests but out of desperation. The IRA move active service units to the continent or the Palestinians opt for terror by Black September because matters are not going well in the chosen arena. Thus expansion and escalation are sought on the cheap: soft targets are rationalized ideologically but new fronts abroad are opened because operations and options have been narrowed by coercion. To avoid the struggle being wound down not only is the arena expanded but also aid and comfort sought—and such aid and comfort comes with a price. Patrons expect returns for the investment: the gunman is seen as purchased ally. When the intervention relies on indigenous support for an alien agenda, the prospects are dim: the persistent failure of Latin American *focos*, the futile spectaculars of Euroterrorists encouraged from abroad, and the scant returns of the traveling salesmen of world revolution. And pursuing their great power interest, the great powers find the imposition of control, the funding of subversion no easy matter—most especially if there is not dream but only interest involved.

Mostly intervention, costly, symbol and sign of power held, power deployed, is state directed against a perceived underground challenge. While any armed struggle generates similar problems, those poised against a real foreign force are especially acute: a real alien occupier, a real monster is always to underground psychological advantage and may offer tactical opportunities as well. For Iran it is easier to target the Great Satan than a heretical moderate Ayatollah, easier to fund subversion than renew the war with Iraq, easier to matter in Lebanon than in Egypt or Algeria. In any case, most armed struggles that persist tend to attract at least some external concern, gain enemies and allies, tend to bleed out of the original battle arena—and this remains true in the postmodern world. Those responsible rarely have sufficient hard data to hand, even experience is apt to reinforce stereotypes and innocence. The British know the Irish but the Irish they know are symbols and assumptions. Somehow after centuries the British still do not really know the Irish nor desire to do so. Minds are not easy to change and even tactical data not easy to find.

New, hard data is rarely presented to those who have opted for action. What is wanted is targets and vindication. In fact there often is no ready market for data on the psychological climate, the reality of the dream, or on the unconventional factors that matter. What is absorbed is reinforcement of expectations, target lists, and tangible directions: the map as terrain. The map sought is guidance for an effective military campaign. That campaign is inevitably based on military assets and so often inappropriate in an asymmetrical struggle but still congenial to those with military force and tangible assets.

Intervention—the extension of power—is a matter for generals, a matter of using force at a distance to achieve vital national interests: and force is military force. Thus the arena is conceived as a battle arena not a matter of perception. Even at that the arena is apt to be alien, the cultural arena strange, the food unappealing, the countryside exotic, the trees without names, the people with strange names, the language different. This is true for those who would aid subversion as well. Those who would foster revolution, however, are apt to dispatch money and arms, most welcome, and advice, often ignored. Thus involvement stays at a distance, a few advisors, an ideologue or two, Regis Debray on his way to Che Guevara wandering in the wilds or a Carlos operating alone spending charisma wrapped in ideology. Those who arrive in any number face the same obstacles as do the allies of the government: strangers in a strange land. At least the alien army can seek seek reassurance from what is familiar: weapons, tactics, doctrine, habits of the vocation.

The military is rarely ready for non-military factors: the Americans had barely heard of the Dominican Republic or Somalia, the Russians in Afghanistan found they did not know their neighbors—any more than after centuries the British actually know the real Irish. In an armed struggle these differences, most of all the culturally impose psychological differences, matter where this is not the case in real war. The German army can be defeated without knowing German or recognizing Wagner, the Japanese without understanding the tea ceremony.

Armies can clash to effect without the intrusion of perceptions and cultures, without factoring in the shifting psychological factors of the arena. Perception and culture are relevant only if as indicators for traditional military decisions: the German armies are likely to do this or that and the Egyptians or the Israelis can be expected to act as they have in the past. Armies expect to deal with armies. The people, their hearts and minds, perceptions, impressions, cultural factors are not congenial even when recognized important. All those who respond to an armed struggle, those who support subversion, those who send troops from a distance to crush subversion, those who watch the available footage on subversion on the evening news, everyone prefers a suitable explanation that encourages a familiar reaction, includes past policies, and offers present predilections. As always few want their perceptions adjusted by reality, their troops given unusual orders, the views of the Irish challenged or the faith of their followers shown false.

VIII

In almost all matters, governments are most comfortable dealing with governments and so responding to the irregular often violates bureaucratic practice. Revolutionary states are administered by bureaucracies as well, have the same inclinations. Armies want to deploy against other armies—and are apt to do so even when the army is irregular, unconventional, hardly to be found. Those underground want to be above ground, yearn to be legitimate and orthodox, a real army and so award their gunmen ranks and issue military proclamations. In an event largely given meaning by perception, all responses are in substantial part predetermined by habit, experience, desire, and need.

Inevitably the conventional insist on the conventions whatever the reality on the ground: Americans distribute weapons that are more dangerous to the users than the targets, the Russians teach tactics ideal for partisan war but not for African guerrillas, the IRA sends off carbombers wearing military berets, the World Trade Center becomes an appropriate target and the starvation of children an accepted tactic. If possible, reality becomes what aspirations imposes—and if not there are justifications and rationalizations to hand, light at the end of the tunnel, and others at fault. There is a constant appeal in responding to an armed struggle to shape reality to convention, to assumption.

What is going on may be complex but the response rarely echoes that complexity. What is true for those actually involved as patrons, allies, or disinterested participants dispatched for good works and good reason is true for most observers. What is wanted is an easy explanation of events that encourages congenial responses. The television audience wants a sound-bite that explains Somalia. The Foreign Office does not want to be informed that a distant ally is murdering children as policy. The Ayatollah wants to hear the evidence that Allah has been the answer.

Yet, even if the armed struggle is interpreted and transformed by the conventions at play, the vital factor is that the conflict remains unconventional. The traditional values of battle—flexibility, concentration, audacity, and secure communications—the virtues recognized by Caesar or Stonewall Jackson, are only a beginning. A Dragonwar is not just another war. The arena is shifting, a matter of the mind as much as geography, always strange. The results are always uncertain since the direction and the intensity of the struggle are uncertain. Frustration is integral to every struggle for everyone involved. The underground lacks the means to act effectively and the center to deploy effectively. How to concentrate against the invisible? How to attack a conspiracy? How to defend the conspiracy without resources, without friends, without respite or firm prospects? What is unconventional about such a campaign is so much falls outside the text. Everyone, the near and the far, those who have guns and those who read two-week old newspapers, must decide what matters. Those with more to risk pay more for any misperception—Che Guevara captured, murdered, his body propped up for photographs in Bolivia, the Americans fleeing the Embassy in Saigon by helicopter in 1975, the rumble of the rubble as the Grand Hotel in Brighton collapsed around the British Conservatives. And Dragonwars mutate and change as do the perceptions of the observers. Nothing to do with an unconventional war is conventional except the determination of all to make perception fit reality.

And if what matters is what matters to the involved, those matters often shift, reality can not always or for long be denied. All the involved accept that there will be a tomorrow. Where once there was hope that the underground could be crushed—always those in power seek a swift and effective means, a tactical quick fix—there is acceptance of protraction. Internment or summary executions will not work. And on the other hand, the liberated zones have not gobbled up the country nor the people risen nor the masses mobilized. If there can be no immediate solution then most—including most of those within the battle area—prefer to pursue the underground by conventional and congenial means, spiced with the techniques but not the dynamics of the underground. And those underground seldom can have preferences, must preserver relying on conviction and will since there are so few assets. And those who intervene are thus given the opportunity to become more deeply involved, become part of the problem. Even those who analyze and observe may become part of the problem—certainly do in the case of the media, often do if engaged in analysis. Those who do no more than watch, watch the news, watch the headlines, watch at best erratically and at a great distance see and so shape one of the phenomena of the times. Each and all react, however haltingly, to a Dragonwar and so create a complex nexus, their reality, one for the gunman, one for the guru, none that satisfies all but all shaped by a violent irregular war, unconventional in concept as much as in action.

Part III

America and the Dragonwars

7

America and the Conventions of War

Inevitably, almost all those who respond to an armed struggle become part of the problem, perhaps part of the solution, but often no more than an inattentive public. The underground provides the problem, moves against a target. Most prime targets are shaped as nations, nation-states, those in control of the arena and their patrons, mentors, or masters. There have been thousands and thousands of subversive conspiracies in this century, The frenetic and faithful have risen against all the old empires, against tradition, against class and order and dominion, against brutal authoritarian regimes, and regularly against many elected governments. Decency is no more a defense against the faith than a general in power. Every system, nearly every state has had rebels, rebels against the crown, gunmen with narrow ambitions and those against history or a universal system.

Swirling about them there have been as well the irregulars and unconventional driven less by a dream than habit and inclination. Some seek vengeance, tribe triumph, or to defend their village but each also presents the center with an unconventional challenge. In fact the center often exists to impose order on the chaos, to deny schismatic dreams, to be a center of gravity—to pose the state against anarchy.

Each state responds as is necessary, as experience dictates, as perceptions indicate, responds not simply as a policy matter or a police matter but also as a nation. The British view of the Irish, the German distaste for disorder, the weight of history on the Italians and legacies of empire or humiliation on many others assure that each response will be different and special. The response is shaped in part by the general nature of any such struggle, by the weapons and the world balance, by all sorts of factors; but also each arena imposes customs and habits, assumptions, a feel for time, and the limits of the possible. Class and caste may dominate the response instead of a national ethos. Often the history and customs of the region will induce states in Latin America or Francophone Africa to react to perceived threats in similar ways. Each reaction, however, if molded by the dynamics of the unconventional, is also shaped by the ethos of the target.

The characteristics of the nation play a major role in any response. The French are apt to act like the French when Corsican bombs are detonated and the Russians like Russians when challenged in Chechnya. The politically correct find national characteristics invidious, largely a matter of prejudice. Social science is apt to find such categories unscientific, generalizations arising from antidotal evidence. Yet, not only do the French continue to act like the French, the various undergrounds also often adapt their strategies just to that effect: imagine their enemy as national stereotype—the Raspberry Reich, Perfidious Albion, the Great Satan, the Zionist Entity. What any astute traveler can recognize about national manners and customs—the national sport, musical preferences, punctuality, and the nature of truth—can also be recognized in war and politics. And the Germans, British, Americans and Italians, just like the colonels of Guatemala or the life-presidents of Africa, are apt to respond to danger in their own special way.

Nations do have special characteristics even if the ethos is more complex that the popular generalizations. And if Germans are disciplined and the Italians dedicated first to the family, then Americans do act like Americans. And each nations does so most especially when threatened, when a crisis looms, when action under pressure is required. This is everywhere the case: danger intensifies the basic characteristics, exaggerates the stereotype. Sometimes this is to advantage in war and sometimes not. The British, the establishment and the everyday, the army recruiters and the dons of Oxbridge, the judges of the Crown courts, have paid to endure the Irish for centuries and still assume their stereotype is if not accurate then adequate: a Celtic people, creative amusing, at times violent, lacking in restraint, in need of tutelage. The Irish seem at times peasant and at others cannibal but always lesser. The British felt that they could cope with the Irish, and were in any case addicted to the psychological benefits of historical responsibility vindicated. The British shaped their own Irish reality that gave rationalization to interests of all sorts without need of change: Fenians bombs or Provo bombs, Irish bombs, a new free state or not, the British knew their Irish. And for much of modern history, the British knew their Jew, those in the Eastern European ghetto, those in the East End and the Rothschild—creative, talented, alien, and like all off-islanders lesser. But in the Palestine Mandate between 1944–1948 the British found a Jew that they did not know, did not like, not easy to intimidate or even to treat as lesser. The armed arrogance of the Irgun and the Stern Group brought out the worst of national traits. Frustrated or humiliated, the British are apt to be very bad enemies indeed—tactically this was the case with the IRA. With the Zionists of Palestine, the British found Jews quite different from all other Jews and all other colonial people: not natives, not in need of tutelage, not impressed with English civility, not grateful or subservient, not proper enemies being victims of the Nazi. The British never really forgave Begin and his gunmen for their denial of the legitimacy of the British presence in the Mandate: for Zionists found the British anti-Semitic, arrogant, and most humiliating for the British army and government, incompetent, inept in their tutelage, a lesser breed of enemy. In time Zionism, the new state of Israel, the army—Zahal—and the military doctrine would find most enemies not unlike the British. Each nation shapes a congenial enemy and none so outraged as when the image cannot be imposed.

The Germans treated their variant of Euroterror as an intolerable affront to the good virtues of the risen German people. They found random disorder in a millenarian cause psychologically distressing and especially so because they recognized that the need for order had once before been exploited. Baader-Meinhof made the German establishment unpleasantly aware that a new German could not be made in a generation, that besides the old guilt hidden by prosperity and civility were the old German weakness for order over law, for authority in their society. On the other hand, during their years of lead and terror, the Italians responded to an undesirable burden, seeking not immediate order but patience to persevere. The Italians did not long for order as did the German establishment or to be mentor as did the British but only to be left alone in the safety of the family. Even the *Brigate rosse* was a family if one engaged in a crusade to strike at the heart of the state—a state that a decade of terror revealed had no heart. The German state has a different history from the Italian and Italians are not Germans and so the two asymmetrical struggles were different and different in part in Italian and German ways.

The British and Irish and Israelis all came out of a special history. They were all different but the necessity to respond to terror, to an armed struggle meant that each reacted similarly in deploying security assets: the British against the Irish, against the Zionists, the Zionist against the Palestinians, the Germans and Italians against their own. All states deploy their power in relation to the perceived seriousness of the challenge, the intensity of the struggle, the agenda of the establishment; but all employ similar techniques and tactics whatever the ethnic component or the historical record.

In a post-modern, post-industrial society the police, the military, the establishment, on the other hand, is tactically apt to pursue order in very similar ways: the German police or the Italian—or the British—are apt to seek the same choke points, use similar assets in similar ways. The tangibles determine their use. Authoritative regimes or fragile ones simply differ in security deployment, in capacity, restraint, and habit. Habits, history, the national ethos make a difference and this difference gives each armed struggle a national characteristic. Some regimes, of course, represent tribe or clan. Some may represent no one but a family, rule not over a nation but rather over acreage defined on a map and so react to threat in even more special ways.

The American way is special and specific. There is an America way of war. And by in large Americans respond to any unorthodox assault by deploying in a manner that reveals the American way. American history is special, more special than most for the state arises from a society shaped by conviction not birth or breeding, a society of opportunity, accomplishment, and a heritage of triumph, triumph over odds, over nature, over history. America is opportunity made manifest, open, optimistic, competent, a land of freedom, filled with citizens by choice who swiftly become American by nature. Almost all the emigrants are converted to the new dream even as they focus on their special agenda: careers, bank accounts, refuge. Those who stay, who do not go home again to rural Greece or an Irish parish, are no longer native if not fully American. They become hyphenated Americans—in many ways a sincere parody of Americans—but their children are absorbed into the main, everyday Americans touched only lightly by the old ways. America is an idea and an ideological society, one without a personally shared history but united on accepting history as written: Washington as father to many whose ancestors in 1776 were children in Lvov or Liverpool or Canton or slaves in Virginia or Georgia. Southerners who lost a war and blacks who were a lost people have been transformed into everyday Americans, often with a special history but this is not burden but goad. Americans thus are apt not to share grievance but success, are endowed with hope, assume the system is or can be fair, just, effective. Americans are Americans, recognizable any place. On a foreign street they are not only visible by language, clothes, and conduct but also by assumption. At home and especially abroad, they display special characteristics. And so do their institutions, attitudes, perceptions, and actions: the American way is real. And the American way is available for export, not a unique gift of race or arena. It is special and general.

Americans are apt to be ahistorical despite the elevation of the past as patriot example. Americans share history texts not history as lived. History is written back to front and contains not grievance but triumph. History is pedagogical not a burden, even disasters at Valley Forge or Gettysburg, Pearl Harbor or Tet become lessons in bravery and perseverance. Americans thus do not feel the weight of history, find it difficult to accept the legacy that so shapes the Irish or the Germans as very relevant to the present or to war. Americans in fact assume that war is aberration, for war has only rarely touched the everyday in America, wars have been elsewhere, war losses have been unpleasant but tolerable, wars have been something to avoid—and avoid more easily because of the isolation of the nation from the quarrels of others. Thus history has been a case-study of development and enterprise interrupted by wars. The frontier was pushed west, the wilds tamed, Manifest Destiny achieved without shaping a war culture for America but rather indicating the capacities of the people, the returns of enterprise, the malleability of reality.

Thus when history imposes on the American consciousness by war or threat of war, by those pursuing an armed struggle or by gunmen or Prussian generals or communist commissars, Americans, the establishment, nearly everyone is not only surprise but also determined that something must be done immediately to return the nation to normal. Americans are possessed by a sense of urgency to impose tranquility. The British army in the nineteenth century evolved to respond to imperial wars—the assumption was that British character could defeat the inevitable rising or insurrection. The British expected insurrection unlike the Americans and had a military doctrine that allowed General Gordon to take his time reaching Khartoum and the nation to accept on his death and defeat that there was no urgency and no advantage in pacifying the Mahdi along the upper Nile.

Americans want far more because Americans assume that the world is peopled by those not unlike Americans. Americans are all colors and creeds, come from anyplace but are as one on their aspirations and agenda. Americans so assume all are as one, have similar aspirations and agenda, are disgruntled or violent because they are prevented by the times, by their history, by an oppressor from emulating the American way. Anyone of these, the rebel or the gunman, is open to conversion to the American way because that way generates the most perfect civil society available, a medium for development and enterprise that rests not on heritage or class but capacity and opportunity for all, all created equal. And if the reality is less than the theory most Americans insist that there is a universality of aspirations: everyone wants to some degree what America has and what America can offer as means to that end.

Americans expected tomorrow to be better than today, feel that past grievances belongs in the past, look forward to a stable, democratic world order and the end of war, conflict, plague, and privation. And if there is to be a war, the Americans assume that reason having failed then power deployed will effect appropriate changes, benefit all: the Nazis defeated and the Germans bastion of democracy, member of NATO, free from an unpleasant past, prosperous and content. At times war may be inevitable, force required when reason and example fail, but then the ensuing war will be won.

In fact the focus on winning has been so great that the purpose of victory is often simply to have no war, to have no enemies, to establish a global society not unlike the American where insurrection is unimaginable. The American reaction to martial threat is to focus on the power of the threat not the history of the arena, to seek an immediate end to the aberration, to seek victory with confidence, and to deploy all that is available.

What is always available is the managerial skills of an free, entrepreneurial nation and the technological accomplishments of an educated people. These must be deployed quickly and to effect to assure a military victory. Management and technology cannot only win wars but also reduce losses, can release the power of the nation. The American way of war then is shaped by the assumption that war is aberrant, that all the involved, allies and enemies, are not unlike Americans, that history and culture matter little and technology and power much, that victory can be achieved by management and technical superiority. War is too irrelevant to waste Americans when machines can sufficient. And once armed resistance is gone, then America can, if need be, export the American way of enterprise: announce a Marshall Plan to save Europe, set up NATO, pay for aid and development and the Peace Corps. The world can be shaped to the American image because most want what America has to offer. And what America has to offer is universally applicable.

Obviously America has worked, worked well, even during a half-century of Cold War with the rise of a permanent and professional national security establishment. America managed the balance of terror, deployed great power for universal advantage. And when that power is deployed the result is the American way of war: urgent employment of force relying on management and technology to impose peace and normality on an aberration. Since all grievances are amenable to accommodation, war is an anomaly that must be waged by overwhelming force—a force within the capacity of the nation. Enemies are if not reasoned, open to conversion, then to be overwhelmed. It is when such armed and dangerous enemies exist but cannot be easily found that the American way of war proves less than effective. The more asymmetrical such a conflict the more difficult for America. And the more intense the struggle the more anxiety at the difficulty of focus becomes. If the challenge can be defined as marginal, then the sense of urgency erodes, the need for a full commitment unnecessary. So the American military has managed to pursue irregular campaigns at the margins of events by allowing the cunning, time, and minimal assets to pacify Haiti or to pursue the Indian wars.

With the consolidation of America's Manifest Destiny along the Pacific basin, the country was seemingly safe from all but isolated revolutionary anarchists, the last of the Apaches, a few mad gunmen. There was no indigestible radical faith, no schismatic nationalism, no grievance for terrorists. Thus Americans without serious domestic subversion were apt to view most unconventional challenges as external and minor arising from the necessity to protect American interests, follow the flag with expeditions. The insurrection in the Philippines or the aspirations of Puerto Rican nationalists were minor legacies just as was the internal deployment of terror and intimidation by the Klu Klux Klan or by those engaged in labor violence or range disputes. These last were all police matters and transient: once progress and prosperity were assured domestic tranquility would be assured.

There was no internal military threat and external dissent could and was met by the dispatch of expeditions that hardly troubled the population or at times the military establishment focused on other matters. The cavalry had taken care of the Indians while the country expanded and industrialized. The Marines could take care of national concerns abroad while Americans focused their concerns on progress and prosperity, on reform perhaps but never on revolution. If there were threat, it was external—to be met by reason, example, and if need be defensive war. And the defensive Cold War was easily served by American characteristics, for what was needed was the preparation for war, the management of confrontation, the development of complex weapon systems rather than open battle. There were battles but never the great battle.

Americans did not want great battles, did not want war, offered a military career that only the opportunities presented by the Cold War and an expanded establishment had general attraction. America recognized that every nation must deploy military force even when protected and isolated. The perceived aggressive of communism made this clear. The expanded military, however, was not only orthodox in nature and doctrine but focused on what Americans did best not what needed to be done. So there evolved an American way of war that tended to ignore the reality of unconventional conflict.

Most Americans wanted to be left in peace, in splendid isolation, wanted no war. Americans also saw no contradiction in the view that their system was for export. The world should be safe for democracy. The nation must set an example, export free trade as well as products, urge freedom on all, set an example. This would not require entangling alliance or military crusades. In 1916 the country had been too proud to fight—or to prepare to fight. War was something far off or taken care of with dispatch by Black Jack Pershing pursuing Pancho Villa. War was alien if grand and irrelevant if not. Yet, for much of the twentieth century, America was engaged in war, real wars and unconventional wars. Wars might be alien but were constant and often not irrelevant at all. There were the two great wars, the Vietnam war, the often forgotten war in Korea, and all the expeditions and excursions and interventions. There have been special operations long forgotten, expeditionary forces remembered only by historians, swoops and raids and air strikes, wars by proxy and war by advisers. America as a world power, ultimately the world power, was constantly engaged in the deployment of force: war was not a sometimes thing, not excursions isolated from power politics but rather a constant. The navy has shown the flag, blockaded Cuba, shelled irregulars, as well as deployed huge fleets. The Marines have been expeditionary warriors, designated peacemakers, national symbols. The army has followed the flag, responded to natural disasters, the resolutions of the United Nations, and the emergencies of the times. The air force overflies to intimidate, has attacked Libya, flown aid to the starving and troops into action. Despite this, much of the American public is hardly aware of the long history of military involvement abroad.

The great wars, the Cold War, changed America. America must perforce be prepared for a great war, must be prepared to deter a nuclear strike, must be prepared to lead the Western alliance even when, after 1989, there was no sure direction to lead that alliance. On the other hand, the necessity to cope with small, nasty bush wars, too often protracted wars, too often unorthodox, had to be accepted: communism must be contained.

These dirty wars were to remain uncongenial except for the specialists and those intrigued by low-intensity combat. For most Americans, in or out of the Pentagon, such campaigns had little glory, seldom allowed national skills to be employed. Battles should be swift, decisive, and absolute: everything Vietnam was not. And there was horror when losses continued to come not in great battle but inflicted by Somali clans, by Lebanese warlords, by Arab fedayeen, or by Islamic fundamentalists. These events were difficult for many Americans to understand: the involved were too alien, their grievances and aspiration unreasonable—and worse they often were beyond reach.

Each summoned up memories of Vietnam where American virtues were transmuted into dross, when the war was highly asymmetrical and high in intensity and most important beyond reason, beyond the capacity of the managers and technicians and generals all assuming that more would at last be sufficient in a war that could not be won or abandoned or even defined. Americans did not like or understand power that could not be seen, was not tangible, or, causes that were beyond reason or the frustration of tangible assets by invisible will. Americans wanted wars that fit their expectations and assumptions: D-Day replayed.

At the end of the century, without a Cold War, the American public wants no foreign affairs at all, wants to forget old wars and especially unconventional responsibilities. Much of the country would like to return to splendid isolation. Since this is not possible, the easy alternative is to cut cost, narrow commitments, focus elsewhere on welfare and crime and tax laws. After a half century of confrontation, America still holds no brief for war, for military force, for adventures abroad, and especially for unconventional tasking that reveals a world quite unlike that of the Pentagon texts or the American experience. Americans like the weapons of war, technological achievements, accept war as a career deploying skills in systems management to a great institutions, but almost no one cares about actually using military power.

Those abroad have had a different perspective on the American agenda. For over half a century America has been a great power, a military power, the preeminent military power. And with the confirmation of American military hegemony after the implosion of the Soviet empire, much of the world regards any American return to isolation improbable and any self-denial of military capacity unlikely. The Americans find hegemony based on force unattractive. Many critics abroad can not accepted this distaste as real. Many simply assume that Washington, the Pentagon, the establishment intends as always to pursue global interests with force if need be. For some, for many, American has been home to the Great Satan or the Imperialist Center of Capitalism or for others in Latin America a gigantic neighbor, often greedy, seldom disinterested. Their are other perceptions. For many America is and has been leader of the free world. America remains the last great hope to quell disorder, prevent famine, bring order to Africa or democracy to Russia.

Thus while some radicals and nationalists have often seen a Fortress America dedicated to capitalist order, other radicals and nationalists have seen a Free American dedicated to open societies and world democracy. Few see America as an isolated enclave and few can divorce America from the capacity to exert power at a distance, deploy force, wage war. Americans may not know this, may assume at best they set political and economic example rather than exert

America and the Conventions of War   241

military force. Few others agree. To the distant America has long been not just a golden city on the hill but also a military player with armies deployed on the plains.

Nearly all American leaders and much of the population have always purported to hate war. War is not the American way. Aggression is not the American way—we defend, deter, contain, have a Defense Department, want only to encourage open and just societies, order, decency and civility. With the end of the Cold War, the army was apt to be considered as a means to control floods, interdict drugs, respond to famine and catastrophe, win friends and influence people, dig wells, rescue citizens trapped amid insurrection, and appear at parades.

The establishment accepts that America cannot return to isolation, should not do so, cannot give up the responsibilities that history and conviction have generated. In a post-modern world without time and space, with an integrated, interrelated transnational world of global corporations, computer networks, a universal cultural of consumption and entertainment, in a world where Mickey Mouse, the dollar, and CNN are ubiquitous, America is going to be involved whatever the sentiment of the people. And involvement, if past be prologue, will include low-intensity conflict in strange places.

America can no longer deploy military force in preparation for Armageddon. And since America has always preferred—as in all serious matters—to treat war as a compelling challenge to be met by the full resources and concentrated services and attention of all, other military missions appear minor, costly, unappealing. America prefers to mobilize the nation's assets—to get to the moon, to defeat the Kaiser, to defeat the Axis, to contain the Soviets—to pursue with vigor and dispatch a policy that deploys the latest technology and concentrates on the immediate. There is little sense of history since Americans share only a text not experience and grievance, little argument for moderation or delay, and an enormous desire to meet, to overcome, and to move ahead to other matters. In war as in peace, Americans want to achieve absolutes, to control, to concentrate and so to triumph, to win forthwith by dint of skill, assets, and focus.

Small wars, insurrection, rebellion are protracted, unorthodox, the foe elusive, the arena as much psychological as tangible. American assets are hard to focus, impact is slow and uncertain, progress difficult to count. This is not what Americans want, what is expected when the national purpose is deployed as military force. Americans want the Marines to win big, win quick, and come home. So do the Marines. So does the Pentagon. Military force is for real war—and all those who run special operations, pursue rebels, seek out terrorists do so at one remove from national concerns and proprieties. Military force must be forceful, orthodox—and even after the long Cold War—limited to need. Force may be needed for war; but still at the end of the century as at the beginning for most Americans, war is an anomaly, an aberration. War requires preparation, high taxes, waste, the establishment of military professionals, who are little understood if not actually suspect. Not only is war a reluctant policy choice but also even preparation and readiness are onerous. Without a clear and present danger, the Americans lack enthusiasm. The Cold War with the prospect of assured nuclear destruction concentrated mind. There was an absolute threat that had to be met absolutely.

The threat persisted for fifty years, eroding American isolation and creating a large professional American military-industrial complex. Even after the collapse of the Soviet empire, the reality of the world power balance required the continuation of that complex if not at the same level of readiness, if not with the same budgets and certainly not with the same demanding mission. Still, the military establishment from the troops in the field to the sub-contractors in the aerospace industry remained a reality but an American reality. The military was very American: the Pentagon both reflected and imposed the American ethic. And the Pentagon and the people still prefer real war, full commitment, elegant weapons, and open battle. This is the American way.

A lack of much historical enthusiasm for a professional military establishment did not mean that American society was pacifist. Martial violence is as much a part of national history as elections or technological triumphs. And although advocates of aggression are few on the ground, Americans have accepted and at times encouraged appropriate expansion: an Open Door in China, the Constitution followed the Flag in Latin American, and the need to make the World Safe for Democracy required entry into the Great War in 1917. With the end of the war against the Axis, the evolution of responsibility meant Washington and the people accepted the need for power deployed, power prepared, for a professional military establishment, constructed without master plan, often in haste but still arising from the premise that peace was ideal, advantageous, and natural.

America and the Conventions of War 243

Americans never have felt imperial, seldom have accepted a civilizing mission even when engaged in imperial adventures. Empires were of the old world not the new. America was that city on the hill offering example, offering the means to peace and prosperity, revealing how to reason together, work together. So most of America's overseas adventures were assumed disinterested or imposed by others, necessary adjuncts of domestic interests. The great responsibility after 1945 was to contain the Soviet world, deter Soviet aggression—and pursue normality despite the existence of a new military-industrial complex. It was politically prudent always to urge a modest military and to limit enthusiasm for global responsibilities. At home and abroad, there were always critics enraged that American imperialism was so denied and so obvious: dollar diplomacy, banana wars, CIA dirty tricks, expeditionary forces, Vietnam, of course, but also Iran, Guatemala, Laos, Nicaragua, and Cuba. And there was as well the export of culture, the far reach of American capitalism, the inundation of the world with American films and American appetites, fashions, and attitudes. The shallow American dream of the good life, everyone could be rich and famous if only briefly, was spread abroad by television role-models, by Hollywood features, and by products and expendables. The world was far from homogenized but wore Harvard sweatshirts, watched *Dallas,* and spent dollars. American example and investment created new Asian offshore neo-colonies, maintained a controlling grip on Latin America, ignored Africa or India because there was insufficient money to be made, so sought out China for the same reason. There might no longer be a Soviet empire but for many there had always been an American empire.

Most Americans still felt their nation disinterested. The Cold War was justification and triumph. The collapse of the Soviet Union vindication—and the call for a new world order of democracy and open societies sensible. In the last decade of the century, American military power was deployed, if at all, to succor the ruined, pursue terror, achieve peace missions, support the United Nations. There would be no great war but the post-modern world could not—alas—do without military capacity, American military capacity. The Gulf War proved this. And all assumed that the military so deployed was defensive, democratic, responsible—with luck transitory—and essentially decent. None of these characteristics were traditional martial virtues and this too was acceptable in Iowa or the Bronx. The American military has always been special, a citizen's army, an *American* force.

If most citizens accepted that military force is defense, needed but expensive, they also feel that the country has shaped that military force into an American image, constant to the nation's ideals. The military may be stepchild but it is not an orphan. In fact the military has become a child of the times, an integral part of American life, conduit to the middle class, to career options, employer and patron, a vast stratum of the land of the free. It is part of the nation. When President Woodrow Wilson led America into war in 1917, he noted the task was peculiar, "it is not an army that we must shape and train for war, it is a nation." Nations, however, shape armies, molding the military long before academy classes, indoctrination, long before uniforms are issued or the regulations read.

The army—the navy, the uniformed military—reflects the central values, the buried agenda of the nation. An army of a great power is in many ways the people assembled, certainly since the rise of national armies with the French revolution. Sometimes, there really is no nation and the army remains alien, fragile, a paid ornament of the elite or the power of a single tribe. Sometimes, the army is a professional institution retained for military purpose with loyalty and efficiency based on appropriate reward and parochial loyalties—the eighteenth century made visible in the French Foreign Legion. An army may be creature of a family or clan, assume as mission the mandate of heaven, respond only to payment, represent in each society various roles and missions. Trotsky pointed out that the army copies society not the reverse. Mostly, however, armies and navies are integral to and extensions of modern national states. Even swept up by great ideologies or profound religious convictions does not mean the volunteers in the crusade shed their heritage: the Russian Red Army was as Russian as Red and the Iranian martyrs who died in Iraqi mine fields for their faith went into war as Persians shouting Shi'ite slogans but shouting in Farsi. Western nations have Western armies and the British tend to produce a special British case reflecting a narrower society. Each army is special. Some armies are so special as not to be recognizable as armies at all but clan warriors or bodyguards. Some have an imposed mission to cherished the national ethos or socialize the masses or simply administer the government. Some develop the economy, see that the trains run on time, or take over the customs. Some armies have no military role at all, exist for display or as symbol of independence and so find occupation as peacekeepers. For his Chinese com-

America and the Conventions of War   245

munists Mao wanted a Chinese army of workers and peasants to win the revolution not a Russian Soviet army of the proletariat to defend the revolution. Each army tends to be a special case as well as a general example. So the American military is American first, reflecting the dynamics of society that had evolved over the century but was still demonstrably American: optimistic, ahistorical, cohesive, independent, often innocent, dedicated to competence and enterprise.

National characteristics are too often caricatures disseminated for special purpose or the stereotypes of the ignorant. Yet, despite the vast internal variety, the varying heritages and accents, the power of the parochial, Italians are Italians, Americans American. Certainly, much of the world has a particular vision of the American, perhaps a brief exposure to a few individuals but largely the result of the cultural impact of the media. There is as well a more elegant stereotype that filters out to the general public from local elite fearful that Shakespeare or Picasso will be overcome by Walt Disney and NASA. The more sophisticated the observers, the more complex the caricature, but the result is an agreed version more related to need than reality, to image than fact. Hence the cautious and the complex, the trained analyst, the historian replete with exceptions or those who can easily recall horrors founded on stereotypes are reluctant to deploy national traits for any purpose.

Despite all this there are national traits. Americans are *sui genesis*—at least in some ways. Americans tend to carry about certain assumptions and attitudes, a nexus of beliefs. Anyone abroad can recognize not only the ugly American but also any American. And the astute as travelers and observers have long written on the American character. And this character is as readily found within the American military as within the nation as a whole. All these national characteristics are as easy to satirize in the military as out. Rambo is a Hollywood artifact and John Wayne was a Hollywood actor but each reflects haltingly actual American hopes and fears. Some American soldiers in real life are crude and cruel, arrogant, racial bigots with guns. And of course, in real life many American soldiers are citizens dedicated to alleviating injustice and poverty, eager to encourage the benefits of democracy and freedom, And most are fair, decent, pragmatic and hardworking, egalitarian: Americans hardly perfect but easily recognized. All are children of the golden door who have inherited the first modern revolution, ingested and transformed in their own image the poor and wretched, the world's discards. All are emigrants to opportunity and many recent arrivals. Some are black and some white, some educated, and some not, most are everyday, all are different and all alike.

Americans seem innocent only in their enthusiasm and in their ignorance of the dark side of history that so blights old nations, parochial tribes, cynics, and survivors. They, like the world, have a dark side, the obverse of their virtues. Often morally arrogant, narrow in vision, simplistic in analysis, materialistic, crude, rude—ugly. So, too, the military but mostly, like the soldiers of every other nation, they are a mix, displaying both sides of their character simultaneously, filled with exceptions and special cases. The American national character determines the way most American wars are fought; for, more often than not, America can impose the American way of war upon any conflict no matter how unconventional—and most wars are neither unconventional nor so special that national character plays a compelling role. Wars are won by the big battalions, by weapons and morale and by generals with luck, won by the application of skill and the weight of metal. All these Americans have had in abundance, skill and arms and luck.

In general Americans seem driven by an urgency to do, to make, to build, to accomplish. Americans seek to act not to survive, not redress old wrongs but to achieve. Growth is assumed, success a goal. Given a problem, Americans deploy for a solution, dedicated to the proposition that all variables can be controlled and the appropriate resources conjured up to meet any challenge. First choice, always, is an immediate and effective technique and especially a technological solution— the tech-fix. And not only do Americans have to hand the necessary tools but also the equally vital managerial skills. As challenges grow grand, more complex, then the application of the appropriate techniques and technologies depends in growing part on managers. Ordered to fly to the moon, Americans will manage. And the result is they shape the future to demand, a future that can be shared by any willing to adopt the American way. For Americans see others as a mirror-image of themselves, operating under the restraints of the same universals, capable of the same kind of accomplishments that will lead to a better man-made world. The problem is not that foreigners are not Americans but that they do not live in a society modeled on that of America.

America and the Conventions of War 247

Anyone can succeed deploying American example—society and history are malleable. In fact, for Americans it is a moral and proper duty to create by example and to export on demand the American way. If the great American prairie could be transformed from a playground for bison to the world's breadbasket by the application of American means, then there is no need for famine in India or Africa. Anyone, anywhere can be taught anything necessary to the good life—and what more is there? Who really wants vengeance alone or merely to do harm or to rewrite history with blood? Certainly not Americans. So all Americans tend to approach life with a singular set of assumptions—and it takes only a generation to make an American, which is what the American way is all about, making the future fit the dream.

This perception of America, of course, is not simply the caricature in television advertisements, children's primers, or political speeches, but an articulated vision of a real world edited by faith. Americans believe in America—the more so when abroad, when amid aliens. True, America is far from perfect. There are inequities. Yet, such are seen as aberrations, seen as passing: injustice is compensated, affirmative action taken, opportunities created, bias legislated away. The wretched will be given hope, jobs found, justice assured, addicts cured, crime reduced, and charity encouraged. The American dream lives. Every American regardless of heritage, race, creed, or origin, sooner rather than later, willingly or no, carries the vision, assumes the future can be molded, should be transformed. And the American way is not parochial, has always been a beacon and a guide. The golden gates open both ways taking in the wretched and unwanted, dross to be transmuted into precious citizens, and releasing an example for others who can go, then, and do likewise.

For Americans the worst of times usually appear as war-related times, even more dreadful in waste and cost than economic depression. War in retrospect may be a time of heroes but even then not worth the cost. War, especially great wars, divert the nation rather than proffer special gifts. Some Americans do march to a military drummer but only a very few advocate the advantages of resort to war. If war does threaten, Americans, often to the last possible moment, hope that conflict can be avoided, then assume that the nation can be organized to cope. Wilson and Roosevelt kept us out of war—for a time—and then mobilized the nation. Truman tried to fight a small war, a police action, in Korea and so managed a most unpopular war. In Vietnam, Washington sought a real war but fought a limited one, limited by commitment, limited by what was asked of the country, and what was assumed at stake. Afterward, those most bitter felt America had failed only because full resources had been withheld—more would have proven ample, always had proven ample once the whole nation was mobilized. So no one, the bitter, the pragmatic, the critics wanted another such war—really any war that was not imposed.

In a real war the nation makes haste to deploy the country's resources in a drive for unconditional victory in a just cause. In the process of waging war, the nation and the military commanders assume that the great industrial and technological dynamo can soon be focused on the national needs, the country managed to victory with massive power on the battlefield. For much of American history, such a victory was always assumed to be the prelude to peace, the signal to dismantle the war machine. There never again would be a need to make the world safe for democracy or a danger of Axis domination. Each war was a war to end war. Regular armies were distasteful and expensive. The preparation for any future war was best postponed or left to a few selected professionals given tight budgets.

> "Although most history books glorify our military accomplishments, a closer examination reveals a disconcerting pattern: unpreparedness at the start of a war, initial failures; reorganizing while fighting; cranking up our industrial base, and ultimately prevailing by wearing down the enemy—by being bigger, not smarter."
> —General David C. Jones, Chairman of the Joint Chiefs of Staff, 1978–1982.

This was the American way, arising out of assumptions, values, and priorities, the way to fight real wars. And anyway, the American way won all the wars, all the battles from San Juan Hill in Cuba to Pork Chop Hill in Korea.

For most the Cold War truly began in June 1950 in Korea. The new confrontation was one that could not be won in battle and could not be abandoned because of the Soviet's strategic and ideological threat. War for Americans changed at mid-century. The great threat of the Soviet Union armed with nuclear weapons imposed protracted commitment, a permanent military-industrial complex. The system could not be dismantled.

There was by no means a straight march onward and upward by the new military. The traditional dismantlement of the war machine had begun at victory in 1945 and the demands of the new, dimly-perceived Cold War were haltingly handled. There was a threat to Greece and Turkey, communist insurrection in the newly independent Philippines—and the Western imperial allies were under attack in Malaya and Indochina. No matter, the army went from 8,266,373 on active duty in 1945 down to 591,487 in 1950. After the Korea war, even though the permanent nature of the new global responsibility was accepted by all but a few isolationists or neutralists, the numbers dropped from 1,107,606 in 1955 to 871,348 in 1960. The same was true throughout the military-industrial complex, a rising and falling graph related to threats, to actual combat, to the necessity for new weapons, for new training or for new directions. Military assistance was $5,744,000,000 in fiscal year 1952, $1,022,200,000 in 1956, and $1,340,000,000 in 1960. What after Korea became patent, was that national defense required an enormous commitment of resources and that the military establishment would remain huge, requiring all sorts of managerial skills. These factors imposed priorities in the preparation for eventual war at variance with those of the past. The new Pentagon, however, saw the major mission of the Cold War as conventional, strategic deterrence incorporated into the military and coupled with regular forces shaped to regular war.

The military took the new—new money, new equipment, new recruits, new tasks, new opportunities, new missions and roles—and created a giant establishment that revealed the nature of the American nature grown grand. The curious dilemma for the new Pentagon power elite was that as their role grew, as their ranks multiplied and their budget soared, as their tasks became complex and global, the prospect of actually waging war receded. No one wanted the three-service strategic nuclear force to be unleashed in some suicidal spasm. No one much could imagine a formal European war between NATO and the Warsaw Pact that would not risk such a spasm. And no one could quite fashion an appropriate mission for the remaining conventional forces denied an opportunity to wage a war against the primary foe. In Korea the problems of limited war, global stability, protracted stalemates, and political calculations indicated that the military would seldom have an opportunity to replay the great wars of the past. Korea at least had been within the American tradition, unprepared at the beginning, mobilizing while at war, set battles, attacks and retreats, and if Washington had persisted, real victory. After Korea the military had to invest much of its assets in structures that must never be used but must be kept usable.

As a global superpower in a dangerous world, America reluctantly accepted the need for a military establishment, one that absorbed millions of citizens, hundreds of billions of dollars a year, shifted domestic priorities with huge research and development projects, vast procurement programs, and broadening career and employment opportunities. America depended on assets that could be counted and did not need to be sly or devious, to resort to the weapons of the weak: perception, psychology or the intangibles. It was not the American way to depend on surprise and deception. Back in 1943, in the midst of the Axis war, the United States Chief of Naval Operations Admiral Ernest J. King turned down a recommendation for midget surprise-attack submarines, "The element of surprise has been dissipated..." "Surprise was the tool...of despair, of have not nations...not for us." Why deploy deception when power would do? Why bother with the irregulars? American fought righteous wars with patriotic troops and proper weapons. American was an arsenal of democracy. And when a few Marines or a scratch expeditionary force had to cope with limited means, the conventional at the center remained focused on the prospect of real war.

Real war required a huge army, vast in size and capacity, if not doctrine and assumptions. Britain might have a doctrine for imperial wars, even the Marines a manual for expeditionary forces, but for most Americans war meant great battles and fleets maneuvered. And if at all possible, such wars were to be avoided. This seemed no longer possible after Korea. In a bipolar world somehow America was at one pole, responsible for the free world, must be armed in peace, prepared for war. If there were to be war, however, it would be an American war, entail armies and fleets and the reality of the nuclear strike force. The Pentagon was focused on the great issues, the real challenges, classical wars or elegant modern military technology, not marginal threats. Such small wars, as had been the case in the past, should be left to small units acting outside the usual tradition or by traditional deployment or better yet left to province of politics or of diplomacy, to the spies or the aid programs.

The Pentagon engrossed with bureaucratic techniques, the research and development of advanced weapons systems, the need for operational response to global tasking, the new military empire ceded the covert world to the new agencies. Some of the soldiers without a war, unhappy with silo duty or Rhine army exercises, were attracted to the challenges and opportunities of communist insurrection, the need for an anti-insurgency doctrine and a counter-guerrilla capacity. They found lessons in British experience and a hero in Magsaysay's friend Colonel Edward Landsdale, especially that of Oren Wingate in Palestine and later Burma. In 1961 they were encouraged by President Kennedy's interests in special forces. This was not true for the Pentagon, new president or no. The admirals and generals let others bear any unconventional burdens, make any irregular sacrifices. They intended to fight even, especially, in Vietnam as regulars. Others ran a war in Laos, subverted Latin American radical regimes, engineered a coup in Teheran, and funded Egyptian colonels. Other departments established false flag agents and front airlines or organized massive aid programs.

Such matters were not relevant to Pentagon concerns—subversion, propaganda, social revolutions, the eradication of disease, the purchase of unions and journals, land reform or commando airstrikes in Indonesia. And when small tasks were assigned, the Pentagon was apt to send in the big battalions to impose regularity on the margins. Vital national interests were seldom involved in the Congo or Lebanon or Guatemala. Few unconventional conflicts intensified sufficiently to involve vital interests that such a real military response was necessary. Advisors and arms could be sent into Greece or to Philippines but no more was needed. The fleet could be directed appropriately and the Marines landed at Beirut in 1958—in a display, temporary, impressive, politically motivated, and so basically unwise. The American military had always had a tendency to divorce the advantage sought from employed force from the primacy of the needs of that force: America fought to win but the generals and admirals were not interested in the spoils, in the nature of victory. While a Winston Churchill sought to maintain the empire, deny the Soviet Union the Balkans, see Germany ruined, the Americans focused on crossing the Rhine, ending resistance, bombing the remaining strategic targets. War was about war not politics by other means.

Mostly in a world grown hostile, the Pentagon was concerned about orthodox American military capacity, the core alliances in Europe and

Asia, and, if need be, transforming the lesser friendly armies into the American image. Ethiopia or Iran or Thailand became the site of money properly spent, new American weapons displayed—and sometimes purchased—and American instructors indoctrinating new officers and promising men. Building new armies in the American image was a comfortable Pentagon mission not unlike the challenges to building a great American military establishment. The Pentagon could manage this—in fact manage any war, manage the Cold War where battle never came, capacity was never tested, and readiness was required not skills won from armed engagements.

The Pentagon by necessity became enmeshed in the management sciences, organization theory, systems analysis, often in the fancies and fantasies of the academic world from game theory through the new directions of the social sciences. The logistics of maintaining a military establishment absorbed an increasing number of personnel. There were in the Pentagon layers of middle-managers, often seeking a career role, often trained in mid-career by social scientists or business school faculties. Officers went off to good universities or bad just as they had once gone to a conventional tour. Officers were apt to find military careers as bankers or computer programmers. The new military managers were familiar with the product of think tanks, with quantitative methodologies, intimate with bought consultants and the skills of the private sector. The wave of the future seemingly lay in the administration of the Pentagon, in the means of preparing for war rather than in fighting careers. Management loomed very large in the evolving American way of war. Much of the military became a service industry, a reflection on the transformation in American industrial and commercial life—communications that had deployed runners in 1945 by the end of the century bounced off satellites. Wars were simulated. Computers were not just carried into battle but employed as crucial in command and control and in employing new weapons.

The growth and permanence of the new military-industrial complex imposed certain requirements on the Pentagon. The role of those involved in actual battle, command in the field, declined. The Pentagon had to managed itself, manage a world strategy, manage weapon systems. So technological complexities increased but battle efficiency did not. The arms race was about arms not used but acquired and deployed. Almost every survey of capacity indicated rising cost to go with rising technological innovation. During the Axis war, a *Gator* class submarine cost $5,500 a ton, next the new and elegant *Trident* class figured out to $1,600,000. The unit cost of tanks went up four percent a year after 1945 and that of other weapons systems averaged twenty percent. Some of the weapons did not work as planned. The M-16 assault rifle had to be redesigned so that it would not jam—too many managers. The Pentagon wanted elaboration, complexity, bells and gongs, wanted to keep in front in the arms race. Other weapons grew too elaborate to fulfill the original mission. Some were canceled. Cost overruns were the norm: on twelve major weapons systems of the 1950s, final costs were 2.3 times greater than original estimates. Defense Secretary Robert McNamara's effort to reduce the different weapons systems produced in the F-111 a plane with too much advanced technology to function in any mission. Unit cost went from an anticipated $3,400,000 a unit with 1,7000 planes produced to $14,900,000 each for the 562 completed by 1976. What was true in 1976 a generation into the Cold war continued to be true a generation later. It cost billions to build a Stealth airplane, untold billions to keep ahead of the Soviets. Weapons cost more and did not deliver assured security. Weapons were for war and war never came, not to test the weapons or to test the military.

The problems of the Pentagon in time seemed systemic rather than the result of individual error, service misjudgment, or poor oversight. The military wanted more, more higher tech weapons and instead got less at great cost. There was a lack of the free play of market forces, lack of singularity of purpose, the erosion of bureaucratic priorities, and generally a lack of war to impose efficiency for combat. The Pentagon grew not only big but also unwieldy. Repeated efforts at reform, reorganization, and realignment changed little: the nature of the Pentagon arose from the nature of America.

The Pentagon in a sense was American without the rules of reality properly imposed, an institution that could shape perception to mission—even if it on occasion meant shaping the mission to perception as was the case in Vietnam. Repeatedly tasked with unconventional missions the military felt aggrieved. No American wanted war and few Americans found more than necessity in engagements in Latin America or crises in Asia. The Defense Department wanted to defend, to be prepared, not to be tasked with uncertain missions permeated with politics and irregularities. At the end of the Cold War, the Pentagon finally was tasked with real war. The Gulf War, however, was more like a field exercise than a military conflict. And the Gulf War was as special as Vietnam and as likely to be repeated.

Other institutions in the national security system had been equally transformed. From the core of the system at the National Security Council out to the ends of influence involving special laboratories, government funded research centers, secondary military contractors, individual consultants, a vast new institution arose. There were over a dozen intelligence agencies, often with secret budgets and at cross purposes. The Pentagon was responsible for an enormous, permanent business, a huge budget, all sorts of staff and personnel, educational systems, airlines, housing, huge bases, a world filled not only with elite combat units but also warehouse inventories, specialists on import-export procedures, foreign area schools, motor pools, off-base housing, hospitals, arms salesmen, pension plans, and the parallel programs and proposals of a great society. Neither budget cuts nor base closures changed much. The Pentagon, like the rest of America, had grown great, absorbed new responsibilities, assumed new roles, taken on vast new tasks.

The Pentagon, like the new service industries, the new business and management specialists, was attracted to American skills, committed to scientific management that had most visibly arrived with President Kennedy's Secretary of Defense, Robert McNamara in 1961, but was inherent in the problems presented by simple growth. Defense had to be managed like a business, using the more advanced operation techniques. All the new techniques and technologies did not transform the American perspective on reality: the nation and the Pentagon remained ahistorical, dedicated to reason but accepting that if there were war then dispatch, enterprise, skill, technology, and conviction would win through forthwith.

Without a great war to indicate the somewhat different priorities than those of the private sector, based on profit, the new managerial means proved particularly appealing. And the new techniques and technologies arose from that aspect of the American character dedicated to managing solutions, counting progress even if the means became more significant than the goal. Martial virtues were hard to define, hard to teach, difficult to quantify. Martial experience, leadership in the battle arena war command, grew remote from Pentagon concerns. There were thousands of officers with advanced management degrees but only a handful with doctorates in military history. Classical strategy was irrelevant to management strategies.

The skilled and specialists were, if anything, increasingly entranced with the numbers, charts, and graphs, the applications faded, once removed, as quantitative results determined policy. McNamara ruled the Pentagon with a slide rule and charts. What could not be counted, would not compute, was lesser, unscientific, outside much of the system. And beyond the system was the American perception of the world that often isolated both the military and the nation from those who relished protracted conflict, were driven by historical grievance, who did not believe in reason or accommodation, who relished violence or were willing to use terror, atrocity, and horror as means to achieve a goal beyond the imagination of those within the Washington Beltway or beyond in the heartland.

The new military establishment, massive, permanent, refined, and technologically elegant, continued to be a reflection of the new America, a long way from the cowboys and pioneers that inspired the few special operations advocates. The American military felt that as the years and campaigns stretched out, as the Cold War became institutionalized, the text for the unconventional could be easily read. And the text only returned the lessons sought, reassured the conventional that the congenial was effective, Clausewitz valid, and tomorrow could be met with today's capacities and assumptions as had yesterday's war.

The military machine was geared to a great war, to repelling the Soviet threat, as global and strategic threat. Yet the threat was shaped so that no direct battle was possible—such a battle might trigger mutually assured destruction. The country and the Pentagon was prepared for a war that could not be won and should not be fought. The posture was enormously costly. Airplanes cost what once could buy an air force. And such airplanes had no proper role in most small wars, in exerting power in the Third World, in interdicting terrorism—to cost out the price for one small sortie on Libya on April 14, 1986, would boggle the imagination. Airplanes, missiles, all the necessities of an elegant modern war had gotten no cheaper over time nor the Pentagon any less concerned with procurement. And there was hardly any need to manage them in battle: all the battles were against the feeble and the frantic—until the Gulf War and that fought at half-speed against a ineffectual regional power. The Gulf War used brute power and hi-tech wonders to defeat Iraq. All the managing and planning that had been the regular military establishment had been effective, if at great cost, and the weapons, purchased at enormous expense, had sometimes worked, and those in charge showed a complete command of logistics. The Gulf War was what the military wanted, what they did best if at cost.

Over time everything military appeared more complex, more costly. The complex rested on a huge bureaucracy that grew unresponsive and parochial. The expensive new Bradley M-2/M-3 fighting vehicle has no real mission—a design blunder—but ways were found to deploy what had been ordered. The vehicle became an armored taxi, useful in the Gulf War when resistance was slight—and after that war what challenge was left that would require more than an armed taxi? The smart bombs that looked so smart on CNN did not look so smart later on paper. Pilots can hardly afford to practice with missiles priced at over a million dollars per shot or even America afford too many $40,000,000 top-of-the-line airplanes. And when cherished high-tech weapons were used, flaws were found if not announced. The search-and-destroy teams relying on the latest technology did not always find anything to destroy in the Iraq desert. Stealth technology worked but the airplanes were fragile, inordinately costly and lacked a real mission—they had a role in doctrine that after 1989 had become irrelevant to reality. The actions in the Gulf War were congenial because the conflict resembled a great field exercise rather than a military campaign. America was superb in deploying force as a distance when given time. And then all the weapons systems and carefully trained troops could be unleashed against a vulnerable and evil enemy. It was the perfect Pentagon war—mistakes or not. It reinforced the system so that whatever else at the end of the century the Pentagon remained orthodox in structure and assumption.

The national security community, the military-industrial complex chose to shape any lessons learned as to what was congenial: and this tendency, if not especially American, has been an integral aspect of doctrine and policy: preferences deployed rather than reality recognized. Americans like the tangible, seek technological solutions, prefer numbers and charts and graphs, what can be touched and organized and managed. In the Gulf War America fought an enemy who did all the orthodox things poorly and so deployed a vulnerable conventional military force and allowed the Allies time to set up their exercise. And Iraq responded without flexibility or a sense of reality when the battle was joined. Americans were delighted with an American war and an Allied victory; almost no Allied casualties, few civil-

America and the Conventions of War 257

ian casualties, technology and organization displayed, the sand table of Iraq transformed into a killing ground of those loyal to evil, and a triumph recognized at the end. The triumph, of course, was shaped not by a coherent agenda but by the military needs of the moment: the Allies won over the Iraqi armies, reversed aggression, but did not go on to exploit victory. What had been wanted was victory and victory won by the assets to hand. What is always wanted is to employ the assets America has to hand, finds congenial.

What better spy than a high-tech satellite that cannot deceive, betray, is not amenable to cultural considerations or tempted by personal desires? Satellites are costly, complex, impressive in their capacity, vital in certain defense areas but of little or no use in others. In fact the returns of a satellite can be twisted, technology deceived. A magician can make an elephant disappear on stage and so too can a deception specialist deceive a machine. In any case the machine can not produce on demand the motives, intentions, and inclinations of an enemy. Some things can not be read off a photograph—or not read until too late when intention becomes missiles or a tank attack. To find out motives, intentions, and inclinations an intelligence system need rely on people: fallible, easy to corrupt, often duplicitous and inaccurate, hard to organize, hard to manage, unable to generate charts and quantifiable data. This data is required if the nature of the arena matters and not if, as in Iraq, power alone will do, elegance is a matter of tactical deployment, and what drives the Iraqi reduced to tank numbers.

The American intelligence system, understandably prefers satellites to co-opted spies and hired agents, prefers the hard data of numbers to the vagaries of the culture gathered by agents. The American intelligence system is inclined to discount the utility of such spies and agents despite the appalling impact of a few such moles on Western security. The British counter-intelligence service at MI 5 became all but the English branch of the KGB. And one marginal CIA employee sold out the entire agency apparatus in Russia. Spies obvious can pay dividends. Yet deception, dirty tricks, lying, cheating, and special operations are, like war, considered aberrant, a risk rather than a means. So the system wants money for satellites but does not really want at the same time to spend money on people. The returns of agents is uncertain, qualitative, covert, corruptive, and deceptive. Advice given by those in the arena, by specialists, by those not driven by

American priorities and assumptions is advice that the establish often does not want. Power alone does not work underground, did not work in Lebanon or Somalia. The quickest route between two points in unconventional wars is often crooked. Underground all action is protracted, vague, and only rarely tangible. The best thing to do is often very little.

What those in power in Washington have been apt to do is not simply dislike such advice, especially if from covert sources but to deny its validity and even the utility of people—too fallible, too alien, too difficult to manage. So even without a Soviet Union, American security requires not insight into the post-modern world, needs spies and specialists and irregulars, but more of the same, more systems, more Stealth bombers, more satellites. The world may have become an arena for the covert, the illicit, the dreadful, and the marginal; but the Pentagon has accepted this reality only reluctantly. For the American establishment, for Americans, the covert, the illicit, the zealots and gunmen found in unconventional conflict are best avoided if possible. And if this is not possible, then the system should deploy the system's assets whether useful or not: the air force asks Rand what to do about terror and the Joint Chiefs offer seminars on terrorism taught by bureaucrats and career officers often innocent of the real world much less the underground.

Still, America had always been able to deploy about the margins without great risk or committing too many assets: dispatch Marines, mount expeditions, supply gunboats to the diplomats and cavalry for the Western settlers. There were the little wars, the Indians wars, the banana wars, the gunboat wars. American troops in the Caribbean and in China, in Mexico, in Russia. During the Cold War American troops were station almost everywhere but inside the Soviet bloc—bases in Greenland, listening posts in Australia, an army in Germany, and carrier tasks forces on call. The small wars, on the other hand, received scant resources, but the military, limited in means, without appropriate doctrine, coped, persisted, and usually prevailed. These campaigns did not need to be managed and organized and could be pursued without technological elegance, and could be pursued beyond career patterns, general interest, and the resources husbanded for conventional war.

Americans did not seem to hate wars that could be fought out of sight, by limited means, for what could be construed as decent purpose. The little wars were often secret wars. American was not like Britain with a doctrine for imperial war and an army that was unfit for conventional battle. The American professional army remained a cadre for great wars—the small wars were incidental. Americans did not long recall the enemy, Aguinaldo in the Philippines or Sandino in Nicaragua. There were handbooks and small force doctrines but all this mostly disappeared when real war transformed the military after 1941. By the time of the Cold War, American had no irregular doctrine, only a neglected and forgotten exposure.

The little wars had been relegated to shelved texts. In the real world small American commitments of military force were in charge of those who learned on the job. The new permanent Pentagon was dominated by those who had experienced great war, the Axis war, and then a real war in Korea. If there had been exposure to unconventional war before Korea—in Greece, in the Philippines, in the Anglo-French imperial experience, the impact had been slight. Mostly the military sought not new tasks or new understanding but to hold on to sufficient assets in a time of scarcity to pursue conventional war. The admirals and generals knew the history of the last war, their war, and soon had the means to fight it again with more assets to hand. The Cold War concentrated minds on the need for a real, peacetime, professional army. Special operations could be run by the intelligence community or would be amenable to orthodox military responses.

Except for a little romantic admiration for Special Forces on the part of John Kennedy, most of the politicians were as orthodox as the generals. The great threat was the Soviet Union, the communist danger, the need for deterrence and defense. Subversion, stealth, deception, terror and insurrection, and all limited wars were un-American. Unconventional conflict was apt to be generated by revolutionaries and best quelled by orthodox means. The texts recounted only the great battles, Gettysburg and Belleau Woods, Tarawa and D-Day.

A few politicians, some specialists, even those of flag rank, urged an unconventional capacity on the Pentagon. Those who advocated these different tactical directions soon had career problems. President Kennedy's interest was tangential and passing. After Vietnam almost no one wanted to contemplate the guerrilla, irregular war, the armed struggle. So most of the time, even during the best of times, the unconventional theorists had a small audience and little influence. The Pentagon could not afford to be interested in hearts and minds when a vast Soviet army waited at the Iron Curtain. Even after 1989 when new missions were sought, the assumption was that terror or crime of plague could be countered without a shift in priorities, no need to invest in irregulars.

There were viable professional reasons to distrust the focus on the unorthodox. For very good professional reasons, most admirals and generals have found the irregular unwelcome, special units too special. The commandos or the rangers erode regular units by recruiting the ambitious and creative. Such units were apt to follow their own agenda not that of the service. To deploy the special required not just toleration but enthusiasm. And professional military people had always found elite units, unorthodox tactics, independence of mind and techniques uncongenial if on occasion necessary.

Generals are conservative. Admirals are conservative. And American generals and admirals had little personal experience with small wars and small units and in any case were tasked to provide the free world with a defense that must rest on elaborate missile delivery systems, big divisions, planes and tanks and guns, not a few flashy men in berets. There was an inclination to ignore that such men in berets and their units might be the only ones actively deployed—the rest, the tanks and missiles and divisions were waiting for war, were deployed for defense, were unused. The conventional, however, had to be in place. The Gulf War was not fought by ranger units nor need African capitals be evacuated by draining military readiness. Real wars needed real armies and other missions could be managed by regulars. So the Pentagon did not want to adjust for change especially when there was no need. No military system likes the special.

In sum, the Pentagon controlled enormous power, huge armies, advanced weaponry, institutionalized skills, and high professionalism all driven by American dynamics. And America in general and the Pentagon in particular liked war to be orthodox and so subject to managerial skill, technological elegance, and the enterprise of the nation. This was not the case when the enemy deployed unconventional means, relied on psychological factors, used subversion, subterfuge, terror, and surprise. No professional army likes the challenge of unconventional conflict. And the Americans in general and in particular were especially opposed to the requirement of small, peripheral wars—too protracted, insufficiently pressing for great commitment, vague in direction, intensity, and even outcome.

When the marginal challenges occurred, only the smallest or most conventional could be managed. When asked to respond to the hostage seizures in Teheran by President Jimmy Carter in 1980, the national security community had to begin from ground zero. No one was ready. The CIA had no assets and little relevant intelligence. The Pentagon had no unit ready to move and had to create one—a force that had no coherence or experience. The unstated hope had been that without such special forces no administration could task the Pentagon with special missions. It was a force that failed because the system was not prepared for the unconventional. When the military system did respond to the unconventional, it did so within service and bureaucratic perimeters: proper titles and offices, proper training and recruitment. The irregular was made regular and the special forces not so special or so effective. The American myth of the frontier, the Indian fighter, the civilian marksman, and the romance of special forces that had appealed to President Kennedy had little to do with Pentagon priorities or American realities during the Cold War.

The military-industrial complex was in place, was real, had real assets to employ. During the Cold War, the military managers reproduced their ranks swollen by grade-creep and career prospects. Their responsibilities fragmented into scores of competing bureaus, each tinkering with part of the process, none responsible for the general good. When wars came they tended to be small and irrelevant or relevant but limited. The result was an institution that in theory deployed power was largely focused on growth, growth managed, growth distributed, growth in systems, in budget, in numbers. This was the American way: more is more, less is never more. All evidence for fifty years indicated American growth in capacity and so influence and control. When the Soviet Union imploded, there was a new world order fashioned to American orders. A billion Chinese became capitalists and democracy had takers in Latin America, Eastern Europe, and even in Africa. If there were neither clear orders nor an agreed new world order, there was at the end of the century prospects that the next century would be secure, prosperous, disturbed at the margin by ethnic quarrels and cultural dissent, but essentially orderly. The Pentagon had to respond to novel agendas, conflicting tasks, ill-defined responsibilities and the down sizing of the new establishment. The military core had not only found the future open to military ambition but also found a central role in the new America, in the new world order.

Yet the Pentagon elite still had little concern with the small, the irregular, or the unconventional. Careers were not furthered by small wars nor were most American interests. Covert action during the Reagan years had revealed the risks of operating beyond conventions. The Iran-Nicaragua debacle run out of the White House basement was ample proof that the irregular was ineffectual, illicit, and unneeded. And this is what the Pentagon, like most orthodox military organizations, had always assumed: the unconventional were marginal. Yet, after 1989, these were the only wars in prospect. The Pentagon found the adjustment difficult, the old mission gone. Thus the Gulf War was vindication, preparation paid off. The war had been won according to the book. The great battle was launched against demoralized and dazed Iraqi forces unwilling to resist and often unable to flee.

This was real war against a real and visible enemy. The high cost of weaponry, the need of a conventional army, the advantages of the Pentagon system had seemingly proved out on the ground in pursuit of vital national interest in alliance with everyone—Syrians and Israelis and the Russians—even the ayatollahs had been confounded. In contrast the militaries' special units, Rangers and Seals and special forces of all sorts, had simply lacked a real mission. They were sent along as extras in the great drama. There were no strange cultural and psychological factors to consider for the war was fought on a sandtable against a monster and his creatures who could and were defeated the American way. Overwhelming force and state-of-the-art weaponry were concentrated and astutely deployed with dispatch.

The military as the Gulf War indicated was finely organized, elegantly armed, ready for any conventional deployment. And the military establishment was also integrated into American society, was no longer simply soldiers waiting for a mission. The military no longer could be assured of an appropriate grand mission and was reluctant to become too involved in unorthodox tasks—irregular missions tend to end ingloriously. Intervention in Somalia was a disaster. Intervention in Bosnia was a risk few had wanted to take. Intervention in Haiti seemingly simply postponed chaos. Action abroad short of real war engendered suspicion within the military establishment even as seminars were set up and courses arranged. In fact most of the politicians and nearly all the voters no long felt the prospect of real war exited. The entire defense system was target for cuts and subject of growing criticism. The voter did not want weapons systems but jobs and good schools and no crime. The intelligence complex wanted more satellites that could not focus on the new enemies—warlords, narcoterrorists, computer thieves, ethnic fanatics, or on legitimate economic competitors. The voters did not really see the need for an intelligence community when the real danger could be found in the streets, unemployment statistics, or the future of the Dow-Jones.

The world might be new, the global order different, the threats transformed from marginal to major but the military establishment was no more enthusiastic about the unconventional, about small wars, about expeditions, peacekeeping, spies and agents and assassins. And neither were Americans who did not want Somalis to starve on television or Serbs to slaughter Croats or Croats slaughter Serbs, but also did not want to become involved. The terrorists and plagues could not be easily countered by generals and admirals. And in this the generals and admirals and voters were as one. The establishment was not alone in seeking to avoid the dangers and small rewards of the unconventional. America does not relish the unconventional. The voters like the military wanted wars to be short, decisive, and conventional so that the nation could again focus on enterprise and development. At the end of the century, Americans and their government simply wanted peace and quiet, the benefits of power without a protracted involvement in violence, subversion, and the Dragonworld that such power assured.

For over fifty years the Americans had insisted on deploying the American way of war for each assigned mission. The American way was exported and taught as sound and effective in all cases. That such a view was parochial, unique, and faulty in the field meant little for always the power to impose American perception on reality mean that America was not so much doomed to repeat the past as to pursue the future with costly but effectual means—as long as will and power existed at the end of the century as they had for centuries past. And such a course had the sympathy of many professional establishments past and present, fit America's priorities and assumptions.

The American way of war after 1945 changed only in size and capacity. The basic assumptions remained. The immediate past was, in fact, a chronicle of experience shaped to assumption, irregularities forced into orthodoxy. Abhorring the dragons, the nation was assumed sufficiently powerful to fight them but wanted to do so in open battle. If grievance could not be accommodated, war must be pursued swiftly in isolation from the perceptions of the enemy, pursued with dispatch, and if possible with a reliance on organization and technology. Wars should be won not protracted, should not involve intangibles and perceptions, require deception and time. If this had not been the case in the past, no matter, it should be in the future. The Pentagon adjusting over time learned little from example, forgot much, isolated internal critics, and ignored unorthodoxy as the plague. The realities of power allowed America to pursue dragons in their own and often costly way. Tangible power mostly worked. And when not available special forces could cope out of the loop. Mostly then irregulars could be dispatched or ignored. The Pentagon simply dispatched the conventional, shaped reality to what was wanted, pursued an asymmetrical war of will against assets as if a replay of D-Day or Gettysburg. The American way of war deployed against the irregular was expensive, often futile, at times humiliating but intrinsic to the nation and the people.

8

Unconventional War, the Cold War, and the American Way of War: Greece to Vietnam, 1945–1966

America's long march through unconventional war was remarkably only in that so much unconventional experience was transformed into such orthodox doctrine. The focus of the military enterprise was—as was, of course, proper—on the orthodox requirements of real war. Unconventional war was assumed—as was, of course, often the case—province of special forces, special interests, and special pleading. Partisans, guerrillas, rebels, and terrorists might be spectacular and bold but were not of vital national interest. It was, indeed, even difficult to define these conflicts that were short of real war: bandits more often than patriots, partisans attached to regulars, commandos without commanders. The orthodox after 1945 were apt to see communist revolution behind most insurrections—there were few insurrections inside the Soviet empire. Not really until the creeping involvement in Vietnam, coupled with the global wave of guerrilla conflict, student unrest, and national liberation struggles, did unconventional war require a policy response. Until then American experience was shaped by habit and Cold War assumptions.

Most Americans had hoped that the postwar world would be peaceful, democratic, the Soviet Union tamed, the Axis punished, and the evils of the old global system eradicated at no great cost. Most hoped there would be no need of great military power and no American role but reconstruction. Those who did not feared not simply the Soviet Union but also the dangers of doing nothing in a still chaotic world.

Mostly it was a matter of degree, the interventionists wanting a larger role, a more active posture, a commitment to the United Nations, to rebuilding Europe, to a global mission and the isolationists as little of this as possible without eroding what had been gained by military means.

Both in 1945 were apt to assume that the vast military establishment could be rapidly dismantled and that global responsibilities could largely be discharged with diplomacy and aid. Everyone wanted to pursue American business as usual. Instead, there were post-war wars and wars not easily ignored or isolated. China was in the midst of a bitter civil struggle that involved American interests. The Zionists of Menachem Begin were in revolt in the Palestine Mandate in an armed struggle that involved American emotions. Elsewhere, the old empires were filled with nationalists who did not want to return to Dutch or French or British tutelage—even in Africa there were new nationalists. Europe was prostrate and the nation's allies, Britain and France, all but destitute. The Russians were increasingly truculent. Yet, America had no clearly defined military mission. Washington encouraged the United Nations, sought to aid the wartime allies, and viewed the Soviet Union with alarm. America felt that insurrectionary violence was most likely to arise from privation: aid and comfort would erode discontent and the apparent charms of communism.

America sought to revive Europe but not to encourage European imperial interests. This was not the American way: that way was made clear on July 4, 1946, when the Philippines became independent. There was no American empire, no American imperial ambitions. There was thus no need for more than a small and necessary army. To unite the services for efficiency's sake and shape a peacetime military was the task President Truman assigned to the new Defense Department secretaries, James V. Forrestal, appointed in 1947, and then replaced by Louis A. Johnson in 1949. By 1949, however, the general hope that economic aid would not require a military component in Europe had not proved out. America's future was tied to Western Europe, to the West, to the free world beyond the Red armies—and those Red armies had made possible a Soviet Eastern European empire, remained beyond the Iron Curtain, huge, dangerous, and sinister. In October 1949, Mao triumphed in China and the communists dominated Eurasia. It was the power of these armies emboldened by communism that threatened vital American interests, not the irregulars at the margins. Yet, it was at the margins that the East-West conflict could find an arena that would not lead to direct confrontation. And at the edges, the unconventional—riots, strikes, arson, guerrillas, agitators with guns, and partisans with Mao or Marx as guide—appeared to threaten America if in strange places.

Greece, 1946–1948: American Conventions and Insurrection

Washington increasingly found evidence of Soviet ambitions, bad faith and greed, duplicity. Moscow seemingly was involved in encouraging disorder all along the fault line that became the Iron Curtain. Greece was typical, an arena assigned to the Western sphere but too tempting for the communists to deny themselves. The British forces in Greece found that once the Germans had withdrawn to the north and the war ended, peace and order did not automatically return. Axis occupation had simply hidden long-lived and incredibly bitter historical and ideological feuds on the nature of the future. The government in Athens could barely control the streets of the capital as the people began to divide, not only on ideological issues but also on more parochial matters. By the end of 1946, the first year of "peace," the country was gripped in a civil war. In Athens the government was supported by the conservatives, the anti-Marxists, the Church and Crown, and beyond the capital tolerated by much of the country as legitimate if not inspirational. In opposition, increasingly in armed opposition were the radical communists and their allies and associates of the Left who hoped that revolution could come to Greece. Beyond the parochial Greek concerns, it was easy in Moscow or London or Washington to see an East-West conflict.

Communism, fueled by the victorious Red Army, had swept through the Balkans and had lapped over the Greek borders. Tito's Yugoslavia and communist Albania encouraged those who in Greece saw revolution as the answer. The shooting started without proper front lines or armies deployed. The Athens center, responsible for a ruined country, poor at best, soon ran through the limited assets available. The British ally, possessed of a victory that had returned little but austerity, could not continue to underwrite an expanding Balkan war. On February 21, 1946, London notified both Athens and Washington that the end had come. The Americans could either take over the responsibility or watch

Greece disappear behind the Iron Curtain. On the same day, February 21, Dean Acheson and Loy Henderson advised Secretary of State George Marshall of the pending disaster and the need to aid the Greek army directly and then address the economic and social problems once order had been restored. Marshall agreed with the need for a response. President Truman soon requested $300,000,000 to support a free people. The United States, as ally, as patron, as participant, had become involved in an irregular Balkan war with unfamiliar and uncertain friends and dedicated and hardened opponents.

The Americans advisors and consultants had a limited command of the Greek arena, the nature of unconventional war, or of the appropriate response to the communist threat. First, those Americans involved in Athens and Washington accepted that the Greek arena was simply the chosen battleground of international communism, so the war was more important than it seemed. The parochial rivalries on the ground, the long local history, the special conditions, the power, influence, and predilection of the players were a secondary factor. The big picture was more crucial than the little. Whatever the context, the war was a war and the task of the United States to see that it was won. And winning wars was an American tradition.

In 1946 on the ground, the Joint United States Military Advisory and Planning Group, JUSMAPG, perceived an ideologically inspired conflict waged by partisans who were supplied and abetted from cross-border sanctuaries. These irregulars were harbingers for an escalation to conventional war in the north that would threaten a Greek government devoid of appropriate resources. The immediate enemy were the partisans—guerrillas, insurgents, the rebels. For the Americans the terms were indistinguishable and irrelevant. The immediate task was to equip, indoctrinate, train, and deploy the conventional Greek army to meet and defeat the threat. Such partisan threats were essentially conventional and could be met by well-balanced military operations that would clear the contaminated zones. The conventional could and would sweep through the irregulars—ideas, hearts and minds, the perceptions of the arena were irrelevant to an effective military response. What was needed was an effective Greek military.

There was at first some concern that America might have to supply sufficient armed force to shield the emerging Greek army until the troops were ready to fight. This was a politically unpalatable option: Americans had finished with war. Advisors, money, supplies, arms, experience, example would have to do—should do. So the Americans set about transforming the ragged Greek army and distributing aid to a devastated country. Not unexpectedly the advisors soon were managing, when possible or unavoidable, an unconventional war.

The American military advisors were largely unburdened with theories of anti-insurgency. JUSMAPG were in the Balkans to help an allied government threatened by communist subversion, communist partisans, and the enmity of the communist bloc. American intended to fashion a Greek army in the American image while at the same time responding to the partisan threat. The task was aided in large part because the Greek government and the Greek military tended to see the threat in American terms. The Greeks, military and civilian, wanted a regular army to pursue real enemies. Everyone assumed that the American way would work in the Balkans if the communists could be held off for a time.

Fortunately for the Greco-American effort, the insurgents were also free of revolutionary-guerrilla theory. The communist party core of the insurrection, the Kommouniskiki Komma Ellados, KKE, had arrived at the barricades with no agreed strategy of revolt. The dominant figure in the KKE, Nikos Zakhariadis, had no knowledge or experience with guerrilla war. Until September 1947, he remained in Athens convinced that the revolution would be won in the industrial cities of the south: the insurrection of the proletariat was ideologically fashionable. In September, however, Zakhariadis moved north and accepted a new strategy: a large area of Greece would be seized and rebel governing authority exerted in the liberated zone. On December 24, 1947, Radio Belgrade announced the creation of a Provisional Democratic Government of Greece led by a variety of radical figures and controlled by the KKE.

Although there were retrospective nods to Mao's theory of guerrilla revolution, the decision to escalate the conflict through creation of an alternative government and a liberated zone was local and, as would be clear in the fullness of time, disastrous. The KKE had opted to fight an irregular war in an isolated region against a regular army. For the Americans and for the Greeks, the enemy had been transformed from covert insurgents into a relatively orthodox force, a revolution into a civil war, and the conflict into an international battle with the communists as aggressors. The American way of war had discovered an appropriate foe.

Soon, in a few years, all was clear: the communists had overreached, the Greeks grew stronger with American backing and despite setbacks defeated the rebels who did not benefit as imagined from the cross-border patrons in Yugoslavia, Albania, and Bulgarian, and ultimately, for all sorts of reasons, were abandoned by the Balkan communists and the Moscow center. All very neatly done, a civil war won by American means.

Not all was as neat on Christmas Day 1947, when two thousand men of the Democratic Army under General Markos moved out of Albania. The Communist Democratic Army possessed small arms, mortars, Skoda 75mm-guns, and was supported from inside Albania with three or four heavy 105mm pieces. The Greek garrison was surprised at the attack, at the heavy fire, at the artillery support, but the government troops held on waiting for reinforcements. These were delayed by the weather and by the loss of a bridge at Bourazani. The Democratic Army captured one of the crucial heights, threatened two, and ignored one—a gross tactical error that even the withdrawal by the Greek National Army from the threatened points did not rectify. As the year dribbled away, the Democratic Army stalled and mounted only minor probes west of the town. Not until early on December 31, did the insurgents mount an attack on the last height Prophet Ilias; but by then it was too late. The Greek command had been far from idle. Supplies had been flown into Konitsa and dropped from the American Dakotas. Reinforcements were brought to Epirus to harass the rebels around the Bourazani bridge. The Greek air force strafed rebel positions with machine-gun fire and rockets. American mortars and automatic rifles were used liberally against insurgent positions. There was constant, heavy artillery support. Some of the insurgent forces, unused to real war, began to withdraw on January 1. Others grudgingly followed. Eventually the Greek National Army reached Konitsa on January 7. The Communist New Democratic Army lost 1,200 men, killed, wounded, missing, deserters, lost tactical surprise, lost any immediate hope of a liberated zone, and lost momentum, the sense of inevitability that had grown over the previous years. Regulars tend to win over irregulars in conventional war—to go orthodox is an enormous risk and had been the Greek communists' first choice.

The Athens government recognized the shift—the stumbling, drawn out, nasty little winter engagement at Konitsa was in a sense a great victory that cost 104 killed and 356 wounded. The fighting had displayed Greece's assets to the full: artillery support, new weapons, air supply and air strikes, coherent response and reinforcements—an all-purpose, balanced response. Even with the 105s in Albania still firing, the communist partisans had been driven back. "The Provisional Democratic Government of Greece" had misjudged the correlation of forces. The KKE was outlawed. The members of the "Provisional Government" were deprived of Greek nationality. The civil service was purged of radicals. The tide had turned seemingly almost at once. In January 1948, the rebels were left in exile arguing whether to continue fighting a civil war or revert to guerrilla tactics without holding ground.

The Greek civil war was far from over and the Greek National Army not then, not ever, a simple American clone; but in time Athens would prevail. The Greek Dragonworld was smashed and the faithful dispersed by straight military power. Hearts and minds came later if at all. The dream faded. Time became an Athens asset. The government was carried to victory and stability by deploying an army that consisted of three corps with armored units, artillery support, motor transport, and effective communications nets. There were air units for reconnaissance and close support. There was a coherent command structure and all the necessary auxiliary services. Roads and bridges had been repaired or built. The telephone system and the radio net was modernized All the infrastructure necessary for the conventional war was put in place. The American money had somewhat eased real and perceived grievances as well as made good some of the devastation of the civil war. The money offered hope. The new roads and working telephones offered evidence of reconstruction and growth. The communists had nothing to offer but sacrifice, theory, and aspiration.

In September 1949, there were only 2,150 rebels left in Greece and by the end of the year 815. On October 9, the communists decided—at last—"to discontinue the armed struggle for the time being." Another day never came. The cost of the war for Greece was greater than the losses of the Axis war and occupation—more killed, more alienated, more executed, more property destroyed, more livestock lost, more homeless, more bitterness—28,000 children were taken into exile by the defeated Greek rebels, many never to return. The KKE splintered leaving five quarreling political heirs in Greece. Most of the rebels lingered on for years in Balkan exile, abandoned by Stalin, tolerated by hosts with no love for Greeks.

So Greece, conservative, democratic, monarchical, tied to the West, poor but developing, moved beyond the civil war to other crises and challenges. The Greek civil war for America had simply been one of the early battles of the Cold War. It was a most minor battle. Even in American policy circles the Greek experience was minor. Other battles would be more visible, the Berlin blockade or the Korean war, and other commitments more crucial, NATO or a secure Japan. Greece had been an incident, a line item. Many of those involved in Greece never reached the center of the Pentagon power circle, Many of their lessons learned went unread by the ahistorical Americans.

In fact most of the general assumptions concerning unconventional war heretofore untested had been reinforced. These would resurface later in slightly more complex configurations, adjusted by Western experience with other guerrillas and other limited wars. Essentially, the Americans came to Greece, saw what they expected, acted out of conviction and available doctrine, won on the ground, and went elsewhere reassured that the American way of war could prove effective in such cases. The Greek pattern fit American historical experience and military doctrine. The American way of Dragonwar had been revealed in Greece:

1. The armed struggle was a rebel-partisan thrust by an international communist conspiracy. The guerrillas were advocates of Stalin or Mao regardless of the special conditions or nature of the arena.
2. The conflict was a direct threat to vital American interests and to the national image. Washington perforce must respond so as to intimidate opponents and encourage allies.
3. The threat was vulnerable to American virtues and assets: military force, money, technologies, techniques. All of these could be managed (i) to create an indigenous defense based on American models, (ii) to destroy the guerrilla threat, and at the same time (iii) to eradicate the more pressing social, economic, and cultural inequities that encouraged rebellion and made conventional anti-insurgency operations difficult. One could simultaneously prepare for war, fight war, and make war unnecessary.
4. Such a campaign would introduce democracy, prosperity, and stability, a moral imperative that validated war.

The core of this model—upon which all else depended—if America were not to send in a shield-force—was the indigenous army. No one then or later could imagine a rebel-partisan threat to the American heartland, so there would always be the problems arising from an allied effort in a foreign arena. This had not mattered in Greece and so it was assumed need not matter elsewhere: the world was much the same—cultures were various but not crucial determinants to policy or programs and so men and women had similar agendas and assumptions. Thus an American solution to a global problem, to communist insurgency, to local insurrection was possible. What worked worked: an army worked.

First, the army was to be, if not a clone, then indoctrinated, trained, equipped, and deployed as Americans. Such forces, even the most irregular, were visibly if not actually extensions of the American expeditionary core—same uniforms, same ranks, same weapons, same assumptions about tactics and techniques. No one gave any thought to variations in roles or missions, in cultural factors, or even local ambitions and priorities.

Second, such a force would depended upon numbers, heavy fire, advanced technologies, combined operations. All these would be effectively managed through swift communications, operational skills, and complex command structures.

Third, the military effort was buttressed with non-military programs that were intended not only to create a necessary infrastructure for anti-insurgency but also to ameliorate existing battle damage and under-development. The people on the ground would benefit not just with military victory but from the process of achieving that victory. The local military, too, should be aware of the non-military tasks necessary in such an arena: winning hearts and minds by good works when time permitted. An effective army must be liked, be popular, be representative of society.

The major military thrust of the campaign was to confront the rebels in great search and sweep operations that would cut them off from their sources and destroy the guerrillas on the ground. Ideally, the whole battle arena would be cut off from the cross-border support, from local allies, from outside agitators, and the communist conspiracy.

It was assumed that the rebels inevitably exploited perceived not real grievances. It was further assumed that the rebels were alien in ideology, maintained in place by force, brutal, ruthless, capable of any horror, without restraint or redeeming virtue, amoral or actively evil. Insurrection was imposed from abroad. The irregulars were only skilled in certain unorthodox ways of war. They were also the spearhead of waiting communist regulars, were proxies of the new Eastern empire. These agents of revolution could, however, be cut off and countered.

Despite the obstacles, American progress could be enumerated by recourse to tangible, visible evidence related to size, sincerity of purpose, a degree of effort, and encapsulated into charts and graphs. Restraint, except as a tool of negotiation, is counterproductive. All available assets from firepower to well diggers should be deployed. All these assets coupled with justice will inevitably triumph over the proxies of an evil empire. Local problems could be eased by diligence, careful management, and most of all perceived interim successes.

Greece was model and example, an almost perfect reinforcement of American assumptions about the nature of the world, about political and military power, about the Cold War, about any contemporary war. Greece proved out the American way, the American vision, the reality of American perception. Greece was saved, transformed, joined NATO. In sum it was not so much that Greece taught the American military certain doctrinal lessons as that the struggle provided an insight into the nature of the American military and the American character. These did not change from exposure to communist insurgency in the Balkans but rather were reinforced.

The Philippines 1946–1954: Unconventional American Responses to Conventional Communist Insurgency

The Cold War did not have an easy starting date. The Greek and Turkish commitment was balanced by the withdrawal of American troops from Korea in June 1949—well after the Berlin blockade and airlift had seized the American public's imagination. Containment lacked inspiration and nuclear strategy seemed divorced from reality if all too real after the Soviet's tested their nuclear bombs. The state of no-war-and-no-peace, permanent defense, a professional army, a military-industrial complex integrated into national purpose, and vital national interests nearly everywhere—a bipolar and hostile world—had fully evolved by the time General Eisenhower replaced Truman in 1953. Truman in nearly eight years had seen the world view of the West transformed, American transformed, the roles and missions of the military transformed not all at once but certainly swiftly.

From V-J Day forward, there was not the peace expected. Even before the end of the war, American allies had to cope with imperial revolts, had to devise various strategies to cope with challenges often beyond coercive capacity or previous experience. There was an entire generation of colonial wars, guerrillas, peasant reformers, tribal revolts, and nationalist insurrections. All were apt to have Cold War implications. The Dutch were driven from Indonesia. The French were hard pressed in Madagascar and Indochina. The British withdrew from Palestine in 1948 and then coped, more or less, with campaigns in Malaya and later in Kenya, Cyprus, and South Arabia. Even the IRA reemerged in 1954. In the same year the French withdrew from Indochina leaving four successor states and great American concern. The Portuguese, the first empire and the last, soldiered on in little, forgotten African wars. The Americans had ended their imperial adventure with the independence of the Philippines in 1946. The Puerto Ricans, except for a few zealots willing to shoot the President, did not want independence. And the small islands and bases were, if considered at all in Washington, assumed not to be an imperial matter. What did matter was the entanglement of the Soviets in so many colonial and developing countries—Latin America was filled with radicals, Laos was under siege in Southeast Asia, and Italy had barely been saved from a communist threat in 1948. Some radicals appeared at first progressive—Nasser and Castro—but mostly America suspected that most revolutions were inspired by Mao and Marx, by the Soviet Union, by those who sought the West as oppressor. In fact the postwar world seemed filled with wars of national liberation, colonial wars, civil wars, a limited conventional war to establish Israel and a tiny guerrilla war to seize Baptista's Cuba.

At least the Greek experience, if in an arena alien to American experience, had evolved in a manner comfortable to doctrine and to expectations. Greece was, however, merely one arena. After the return of the American military to the Philippines. Washington gradually discovered the communist Hukbalahap movement, the Huks, feed on rural discontent, official corruption and malfeasance, and the greed of unsavory former collaborators. The opportunities of a transition period slipped into insurrection. As in China and Malaya, a rural communist insurgency inimical to Western interests arose. The threat was taken as a facet of the Cold War rather than a simple Philippine problem—local grievances might be valid but the communists were the agents of world empire. Greek history and Greek perceptions had not mattered once there were money and arms, leadership and will, tangibles organized and deployed. This was the American way and the way most of those concerned with the Philippines viewed the Huks campaign.

The Huks, communist or no, arose from generations of land agitation. One of their most effective military commanders, Luis Taruc, later spoke of the plight of the peasants, "for centuries, land for the landless, had been the peasants cry." In April 1946, Manuel Roxas, who had cooperated with the Japanese and could readily be seen as an American puppet, was elected Philippine president. His predecessor's effort either to terrorize the Huks or to placate them had failed by 1946. The Americans had made a series of mistakes, including the proposed Philippine Trade Act of 1946 perceived in the islands as a colonial maneuver. The fact that on July 4, 1946, independence had been proclaimed as authorized by the United States Congress in 1934, had eased Roxas's problems somewhat; but his efforts to apply the "mailed fist" to the Huks had failed. The security situation deteriorated, despite a variety of rebel weaknesses.

The Huks flourished. The Americans largely stood idly by, unconcerned with the Philippine situation. The priorities and urgencies of the Cold War were not yet in full play: the Philippines was independent and so beyond both control and often advice. And the Americans had no advice, no real grasp of the problem. The 1949 Philippine presidential elections offered little hope of change merely a choice between the former collaborationist president under the Japanese, Jose Laurel, and a traditional, corrupt, and discredited Liberal leader, Manuel Quirino. No one could win but the Huks. And so the Huks with 12,000 armed volunteers and 100,000 active support members anticipated further decay at the center—and by election time Mao's example was potent. All China was communist. The communist in Malaya were engaged. And the Philippine election was corrupt. Violence and intimidation were widespread. Quirino won the office but no victory.

For the Huks, still small, still isolated in several island strong points, the object conditions for escalation and success had never appeared more promising: the center was corrupt, the peasants hungry for reform, the Americans distracted. Most important there was no alternative, no reform party, no prospect of American aid and comfort, only the feeling that the tides of history ran with the peasants and the poor, with the Huks.

In the spring of 1950, the tide shifted. The very success of the Huks attracted American attention. After Mao's victory in China, Washington was increasingly aware of the danger of agrarian reformers under red banners. The Huks were not a solution but a problem. And Quirino lacked prestige, program, viability, and vision. Still, he appointed Ramon Magsaysay as Secretary of Defense—and so everything changed. Magsaysay was the key for he had both the talents necessary at the moment and a program to thwart the Huks. The American concern that something be done had intensified. The North Korean invasion of June 1950 concentrated military minds. The Philippines needed to be protected, stable, secure. The American bases at Subic Bay and Clark were vital national interests, especially so in a Pacific too open to communist ambition.

The Central Intelligence Agency opened a station mid-year and the small Joint United States Military Assistance Group, JUSMAG, was reinforced. Unlike Greece, the Americans did not set forth to create a new Philippine army; instead JUSMAG accepted Magsaysay's priority that a new image was necessary and a new concern with peasant ambitions. What was wanted was a new role for the Philippine army given their mission. Magsaysay wanted an army with a role as social protector fashioned from the best of the armed forces. Most important he was in a position to impose his strategy on the military, on the country.

Magsaysay proved a charismatic figure, constantly on the move, reassuring suspicious or alienated peasants, hiking through the contested regions, constantly visible. At the same time the army was reformed, the brutal and exploitive soldiers punished, new Battalion Combat Teams created. Two unorthodox Americans, neither an area specialist, both shrewd and practical, played a key role in the Magsaysay strategy. Paul Linebarger directed an intensive psychological warfare campaign based on his knowledge of Mao's people's wars. Colonel Edward G. Lansdale was Magsaysay's personal advisor and inspired a variety of parallel programs like the Economic Development Corps that stole the Huk's land reform programs. Linebarger and Lansdale grasped the reality of Philippine perception and sought to deploy tangible assets in the proper arena. The real war was intended to shrink the Huks' capacity and control and incidentally erode the dream. The special programs were aimed at the hearts and minds of the uncommitted, the aggrieved, and forgotten—many of the military programs did not make the war easier to pursue but rather the Huks unnecessary. The military was an aspect of the anti-communist strategy not, as in Greece, the machine. Roads were constructed because roads were needed not to move tanks against an communist army—with roads tanks might not even be needed.

In April 1953, Magsaysay was elected president of the Philippines on the Nationalist ticket. New land reform legislation was passed. American economic aid improved the Philippine infrastructure. Grievance eroded. The power of the dream was limited to the first generation. The Huk bands dissolved. A number of army units were deployed in civic action programs. Instead of a protracted war, the war sputtered out. Unlike Malaya most of the guerrillas did not withdraw into the hills for another day but accepted reality, the failure of the communist dream. In May 1954, Luis Taruc surrendered: 9,695 Huks had died, 4,269 had been captured, and 15,866 had surrendered. The insurrection was over.

Magsaysay and the Americans had evolved a mix of counter-insurgency techniques effective in the field. There had been no replay of Greece. In fact the Philippine experience tended to parallel past American irregular campaigns, limited in scope, limited in assets, isolated from major concerns in Washington, at times relearning the lessons taught at the beginning of the century by the Philippine rebel Aguinaldo. Despite the Huk assets, the advantages of the covert in the wilds, the corruption at the center, the armed struggle had failed and failed without a protracted struggle once Magsaysay appeared.

For those concerned with such matters, the experience had little real doctrinal effect and almost none on the American way of war. The Philippine "lessons" were translated into elaborations rather than axioms. The conventions exposed in Greece remained basic. Americans could, however, see some useful Philippine byproducts to adapt in any future irregular war. The importance of a charismatic, efficient, honest leader, the crucial dynamo, could hardly be missed nor the importance of psychological warfare and civil action programs. All should play a part in any American response to insurrections fueled by the international communist conspiracy.

The relatively minor role of combat was overlooked: no huge searches and sweeps, no conventional deployment of an all-round army with heavy fire support, no big battalions. All these had existed in Greece because the communists had chosen irregular war. In the Philippines the Huks had neither the assets nor the inclinations to deploy conventional formations and tactics and so the military role of the regular Philippine army had been adjusted. In Washington the adjustment was less noted than the results: the army had won, the Huks lost. And the process and so the result had been a mutual accomplishment—allies in war, allies in peace, allies in victory.

The potential problem of a truculent ally was avoided because of American prestige on the Philippine scene. American inclinations and attitudes were accepted as normal. American domination of all Philippine perception was real if not recognized in Washington and often even in Manila. As well for the Americans, the arena was a relatively familiar one. There were even certain historical assets. Americans were generally popular in the islands, had shared both defeat and victory in the Japanese war, had enthusiastically promised independence long before the era of decolonization, and were not, despite initial mistakes, seen as puppet masters. America was admired, American culture and habits emulated, American advice easy to take.

Unlike Greece the everyday people had a role in the Philippines. Their perception was important and arose in part from the reforms and programs initiated to compete with the aspirations of the Huks. Their hearts and minds mattered—in Greece what had mattered were the military tangibles deployed. And most import their hearts and minds could be reached. Lansdale became an authority on all insurrections using the lessons found in the Huk insurrection. What worked in the Philippines could work anyplace: American organizational skills and priorities were universally applicable. What worked in the Philippines could work in Laos and Vietnam. This was not an America assumption alone. Experience is apt to teach what is wanted and congenial. The British, too, would assume that the Malaya campaign was a universal and so that experience applicable in Vietnam. The French would take their lessons learned from Indochina—what not to do and what should be done—to Algeria as basis for their new *la guerre revolutionnaire*. History was assumed filled with lessons, experience made the military wiser. In the case of insurrections and small wars, an entire generation of imperial soldiers was exposed to the national liberation struggles. Tactical skills grew without somehow transforming the strategic results.

The British military theorist Frank Kitson would become famous for his tactical lessons on low-intensity conflict learned from Kenya, Arabia, and Cyprus and applied after a time in Northern Ireland if to no great effect. Elegant tactics, somehow, had not saved the empire but this did not trouble many theorists and soldiers. What was assumed without great thought was the proposition that lessons existed— that there was a strategic plan to be found that mixed the proper tactical and technical means and the programs and priorities of development. There was to be found a means to win low-intensity conflicts no matter the grievance or the enemy. In any such strategy leadership was necessary, a man. In the Philippines there was such a man. The greatest impact was made by Magsaysay in what was largely a Philippine response to a Philippine insurrection. What was needed elsewhere was a Magsaysay, an effective army, and a few Lansdale programs. If need be an army could supply leadership, doctrine, tangible assets and oversight for special programs and special operations. Yet for the American military neither the Huk experience nor the Greek war were a major factor within Pentagon considerations. These had been small wars but dominated by regular forces. Irregular campaigns like the cavalry engaged with the Indians could be managed at the margins. Mostly only a few were interested in the margins. America had different priorities and so too the Pentagon.

The Cold War had meant a constant effort to contain the real Soviet threat, the Red Armies. Countered by the North Atlantic Treaty Organization, formed in August 1949. In addition a general defense was needed not simply around the margins but in the heartland of the free world, a defense against espionage and perfidy, against dirty tricks, sponsored subversion, and all the seditious and duplicitous means deployed by a revolutionary ideology. And for containment and deterrence to work there had to be an investment in nuclear strategic capacity. These were the vital foci of national security concern. When the Korean war came in June 1950, it was not the wrong war in the wrong place but rather a regular war in an unexpected place. The nation's vital interests were assumed at stake and the regulars were dispatched. America required a modern conventional army capable of being deployed at a distance. As always caught unprepared, the Pentagon had to rebuild an army while fighting a war—a war that began as a replay of recent tactics and techniques and ended with the trenches of a previous war. Partisans—or perceptions and propaganda and politics—played little part. Korea was a real war and overshadowed any lessons that might have be added to doctrine from the Greek and Philippine experience.

It was not a popular war because the arena was limited, the commitment limited, the power of America not fully deployed; but it was a war easy for the military to pursue. Nothing need be known of Korean culture. An army had to be deployed not a pacification program. Bridges were constructed so tanks could get to the front and once there deployed as doctrine and conditions demanded. The locals could use such bridge as a fringe benefit of real war. So Korea was an orthodox war if limited. Americans did not like limited wars but accepted Korea as necessary if unpleasant. Few in or out of the military liked police actions, if there were to be war let it be real war. So many still wanted full deployment. Many found deterrence and containment emotionally unsatisfactory.

The Soviet empire could not be brought to battle because of the nuclear stalemate. Increasingly elegant high-tech defense and delivery systems prohibited real war even as they dominated the budget, military planning, and strategic doctrine. The complexities of nuclear strategy that generated debate, vast weapons systems, intricate plans for disarmament or arms control, fears of Armageddon, and absorbed much of the defense budget took place beyond most military and all popular experience. It was a world of mutually assured destruction, weapons systems as counters, academic simulations, arcane policy options offered by nuclear engineers—and it was a world without war. The major military task of the West, of NATO and then SEATO, the South-East Asia Treaty Organization, and America was to contain the Soviet threat, to prepare for conventional war.

The wars that did occur were classical between regional powers with conventional resources. There were as well the very small wars: irregular campaigns, imperial revolts, coups and insurrections and pogroms. Nothing in the Cold War greatly challenged American assumptions about reality or about war. The Pentagon was content to allow others in and out of Washington to be involved in irregular defenses, unconventional initiatives. America's truly vital interests were most served by strategic deterrence, by NATO and SEATO, by conventions and systems. The European experience with imperial revolts—Palestine and Kenya, Indonesia and Indochina—simply indicated that empires were obsolete, their cost was too high, certainly too high for Britain or France or Portugal. Such insurrections were hardly novel nor assured path into the future. Even in Latin America expeditionary diplomacy appeared obsolete—there was a new Organization of American States, hopes for an alliance for progress, no need of Marine excursions or gunboats. Time would indicate that expeditionary missions would still have a role. There would be unwelcome for the military. The Bay of Pigs was a debacle but one shaped by the intelligence community not the Pentagon. If there were to be intervention, the military would offer the orthodox as the descent on the Dominican Republic would indicate.

In any case small wars deploying unconventional means were not very important. Even when such conflicts were inspired by the communists, most were irrelevant or marginal to American interests. And most armed struggles, as the Greece and Philippines and later in the Dominican Republic campaigns indicated, were amenable to conventional doctrine enhanced with special programs, special operations, and democratic ideals.

Vietnam: The Long Prologue, 1945–1961

The orthodox Americans assumptions seemed quite valid in the Indochina case in 1954. The communists disguised as nationalists were a danger but one best met by Washington unconventionally as in Laos or by the dispatch of aid and comfort. For years American funds were channeled to the French effort. The French had simply lacked their own resources. Once tangible resources were in place subversion and guerrilla war would be amenable to coercion. These tangible resources dispatched to the anti-communist forces of Saigon had become a constant in American Cold War deployment. There was, thus, a nonspecific beginning of the American commitment in Indochina. Most of those with responsibility in Washington and all the public hardly recognized the names of the arena. Except to specialists and strategists Southeast Asia was not very important in a world emerging from a great war. Not until the intensity and drama of real war at Dien Bien Phu in 1954 did Indochina focus American public interest or national security concern.

In Indochina the war escalated, spread to Laos, threatened Cambodia, all Southeast Asia, and drained the French of limited resources and all enthusiasm. In Washington the war was easily slotted into Cold War perceptions, an Asian move by the Russians and Chinese. Few in Washington knew of the historic Chinese-Vietnamese rivalry—few knew any Vietnamese history. It was assumed that a grasp of the arena was not vital when the strategic implications of a French defeat were obvious. Only China and Russia would gain and so the West would lose. Incrementally the Eisenhower administration in office in January 1953 became seized on the war, often teetering on active intervention, even contemplating ever so briefly, ever so tentatively, the nuclear option to ward off a French defeat at Dien Bien Phu. When Dien Bien Phu did fall on May 7, 1954, Washington accepted that if anything were to be salvaged and the communist advance prevented, the Asian dominoes protected, then America would have to become more closely involved.

The Geneva peace talks simply divided the country until general elections could be held: no one expected fair election and everyone involved assumed that effective force would decide the future of the south. The American Saigon Military Mission had already been established early in 1954 to assist the government in Saigon in countering communist insurgency. On June 1, 1954, the new Chief of the Saigon Military Mission Colonel Lansdale arrived in Saigon. There would be conventional aid and training soon and special programs and operations, a mix of Greece and the Philippines. America had become involved in a war on the margins.

The remaining years of the Eisenhower administration saw the drawing power of Indochina and Vietnam. It was not a happy involvement. There was little stability, no coherent government, and local forces little understood, uncongenial, and elusive. After a spasm of conspiracy, betrayal, and gunfights in the Saigon streets, Washington backed the strong man, Prime Minister Ngo Dinh Diem. Coherence and control were needed. In Vietnam democracy, enterprise, and enthusiasm could be encouraged and rewarded. The Americans began to train the Army of the Republic of Vietnam, ARVN, as they had the Greeks. The revitalized army was to be an orthodox, full-scale, all-round military force using American weapons, American doctrine, and American experience. Special operations, special programs, civilian development were a lesser priority but not neglected.

By June 1, 1954, there were 792 Americans assigned to the U.S. Military Assistance and Advisory Group (MAAG) and the Temporary Equipment Recovery Mission (TERM). There were 555 Americans assigned to the United States Economic Mission (USECOM). And there were consultants and specialists to organize an effective infrastructure. Fifty scholars and public administration experts under Dr. Welsey Fishel of Michigan State University arrived in 1955 to reorganize the Diem government. A team of social scientists began to examine the nature of the Vietnamese village as a first step in applying American skills to Vietnamese problems. There were Special Forces units in the country, an enlarged embassy staff, more United States

Information Agency personnel, more operatives and analysts from the CIA, more Americans of all sorts and conditions, all determined to build democracy under Diem, to thwart the communists' aggressive designs.

The number of Americans in country crept up and so did the active Viet Cong. In October 1957, thirteen Americans were wounded when bombs went off in Saigon, the first announced American casualties. By 1959 Hanoi has taken over the direction of the Viet Cong insurgency in the South. Aid from Hanoi was on the way. Small-scale infiltration began along a network of north-south trails. As time passed there were more trails, more transients, more guerrillas. In Hanoi the communist leadership assumed that unity could only be achieved by recourse to force, a mix of the armed struggle, subversion, and irregular war. In a sense the view from Hanoi was the reverse of that from Washington: the struggle would win the hearts and minds of the workers and peasants by example—build a nation by the example of the faith displayed. The Americans assumed that not war but the benefits of peace would create a viable Saigon establishment, a democratic and productive Vietnam.

To prevent the effective impact of the new American aid but most important to continue the process of liberation, Hanoi continued to filter light infantry, irregulars, and party cadres down the trails. An armed struggle would unite the country. War as nation builder instead of war as nation defender. History moved through the conflict of opposites not through conciliation or the consolidation of the present. By 1960, an American election year, the Vietnam crisis was but one of many crises. In Washington to Eisenhower, the situation in Laos was more worrying than the problems facing Saigon. Laos appeared most vulnerable. In any case Southeast Asia could no longer be considered either obscure or another's responsibility. Vietnam was a vital Cold War arena. How vital was another matter. Only the most optimistic social scientists and eager special operations advocates saw Vietnam as opportunity. Most in the national security establishment saw responsibility, cost and danger, a necessary investment to protect the rest of "Free Asia" and so the "Free World." From the moment Colonel Lansdale arrived in Vietnam or before, the only real question would be whether an irregular, covert, little war could be run out of Saigon, largely ignored in Washington as had been the case in the Philippines, or whether the permanent Pentagon would wheel up a grander commitment operation.

The Indochina arena displayed various prospects. Laos could, it developed, be managed as a marginal matter. For a long time Cambodia could be ignored. Vietnam proved more troublesome. In the Philippines Magsaysay ran his own war abetted by Lansdale; but in Saigon Diem—despite the efforts of his publicists—was less independent, less prepossessing, less attractive, less effective, more a part of the problem than the key to the solution. And the arena, unlike the Philippines, was not familiar to the Americans. Washington had no credit, no historical friends, and almost no grasp of the local realities. Vietnam was not a Greece, a friendly government in need of conventional resources, or the Philippines, a friendly government in need of encouragement.

To the north for a decade, the American response to Laos was unconventional. Laos was not as vital, not as visible, and possessed hardly any infrastructure to absorb conventional aid. And what was perceived crucial in Laos was not winning but rather not losing. What evolved was a little war run by operatives and proxies at one remove. Laos like Cambodia was neglected and so pragmatically dealt with by agents or diplomats. Laos could be neglected by everyone because the future would be decided in South Vietnam—everyone agreed to that. For the Americans Laos was not a grand success but not a failure either. It was an effective flanking action that was kept from growing grand and therefore vital by the bureaucratic and political wile of the involved. What was special was that the war was protracted, irresolutely pursued by limited means without a commitment to swift and decisive victory. This was not the American way but the way of the pragmatic Americans in county. So in a sense Laos was in the American tradition of Indian wars and expeditionary forces, those on the margins allowed to pursue at length an elusive enemy. In Laos interests were perceived marginal.

This was not the case with Vietnam. Vietnam was seen as the first domino of Asia—if toppled a whole hemisphere of weak and vulnerable regimes, Thailand and Malaya and perhaps Indonesia, would be acutely vulnerable. Vietnam was important, not a matter to be addressed by a few special operators and agents. On the record, on the classified record certainly, American agents and operators had a spotty record, some success in Iran and Guatemala but incredibly naive blunders related to Cuba.[1] Hearts and minds were all very well but Hanoi was investing troops as well as subversive ideas. So the refined American model of communist aggression was rolled into place almost from the first.

This strategic analysis explained so much: the nature of the threat in global terms, in ideological terms, in regional terms, the responsibilities of America, and the appropriate defense for vital interests. In time Vietnam would prove a special and general case, permit both the irregulars and the permanent Pentagon to display honed skills, techniques and tactics and export cherished strategies. It almost seemed that Vietnam supplied an appropriate need, the right burden to be borne for the new generation that had emerged with the election of John F. Kennedy in 1960. Critics, new isolationists, young radicals aside, after the departure of the French, the American national security apparatus judged Vietnam important, vital, and whatever the costs and dangers of escalation, a necessary commitment. Vietnam could be arena for the entire spectrum of American skills, capacities and technologies of war if limited by other priorities. Vietnam was real unlike nuclear strategy or the deployments of deterrence.

There were always critics and delayers in America, those with different priorities or historical memory, those who had doubts about new directions and old myths. Even President Eisenhower in his final summary warned of the dangers of the new military-industrial complex. And always at the edges were those who preferred military isolationism while preaching global involvement. The new internationalism was if bipartisan also querulous, divisive, argumentative, included those urging crusades and others assigning blame. Still the Cold War made everyone an activist of some sort. The Republicans from Secretary of State John Foster Dulles over to the professional anticommunists Senator Joseph McCarthy and Senator Richard Nixon accepted America as a world power. Nearly everyone wanted the nation at peace, the economy sound, labor content, the suburbs filled and society civil. Most assumed that this was possible and possible without stinting on international responsibilities to defend the free world. American could organize both war and peace.

A key testing ground of American resolve was seen as Vietnam where communist aggression must be countered. President Kennedy articulated the assumptions of most Americans:

> For we are opposed around the world by a monolithic conspiracy that relies primarily on covert means for expanding its sphere of influence—in infiltration instead of invasion, on subversion instead of elections, on intimidation instead of free

choice, on guerrillas by night instead of armies by day. It is a system which has conscripted vast human and material resources into the building of a tightly knit, highly efficient machine.[2]

To concede anything to the Eurasian empire would only endanger all the West. The Saigon challenge was part of a global conspiracy, central to the century, a grand alliance opposed to freedom and justice. Local insurrections, nationalist rebels, radical gunmen whatever their banners were in fact supplied out of the communist heartland and must in every case be met on the borders of the West. In the case of Vietnam, a small country under threat from a communist insurgency, there seemed no special reason why a limited American commitment would not prove effective. There might not be a real war like Korea but American aid could close down an irregular war, supply all that was needed. Saigon needed aid and comfort, advice, perhaps advisors, arms for the men. All this could be done by orthodox means, by the Pentagon, by Washington. Even the most disheartening American analysis detailing the unappealing nature of the arena seemingly always ended with exhortations to a commitment that would assure success, a solution. It was assumed that such a solution, such a commitment, unlike most cherished American military deployments could be so limited as not to infringe on the agenda and priorities an America at peace.

From the first there were those who pointed out that Vietnam was going to be a special challenge not easily met. The final cost was uncertain. Major General Thomas J. H. Trapnall Junior, former Chief of the Military Assistance Advisory Group Indochina, submitted a report on May 3, 1954, that indicated the nature of the French dilemma and so that of any successor:

> The battle of Indochina is an armed revolution... a savage conflict fought in fantastic country in which the battle may be waged one day in waist-deep muddy rice paddies or later in an impenetrable mountainous jungle. The sun saps the vitality of friend and foe alike, but particularly the European soldier. Torrential monsoon rains turn the delta battle-ground into a vast swamp which no conventional vehicle can successfully negotiate... there is no popular will to win on the part of the Vietnamese... the leader of the Rebels is more popular than the Vietnamese Chief of State... a large segment of the population seeks to expel the French at any price, possibly at the cost of extinction as a new nation... This is a war which has no easy and immediate solution, a politico-military chess game in which the players sit thousands of miles distant—in Paris, Washington, Peking and Moscow...
>
> A strictly military solution to the war in Indochina is not possible... The Viet Minh... are fighting a clever war of attrition, without chance of a major military

victory, but apparently feeling that time is working in their favor and that French and U.S. public opinion will force eventual negotiation.

The key to this problem is a strong and effective Nationalist army with the support of the Populist behind it . . . I recommend that . . . a full-scale U.S. training mission be established . . . victory in Indochina is an international rather than a local matter, and essentially political as well as military.[3]

What General Trapnall felt was the crucial key, the fix, was the creation of an army that would have the confidence of the people to protect them from communist terror—the complete "victory will be in sight." Even the alien nature of the arena and the unappealing fact of limited war would be no obstacle. As Professor Henry Kissinger pointed out soon after in 1957 in *Nuclear Weapons and Foreign Policy:*

. . . limited war has become the form of conflict which enables us to derive the greatest strategic advantage from our industrial potential. It is the best means for achieving a continuous drain of our opponent's resources without exhausting both sides . . . the argument that limited war may turn into a contest of attrition is in fact an argument in favor of a strategy of limited war. A war of attrition is the one war the Soviet bloc could not win.[4]

Thus even if American faced a long war in Vietnam, both the military and strategic tools existed to succeed where the French had failed. America had the appropriate resources—an endless supply of tangible assets—to win, to turn Saigon around, to fashion an army that would in alliance pursue a war of attrition to inevitable victory. Sitting thousands of miles away in Washington, a generation of American leaders would so believe while in Vietnam for a time, each new levee of reinforcements found what they had been led to expect, an opportunity, the prospect of victory in Indochina. No one factored a dream into the equation.

The foundation of the American analysis of the Vietnam conflict had existed long before Kissinger's book or Trapnall's analysis, before Korea or the Philippines, had in fact been carried into Greece, before the anti-insurgency texts had been written or an anti-communist doctrine refined. American analysis arose from basic American assumptions about the nature of the world and the meaning of global politics at mid-century. The world was shaped in an American image.

In Greece all these assumptions seemingly proved true and policies based on such analysis proved effective. The communist aggressors had played the appropriate part, the Greek ally had conformed to expectations, and in time there had been victory. In the Philippines once Magsaysay was in place, the response to an unconventional conflict, honed down and refined by local needs, proved just as successful. In Korea except as terrain the arena hardly mattered. The Koreans were a vital militarily ally but the greatest asset was American military power, conventional American military power sufficiently effective to reverse surprise and even the enormous manpower of China and with Korean aid hold fast at the 38th Parallel. And like the Philippines and Greece—and the events elsewhere in the Cold War—each exposure reinforced American assumptions, missions, and roles. Later dirty tricks did not work in Cuba and the orthodox expedition in the Dominican Republic did. Again the unconventional was apt to be deemed the province of intelligence and the military could slot irregular missions into a minor role as was proper and congenial.

This meant that there was a role for everyone: the conventional who would fashion an orthodox army to protect the people from communist aggression and the unconventional who urged special operations. There was a role for the social scientists and development experts, the specialists in public health or cost accounting or irrigation, all those who deployed to win the hearts and minds of the people. There were new tanks and new uniforms for the new Vietnamese army. There was new money and civic action programs dedicated not simply to improving an infrastructure for military benefits but also to eroding real and perceived grievances. There was a place for Colonel Lansdale. There were places for the new special forces units and their very limited but orthodox dirty tricks. There was the more important role for the sergeants and captains assigned to training, the staff officers. There was a place for intelligence analysts. And there was the American dream. All this would assure, impose a solution.

No one could imagine a challenge without solution even and especially the generals who read Trapnall's recommendations and skipped lightly over French problems. Americans were not French. No one gave thought that the locals' perceptions of reality might be different their dream different. American universals were universal. Few Americans imagined that their nation's motives might be questioned, especially by those who were to benefit from American sacrifices. What was needed was haste to respond to the challenge. Rather than introspection, long views, alternative perspectives, what was asked then and later was more: more personnel, more instructors, more troops and equipment, a deeper commitment, further technological assets, more programs and plans and assets, more things. And the more that arrived, then the more management skills were required and available. More military technology could always be deployed, more programs and projects initiated, more statistics gathered, collated, and totaled. And so more managers were imported to manage the escalating commitment. Six to eight support managers were required to keep one American fighting. The American way was everywhere to be found. The Pentagon represented American society, a managerial society that generated enormous wealth, variegated products and systems, and the most elegant skills.

The American generals, like the area specialists, the visiting observers, the agents and analysts, saw in Vietnam what they anticipated seeing. There was a state, a nation, a people threatened by internal subversion and external invasion, a democratic country with a prime minister, an army, bureaucrats and offices and administrators in suites. There were generals and governors, voters and citizens. Vietnam was, of course, different from America with mysterious Buddhist cults and strange tongues and curious habits but still a recognizable arena. In reality America's Vietnam was a cardboard world cut out in Washington and set up in Vietnam.

There was no state, no nation, but only the southern residue left after the collapse of French colonial power. There was no cohesion, no loyalty to the center, no center, no dream. There was not a united people but rather peasants with village views and the most parochial loyalties. There were those in the towns and cities focused on the moment and their own small lives. The "people" did not feel especially threatened by subversion or invasion or Hanoi, only by those who would tax the land, draft the children, carry guns, cause eddies in the even flow of time. The "enemy" had shamed the French, a cause for modest celebration. Many of the young and ambitious, the idealists touched by the nationalist dream and the common good were attracted by the commitment of the victors of Hanoi. Most of the others, the peasants and the workers, the poor and limited, were apt to assign the communists a traditional role, for Hanoi might even be the beneficiary of the mandate of heaven, possessed of legitimacy not earned by the bureaucrats of Saigon.

In the capital of the new Vietnamese state, there was no real prime minister but only Diem, a curious Catholic mandarin with a bizarre family, who possessed no traditional legitimacy, no charisma, but only a bodyguard of lies and gunmen. Diem was a self-created dictator who could harm but not dictate. He was a potential danger to stability, order, and tradition and in time murdered by his military creatures. These in turn paraded through the office decked with titles and uniforms but without authority or legitimacy.

There was no army only armed men in borrowed uniforms, going through military motions for salary, loyal to interest, a crowd with American guns. ARVN at times for uncertain reasons acted as an army and at others was cover for private purpose. Those dedicated and brave were lost amid the local and personal priorities of an army that fought, if at all, for fear of worse, an army without faith, without a mission, with a role in Vietnam quite unrelated to American assumption.

Hanoi was center of a different reality. The North was evolving from the confluence of anti-colonial pride, Marxist-Leninist ideals, the shared war against the French, and the fusing of historical attitudes and existing local practice. The mandate of heaven was passing to the new revolutionary mandarins whose banners were ribboned with victory, with the words Dien Bien Phu, with promises of more to come. The power of the new nationalist-communist dream was such to balance the tangible assets of Saigon and the Americans. In a conflict that deployed the irregular forces of Hanoi against the American-configured, orthodox forces of Saigon, the centers of control were also asymmetrical: the illusion of legitimacy in Saigon under threat from the newly mandated zealots of Hanoi. None of this fit the counter-insurgency texts, the American experiences, or the expectations and the assumptions of Washington.

The Escalation of War in Vietnam, 1961–1968

> *"The trouble with our policy in Vietnam has been that we guessed wrong with respect to what the North Vietnamese reaction would be. We anticipated that they would respond like reasonable people."*
> —Paul Warnke, Former Assistant Secretary of State

America created the Vietnam war in its own image. Much of the Pentagon establishment, the military-analytical complex that organized, managed, and reviewed the nation's defense, went to Vietnam and, if in a different climate, organized, managed, and reviewed untroubled by the reality of war. Quantification introduced by Secretary Robert McNamara was fashionable. Systems were analyzed and planning stressed. The best and the brightest were involved. The Pentagon is the most American of institutions and so apt to reflect the tenor of the times. Those on the ground who engaged in real combat often seemed but the tip of the great American presence. Increasingly those on the ground, in conflict were regular American forces, drafted and dispatched for a year but only a few to fight. These, the infantry, the Marines on the ground, the Special Forces, the choppers in and out, were at the margins of the many, the commanders and organizers and suppliers, the analysts and explainers.

In fact the war often seemed to be at the margins far from the bases, officer clubs, exotic night life of Saigon, or the usual camp routines. It was a war at the edges against savages fought for tokens, paid for in body bags and returning only certain career advantages to a few and great risk to the many directly involved. Skill, persistence, firepower, the advantages of technology, the weight of metal often generated tactical success in the field but without somehow achieving victory, in fact without preventing escalation by the enemy. So American deployed more. The intensity of the war was fed by a rising American commitment that increasingly dominated the efforts of Saigon and the ARVN. The more that the local commanders received the more they found that they needed. The more the Americans deployed the greater the defiance and so resistance became.

In the country at large, in Washington and in headquarters in Saigon, the general assumption was that with so many tangible counters to play, with so great a material advantage, with so many skills and such dedication, American military power would sooner or later triumph. Saigon and ARVN had to agree. And those who fought the war were seldom in a position to disagree. Most Americans came, spent a year in-country, and then returned to the real world. The career officers on the way up directed the system in place as doctrine required. The professionals rarely questioned the profession. Those regulars who accepted great risks expected the risks of war and as professionals did not risk too much professional dissent. Those engaged in the air who too paid a price adapted to the special conditions. The purpose of an engaged army is not to quibble over doctrine but to pursue war.

The Pentagon was apt to see the big picture, a matter of maps over terrain. Those touched by the actual horror and terror of war were a small minority, often a changing minority, seldom with influence or much visibility—and those more articulated and professional, even and often especially when at risk, were most likely to accept the system as displayed. So there was no dissent, no effective evidence presented to the organizers and managers of the reality of the war. When dissent came it was domestic, political, students and agitators who could be discounted by the military, by the national security establishment, by those in charge of the war.

The war was defined by the establishment. There was a government in Saigon. There was a local army, increasingly well armed, well trained. The enemy was external, alien, evil, and without a real Vietnamese constituency. The cause was just. Involvement was duty not merely self-interest—promises to keep, responsibilities to honor. These assumptions were reality for the Americans. Washington could see what others could not. American assets were maneuvered on a grand scale as doctrine demanded. There were great sweeps of search and destroy, strategic sites occupied, constant patrols that did not hold ground, massive air strikes, artillery bombardments, more firepower so that fewer men would be at risk. The terrain was reconfigured, forests destroyed, provinces cratered, the geography made to fit the war effort. At the same time society was adjusted, American civilians, contract workers, government bureaucrats, intelligence and military people sought to win the hearts and minds of the population, counter communist subversion, make the country safe for democracy. And away from combat most who came to Vietnam kept to the traditional round of office with proper hours and schedules to meet. Many found the war merely an intrusion on construction schemes or public health programs—if at times a violent intrusion.

Despite the American presumptions, despite the enormous investment, each year seemed more crucial than the last. Each year revealed a crisis more complex, more intractable and each year noted a growing American commitment. By the end of 1961, 2,067 Americans were in MAAG with a total of 3,200 military personnel in the country—5,576 six months later in June 1962 and by December 31, 11,300. The statistics of counter-insurgency showed rising curves of good and bad—the number of strategic hamlets built or the mileage of roads open to traffic, VC weapons captured, VC attacks, estimated VC strength. And easy translation was a war that was increasing in scope and intensity in direct proportion to the American commitment.

In October, General Maxwell Taylor and Secretary of Defense McNamara visited the country and reported back to President Kennedy: "The military campaign has made great progress and continues to progress." There was more power in place. The State Department had different numbers, different conclusions: "Statistics on the insurgency in South Vietnam, although neither thoroughly trustworthy nor entirely satisfactory as criteria, indicate an unfavorable shift in the military balance." None in Washington could imagine quitting.

The failure of the Saigon center produced violent change on November 1, 1963, when the more ambitious and fearful generals finally risked a coup overturning Diem. He and his brother Ngo Dinh Nhu were murdered while seeking sanctuary in a Catholic church. Another brother, Ngo Dinh Con, a warlord in central Vietnam, was taken to Saigon and executed. Ngo's wife, Madame Nhu, labeled the Dragon Lady by American journalists, was out of the country and escaped. Another brother Archbishop Ngo Dinh Thuc was safe in Rome. It was the end of a dynasty but no mandate passed to others—none had existed. Vietnam entered an era of military presidents and constant conspiracy tolerated for lack of an alternative by the American embassy. Vietnam was not at all like Greece.

The American embassy became power dynamo of a war that fed on the American resources that continued to pour into the country after Kennedy's assassination in November 1963. By the end of the year, there were 16,300 military personnel in the country and cumulative American casualties had reached 120 killed and 492 wounded. This American presence was a heavy burden for the fragile and artificial institutions of Saigon to bear along with the levees of advisors, project leaders, specialists, consultants, transients, and journalists. America always had more resources available if needed—and for a decade Vietnam's needs had gradually grown, been perceived as vital to American interests, to the free world. Where once Laos had loomed large or the Chinese nationalist islands of Quemoy and Matsu or even Berlin, now the choke point was Vietnam.

As soon as the trauma of Kennedy's assassination had passed, the alarm signals from Vietnam began to distract the new administration of Lyndon Johnson. Even though someplace back in the recent past, the American involvement might have passed the point of no return thus assuring further escalation, the American military presence had remained relatively small—far greater than in Greece or the Philippines but still limited. It might have been possible to wind down the military presence, not easy but possible, except this was not even considered an option in Washington. Yet in 1964, it was apparent if not admitted that Saigon was not viable without American support.

There were those who insisted that support did not have to be dominated by military concerns. Laos had been managed almost quietly on the spot by CIA agents and advisors, managed without ever quite reaching the top of the crisis hit list, thus requiring an absolute and hurried solution. The Philippines had been managed by the locals deploying requested American assets. Magsaysay's charisma, army reform, civic action, old fashioned public relations and new psyops techniques had done the deed. Greece had not required conventional American troops.

In 1964 there was no set American policy, no relevant recent experience, no general doctrine that required a military escalation involving American troops as a major component of the Vietnam war. In Washington for President Johnson and his advisors any other alternative appeared a non-starter: only the military could stop the rot. Worse, the Saigon regime seemed to have little resiliency, declining assets, lacked will. The government was unsavory and the army unreliable and the countryside hardly safe. Everything seemingly depended upon creating a real army to repel unconventional attacks, to repel subversion, to counter the communists, and nationalists committed to a dream almost no one in Washington credited as real or valid. What else could be done but nothing? And nothing would reveal American as a paper tiger, no sound ally. Logic insisted on an expanding military commitment, had so insisted for so long that no one in 1964 could recall a point of no return. Subsequent efforts focused on finding some single fork of decision, the great divide before inevitable escalation, a golden moment lost. Yet, the nature of American analysis, the unrecognized assumptions about reality, about power provide a single confluence, a long stream that may have churned over dams of decision but tended always to seek the proper level. Once a military solution had been approved, there was no way back—and such a solution had always been assumed.

Once General William Westmoreland moved into Vietnam in April 1964 as Commander United States Military Assistance Command, Vietnam, the war was real and the war was conventional. Westmoreland as American pro-consul and imperial governor in association with Ambassador Henry Cabot Lodge was determined on imposing a military solution, an American solution, American reality. Congress had accepted the American commitment with the Gulf of Tonkin Resolution arising from a muddled naval incident in August 1964. The bombing of the North became policy. Johnson was re-elected by a landslide in November. The military descent on the Dominican Republic to frustrate a radical take-over had indicated the rewards of orthodox military intervention. In the Caribbean, America was no paper tiger, should not be in Asia. On December 31, 1964, there were 23,000 United States military personnel in Vietnam, up by 7,000 in a year; and in the following year came the flood. The American way of war came to Vietnam. The American military in Vietnam grew in 1965 to 184,000; in 1966 to 389,000; in 1967 to 463,000; in 1968 to 495,000; and in 1969 to 541,000.

In Vietnam the Pentagon found a challenge that could be shaped to existing assets and approaches. The war could be managed. The war required all the organizational skills of an enormous bureaucracy to permit combat troops to operate. As combat intensified, absorbing more front line troops and weapons, the Pentagon could easily keep pace, put managers in place, construct the appropriate infrastructure, see that the essentials, the things necessary to combat, were ordered, constructed, shipped, off-loaded, unpacked, erected, counted and cleaned, maintained. This is what the Pentagon did best—manage, manage all the techniques and tactics of a modern war and count the results. The military machine worked as anticipated except that by 1968 the United States had lost too many killed or wounded and not won the war. Vital interests were still at risk but the Pentagon had not managed a victory only a protracted war without prospect of resolution. During Johnson's administration the limited war anticipated by Kissinger, the war of attrition, expanded in response to American commitment. For five years the answers to unconventional war had been escalation—and the Pentagon still felt that persistence, productivity, greater investment would produce the anticipated returns. What was needed was more.

Despite the imposed restrictions of a "limited" war, by 1969 three million tons of high explosive had been used to bombard the enemy. Two million had been expended on all fronts by America in the Axis war. By September 1969, America had suffered 44,798 killed, the South Vietnamese 97,738, and the Viet Cong and North Vietnamese an estimated 547,000—and at that each enemy killed cost $400,000 (including 75 bombs and 150 artillery shells each). In World War I, 17,000 rounds had produced an enemy casualty, in World War II, 25,000, and in Korea 50,000. In Vietnam the number was 300,000—more as less. The military answer was more firepower, more artillery, more airstrikes, more targets North and South, and more sweeps and searches. And when more failed, those frustrated that reality did not conform to expectation were apt to feel deprived, limited, unsupported by the administration or the public.

In Vietnam only the Americans felt limited. The Vietnamese could see the Americans employing untold assets. The American presence was founded on a base of heavy-construction projects—six deep-water ports, eight shallow-draft ports, eight jet air bases, eighty auxiliary airfields, pipelines, roads, barracks, sewers, and power lines. At the peak of construction, United States contractors used sufficient concrete every thirty days to surface the entire New Jersey Turnpike. All the indicators during the Johnson administration were rising: troops committed, R & R tours, missions flown, incidents of plague and venereal disease, assigned journalists, and linear feet of filled file drawers space containing intelligence on the enemy. All of this vast array, this wealth, all these things were deployed against irregulars armed with light weapons and an implacable will. It was the epitome of an asymmetrical struggle of will frustrating tangible power.

In Vietnam the country, the people, the countryside, the culture, the ecological structure, the institutions could not stand the weight of American intervention. The peasants fled from the countryside free-fire zones—and agricultural production dropped. The cities became jammed: urbanization by terror. Some social scientists said this process simply accelerated inevitable modernization. For the rural refugees the city was a wretched existence not a means to development. The old village ties frayed. The family disintegrated. Corruption became rampant. The government was a facade for private interests. The economy was an adjunct to American concerns, a service industry for war. Hell was no longer confined to a very small place but had spread over the country inflicting casualties far from the fighting. There were not, never really had been, front lines or even lines. Anyone in any place could be the next casualty, the next target. Yet much of the combat took place in empty zones far from the cities, marginal land often deserted by the local villagers.

In the North under constant if restricted American air attack, nearly every building was a target for there were a limited number of promising, modern target buildings. By February 1967, 391 schools had been destroyed, eight churches, and thirty pagodas. All were grist to the production line of war, all statistics in a techno-war that permitted, encouraged, the American military to escalate, deploy more, destroy more, achieve every assigned goal. The most obvious, visible indicator of the conventional power of America for most observers was simply firepower. American doctrine insisted that firepower directed against the suspect was crucial, saving of American lives, effective against any target. American artillery laid down barrages. Mortars and machine guns on the ground, nearer the action site, fired constantly. Millions of dollars of ammunition were expanded routinely to sweep a corner of a field, shred a few trees. The sky was an open medium for layers of warplanes from the B-52s on top down through the helicopter gunships. Repeated airstrikes hit a hillside that might or might not contain elephants used along the Ho Chi Minh trail, an enormous investment of costly weapons, a pyrotechnic display for awed hill tribes. American troops moved forward behind curtains of fire, dug into hills under umbrellas of air support, carried on the ground six times the firepower of World War II soldiers.

The American problem remained that in an asymmetrical conflict the conventional assets could only be directed against largely irrelevant targets not against the guerrilla core, and certainly not against their great asset—the dream that expressed itself in an outward form as a guerrilla-partisan struggle. The Viet Cong and the North Vietnamese deployed all the tactics and techniques not only of unconventional war but also irregular war, maneuvered regiments as well as gunmen. There were assassins and pirates, subversives and heavy mortars, bombs in cafes and mines on the roads, and full scale assaults on firebases. Intensity ebbed and flowed. There was no conventional guerrilla war to fit the conventions of anti-insurgency and no war that later analysts, however praised and popular, truly understood. No single textbook example fit. The dream of the underground did not compute. And the communists determined to persist until the will won concentrated on any vulnerability, accepted great risks and high casualties, persisted because the commitment to persist had abiding purpose. American power was directed on the visible and vulnerable.

Hai Thank was a small, mostly Catholic fishing village on the coast of the Gulf of Tonkin, cut off from the main road, little more than a scattering of wooden houses and a few fishing boats pulled up on the beach—not a port, not adjacent to any military facility, obscure, unknown but to its own. An American targeting officer with air surveys, lists and charts, a check-list for the Target Element Summary that included "buildings" and "Water Vehicles" listed Hai Thank. And so the huts and junks of the village were checked off as legitimate targets. The fishing village endured 1,620 bombs, 650 rockets, twelve missiles, and forty-one machine-gun attacks. There were 162 homes burned and 644 bombed out, eighteen junks completely destroyed, and seventy-six damaged. Ninety inhabitants were killed, seven of these burnt to death. Fourteen villagers disappeared at sea. Eleven children younger than ten-years-old were killed. Five new-born babies killed. The village was smashed because it was a target. The system required targets. Targets were to be smashed to force reasonable men to modify their behavior, to accept the reality of power.

By any standard America should have been winning. War is cruel but cruelty ought to impose a decision. Yet, despite official optimism everyone in command in Saigon and Washington still accepted that more was needed, not something else but more. What was wanted, what was asked, what was dispatched was simply more, more troops, more targets, more. All the skills, the technological fixes, the massive focus of the machine, and most of all the successes in the field, the searches and sweeps, the Viet Cong trapped and killed, the regular North Vietnamese units shredded by classic, case-book battles still did not bring victory closer. The Americans crushed the tangible when the intangible was their enemy. Their Vietnamese opponents underground persisted, relying on will over the tangibles of power.

The Underground War of Cu Chi

The Americans were engaged not so much in the wrong war as another war only tangentially related to that of their opponents. And as a result, great power could not be brought to bear even while inflicting casualties, winning ground, dominating the visible arena. The Viet Cong, the North Vietnamese, the enemy fought their other war across a spectrum of unconventional foci for different purpose, on a protracted time scale, out of sight, out of reach much of the time. The purpose was to persist until their will triumphed, persist, impose a cost, inspire continued commitment, keep the ideal vital. The military assets of the United States could not be deployed against the enemy's perceptions but only against their cadres—and so the Americans were frustrated. The people could not easily or often at all be persuaded that the future, their future, could be determined by American aspiration. Killing did not engender victory. All the villages could not be destroyed to save them.

The Americans could not image that their own persistence would not have effect, that the American experience and past success was not relevant. The higher up in the command pyramid the less willing were American commanders to adjust their perceptions. They preferred their own reality, their own charts and graphs, their own reasons. They knew what they knew and they did not want to know about an intractable, novel, and unpalatable alternative reality that was all about them in Vietnam. That alternative reality was not tangible: perceptions are not easy to define much less to observe. Yet, beyond the airstrips and fire bases, there was another world—and in one case that world, the underground of the zealots was real and tangible as well as symbolic and invisible: the tunnels of Cu Chi.

Beginning with the campaigns against the French, the Viet Minh, later Viet Cong rebels had begun to excavate underground storage and hiding places scattered west and north of Saigon along Highways 1 and 13 all the way up to the Cambodian border. No single commander foresaw a tunnel complex nor was there any unifying construction plan. Here and there protection, security, secrecy was available simply for the digging. Gradually a tunnel and bunker complex emerged adapted to local usage especially in the Cu Chi district to the west of the strong Viet Cong Iron Triangle. This was a pyramid-shaped area pointing down toward Saigon from a line between Ben Such and Ben Cat. There was no central scheme. There was no doctrine of tunnel warfare, no ideological rationalization for creating holes in the ground for those supposedly most elusive of all modern military forces—the modern guerrillas.

The Viet Cong simply found that in the area north of Saigon they could function more effectively underground. So they dug holes in the ground. Some holes and tunnels were already there. Soon, bit by bit, vast systems spread out and linked. Diggers working year after year, hollowed out living quarters, storage caves, underground ordinance factories, hospitals, headquarters, command bunkers, and long communications shafts out to villages and on to district and provisional centers. False tunnels and dead-falls were built as were sniper traps, spy holes, secret exits, underground ambush sites. The diggers kept digging. The Americans came, spent a year in-country and left. The strategists from Washington came to Saigon, visited up country, saw what they saw and returned to offices inside the beltway. And the diggers kept digging, unnoticed, unknown.

The Viet Cong, once the pressure built up against them after the Americans entered the war in force in 1965, took the war underground. The cadres hid out in their caverns to escape sweeps and searches and then to emerge and disappear mysteriously through camouflaged exits. They fired from spider holes and disappeared down escape hatches. ARVN or American patrols swept over the tunnel complex unaware and then found the enemy not in the front but in the rear. The French had hardly been aware of the tunnels. The ARVN troops, rarely very aggressive, had seldom been disposed to take the war underground. So the wary ARVN soldier passed the visible entry holes without enthusiasm. When the American patrols arrived, the soldiers found the countryside sufficiently alien without looking beneath their feet for another war. They, too, were apt to pass on, seeking guerrillas where the manual indicated.

In January 1966, General Westmoreland in Saigon ordered a huge sweep by American and ARVN units through Viet Cong strongholds in the Ho Bo woods and other parts of the Cu Chi district. The sweeps were to involve overwhelming forces and follow massive raids by B-52s. In a clear-and-secure operation the area would be made safe for the proposed establishment of massive American base camps ringing Saigon in a permanent pacification line. Involved were the 1st Infantry Division, the Big Red One, and the Hawaii-based 25th, Tropical Lightning. The massive sweep in 1966 cleared most of the area of Viet Cong. In the process, the Americans found beneath the ground evidence of a weird, underground city. American combat intelligence indicated that the complex was huge, reached to Saigon. Unlike ARVN with a leadership dedicated to other priorities and soldiers lacking will, the Americans involved felt challenged by their discovery.

The Viet Cong had been found to be well armed, determined, dedicated, ruthless, effective, and most unexpected of all underground. Above ground the irregulars and guerrillas had been elusive as a result of tactical pressures. Below ground they were invisible. Westmoreland assumed all of the guerrillas were living on borrowed time until the full weight of the American machine could be focused. The whole exercise was to focus great power on the irregulars. If some were in tunnels, truly underground, they were assumed even more vulnerable. So Saigon and Washington were content. Westmoreland's grand plan was operational. The sweep had swept the Viet Cong out of the arena and the American divisions were in control.

The 25th Tropical Lightning, twenty thousand troops and all their technological accouterments and accompanying baggage, moved into the new base camp beside Cu Chi town very near the Viet Cong tunnel complex. The huge camp area was protected by a heavy perimeter defense, mine-fields, barbed-wire entanglements, sensor devices, floodlights, and reinforced dugouts. Out beyond the wire earth-moving equipment bulldozed a huge arc of farmland to create an open field of fire. An area five thousand meters in diameter could be covered with harassing and interdicting fire from the artillery, all dug in with overhead cover. The usual complex of mortars, machine-guns, light weapons, bunkers, strong points, existed as planned. From a secure base the cleared zone could be extended, the guerrillas scattered. In the meantime, the camp had to be hardened and extended. Once the base was completed the enemy could be pursued.

Behind the ring of defenses protected by air-support, helicopter gunships, and even B-52 strikes, the base camp took shape. The 25th had brought precut tent and latrine kits to make do until wooden huts and steel offices could be erected. These wooden huts and prefabricated offices were replaced in turn by more permanent facilities. The offices were air-conditioned. There were ice-machine plants and walk-in refrigerators, ten-kilowatt generator sets, folding beds, filing cabinets, proper water supply, helicopter pads, and clubs for the officers, clubs for the NCOs and enlisted men. There was a radio station. There were barber shops, sports fields, swimming pools, chapels, and motor pools. There was a miniature golf course and repair shops, maintenance facilities, storage dumps. There were typewriters, calculators, copiers, and the endless flow of paper, intelligence reports, rosters and lists, sports schedules, statistics, pay-rolls, charts and graphs, supply lists, grocery lists. To cope with the vast flood of data, were the computers, UNIVAC 1005 and NCR 500 in specially designed trailers. The machines would digest the war bytes and quantify progress.

Except for the sound and light show of distant battle, the Lightning Flash could have been sited stateside, cool offices, crisp uniforms, formalities, airs and graces, a managerial bureaucracy transported beyond the Pacific Ocean to run a war by the numbers, by the manuals, according to the doctrine of the day. No visitor could help but be impressed by the enormous power such as the exercise indicated. Any opposition by irregulars armed with red books and light weapons seemed futile. At enormous cost an orthodox army had been set down on a tactical site swept clean—everything, even the water, had been imported from the United States, transforming the countryside into American reality.

In fact the camp was almost on top of the Cu Chi tunnels. No one had foreseen tunnels. Indeed, the American recognition that there might be a tunnel threat came slowly and then piecemeal. Often even chasing Viet Cong snipers into their holes still gave commanders no idea of the further reaches of the underground complex. Americans went into the tunnels and came back out again none the wiser. Only gradually as the intelligence trickled through from combat experience did the reality of the tunnels take shape. Then American ferrets, pistol in one hand and flashlight in the other, began pushing down the shafts into the dark underground side of the war, not just in and out but along the narrow, dark paths at great risk. A premium was placed on great courage and slim build as the American ferrets, first in isolation and then in more coordinated probes, began to put together a map of the other world.

Creeping through a breached communication tunnel along a narrow, dank passage, twisting past niches and over trapdoors, moving up and down, always at risk, the ferret would unexpectedly emerge into huge rooms. One was a factory, fifteen feet high, where artillery pieces were assembled, mortars made, each shell crafted from scavenged tin cans and filled with explosives taken from undetonated American bombs and shells. One cavern contained two forty-year old 105 artillery pieces, burnished, oiled, ready to fire, guerrilla guns on hold for better days. Elsewhere there was a deep room, thirty feet long and six feet high that had been transformed into a propaganda center. The facility complete with a large printing press, newsprint storage, typesetting devices, dyes, paper in great rolls, and make-up tables. Even more impressive one access shaft led into a Viet Cong command center that consisted of an intact, functioning M-48 tank, stolen from ARVN north of Lai Khe in 1966, buried for three years before being "recaptured" by American tunnel ferrets. The tank batteries worked; the lights functioned; the radio was in use. No guerrilla in an armed struggle ever has enough, always wants more arms so in this case the Viet Cong had armor: buried armor, waiting for better days, symbol and reality.

The Viet Cong were often forced to exist underground in the most miserable conditions without a chance to cook, to stand, often to breathe properly for days at a time, later for weeks. Months, even years passed with only a few chances to spend time above ground in an everyday world growing alien. The rats and snakes, mites, spiders, the special tunnel parasites remained a threat and a reality. There was the danger of infection in the dirt holes, the risks of cave-in and flooding. There was the lack of light, fresh air, cooked food. Increasingly there was the war overhead, the crash and rumble of shells, passing American armor shaking the ground, and the enormous thud of B-52 bombs. Lights flickered, tunnels filled with dust, shafts simply collapsed.

Then the war came underground. Those beneath the ground there were increasingly in danger of American ferrets, of American gas pumped through the system, and of American ambushes set at exit holes. Always there was claustrophobia, the weight of the earth above the nasty bore holes, the realization that there was no end in sight—the faithful must live like animals in the dark, in fear, in misery. For the Viet Cong in the Cu Chi tunnels life was brutal, harsh, crude: "what hard days, one has to stay in (the) tunnel, eat cold rice with salt, drink unboiled water."[5] For the dedicated and determined, those committed to their dream, to the army, in constant danger, miserable, fearful of the day, uncertain of tomorrow, called to sacrifice in a war already prolonged for decades, the tunnel time was the dark before the sunrise. In the dark and in danger, service was special, released each cadre into the whole, strategic effort. Life mattered most under threat, intense, vital, dangerous and exhilarating. In the tunnel amid the war "one is free and feels at ease."[6]

The local American commanders had realized that the men faced a new kind of warfare. Their ferrets began to learn on the job, applying available tools. CS-gas grenades detonated in tunnels would leave lingering irritants on the walls to harass the Viet Cong. A Sears and Roebuck orchard blower could be used to pump acetylene gas through the complex that when ignited burned out the Viet Cong or sucked up the underground oxygen. Mainly, the new tunnel ferrets depended on their own devices, their own learned skills and a few comfortable weapons, a regular flashlight, and proper clothing. Only the slender and agile could move through tunnels built for the small, agile, and lithe. Only the very bold and daring chose to push on into the dark maze beyond the familiar, beyond technological aids and the power of things. In a new and unexpected variant of unconventional war, the regular was adapted to the irregular. This was a war fought on the margins of experience by those independent of theory and command—special forces shaped by a special mission that could not be covered by existing doctrine. Without a text, the Americans responded as Americans, as Americans at the margins had responded to the Apache.

Those Americans further back from the tunnel mouths sought, as always, a more effective, more technologically crafted solution. The commanders wanted a fix to the tunnel problem that would put fewer lives at risk by using techniques to replace or at least supplement courage. In 1962 with the growth of interest in anti-insurgency epitomized by President Kennedy's support of special forces, a Limited Warfare Laboratory, LWL, had been established in Maryland to produce technological solutions to special problems. The experts produced leech repellents, foliage penetrating radar, and high-calorie meals in small packets. By 1967 ninety percent of the products were related to the Vietnam campaigns. Tunnel technology seemed no stranger than many earlier Vietnam problems.

On August 7, 1966, LWL shipped out a Tunnel Exploration Kit, TEK, for testing. Ingenuous, theoretically sound, the kit like much of the American technological deployment in Vietnam did not adapt readily to reality. The cap-headlamp simply provided the Viet Cong with a target and the apparatus tended to bang on the tunnel roof. The communications set with a novel mouth-operated switch did not communicate effectively with the surface, the trailing wires unwinding behind the user only hampered movement. The special tunnel gun, a .38 with a 4" barrel, silencer, and hi-intensity aim-light, covered all the angles except that it lacked stopping power. In sum the kit was a killer but not of the Viet Cong and so the ferrets went on with their flashlights and .45s. No one used the kit.

The same fate met the Tunnel Explorer Locator and Communications System, TELACS, that neither located nor communicated under real conditions underground, leaving the tunnel rat on his own, as usual. At least a conventional flame-thrower did burn off the vegetation and reveal tunnel mouths; but a new Smith & Weston .44 Magnum, firing a 15-pellet bullet almost silently, like the .38, failed to kill the target. LWL took the .44 Magnum back for adjustments and introduced a novel and potent four-segmented cartridge for the Tunnel Weapon that worked but apparently violated the Geneva Convention and was withdrawn. The Tunnel Weapon like the TELACS or TEK became a historical military artifact, martial technology awry, inapplicable for tunnel warfare where the ferrets were content with the means to hand. What worked on the margins worked—what ought to work, what was elegantly designed and ideological correct, was often a danger.

There was no quick fix to the Cu Chi tunnels and in many quarters no great enthusiasm for cleansing the entire complex. If there were not an appropriate solution, there was not a congenial problem—in fact the problem had best be ignored. The American commanders back in the base camps found that running American tunnel ferrets, one at a time, underground, without need of complex weapons systems, vulnerable to any Viet Cong mole, was not especially appealing. Few could really believe the size of the tunnel complex or felt any urgency in clearing the underground world of Cu Chi. It was bad enough to come to battle with the elusive Viet Cong guerrilla above ground without complicating matters by giving a high priority to small, invisible, underground maneuvers. So the 25th concentrated on the big picture above ground and in so doing called in even more awesome fire power. And American military capacity was not without effect, numbers count and destruction shapes battles. Constant artillery fire out of the base camp, demanded by doctrine and inclination, made life above ground difficult. The Viet Cong were no longer above ground. And underground even direct hits by most bombs and shells seldom collapsed the tunnels although exits were buried. What did batter the actual tunnel complex was the impact of the B-52 raids on the area. Each B-52 carried dozens of conventional 750-pound bombs that fell without warning. The thirty-ton load would suddenly crash down across the countryside leaving a mile-long set of huge craters. There were then no tunnels left, only a slab of churned and acrid earth cutting across the complex.

Coupled with the bombing and shelling, the American army sweeps across the countryside gradually depopulated the area. Each year one million civilians moved away from the war in the countryside and into the cities drying up the ocean used by the guerrilla fish, making the life of the tunnel Viet Cong more difficult. Villages were deserted, destroyed, the fields bull-dozed flat, the open zones shelled and bombed repeatedly. Tunnels collapsed. The exits were sealed. In the remaining tunnels, miles and miles and miles of tunnels, the Viet Cong persisted, waited out the sweeps, the B-52 raids, the shelling, countered the persistent American ferrets. They set up scorpion boxes and bee-hive traps, released vipers in the tunnels, built water traps to absorb the gas, protected their dormitories with hornet nests, moved back into cleared tunnels, used the side of B-52 bomb craters for caves, popped out to fire a few shots, went to ground when attacked. There seemed no end of tunnels, no end of an enemy somehow beyond the reach of modern militarily power deployed without challenge.

Any ferret could have explained to General Westmoreland and the rest, if asked, if sufficiently articulate, that search-and-destroy was a failure: the Americans did not ambush only reacted and the great panoply of fire and steel did nothing to prevent the next ambush. And in the tunnels, creeping through the dark, no American could deny the commitment, the capacity, the endless patience that existed around the next bend, in the next level, in the resolution and resiliency of the opponent. In the tunnel arena American power could not be deployed to eliminate resistance: all the killing could not erode the capacity of the enemy to persist. This was true for the country at large. The will to persist was beyond reach of American power.

The very few Americans involved in the tunnel war were exposed to the tangible reality of the underground as well as the will of the guerrilla. The Americans had patriotism not faith, commitment not a dream, and the outside world waiting at the end of a year. Skilled, brave, dedicated, increasingly professional, they were deployed against forces they neither understood nor could easily image. If only power mattered, this would not matter. If power could not erode their enemy's assets, then the nature of their opponent did matter. And even the most innocent American ferret focused on the mission, eager to win, increasingly capable, recognized that the tunnel complex was an outward symbol of an alien faith: their enemy was beyond doctrine—if not always beyond their own skill and persistence. Such skill and persistence, however, could not be deployed in mass.

The Vietnamese enemy had faith, a binding commitment, and a

belief in the direction of history. What was history to an American ferret? Bravery, daring, loyalty to the unit, overarching patriotism were in the service of efficiency, made the ferrets feared, but at the end of a year each left and the enemy in the tunnels stayed on emboldened and maintained by their faith and replaced by the next levee. And what was such faith to Westmoreland or his generals? They simply used what they had against the others—the orthodox tools of a regular army, the professionals and the trained, the great units, the communication nets, the weapons systems and the doctrine of the day. And there was so much to hand that power *must* have effect. The ferrets knew better: you could only win one by one and there was always one more tunnel, one more VC around the corner, one more in the next level. The advanced weapons helped, the B-52s killed and intimidated, bravery and skill counted, but in the tunnel power evened out. Will mattered and winning was surviving—the Americans need only survived for a year while their enemy waited on time and history.

In the strange asymmetrical Dragonwar in Vietnam, the tunnels of Cu Chi presented the most telling example of the two worlds at war. They were worlds that rarely met, rarely came to recognizable battle, each sought to impose their reality, their own vision, so they could pursue their own war. In the tunnels a kind of reality could be imposed by both while above ground reality was in perception. The Americans saw power deployed and their enemy history moving.

* * *

In December 1966, after the 25th had been at Cu Chi a year, Bob Hope and his troupe flew into Vietnam to entertain the troops at Christmas time, bringing another bit of home to the battle. Except for those few men actually in the bush on combat patrol, the grunts on the line or at times those units engaged in massive sweep operations, not too many Americans were actually involved in "battles." In a war without lines, the 25th Division was at least at the front if most of the men were engaged in service industries: loading and unloading, guarding and parading, filling in forms, most simply getting through their 365 days. Bob Hope did not so much bring a bit of home as arrive at a Vietnam home for Americans who at times were engaged in combat. And so Hope's entourage was at home: the thirteen show girls of the Gold-diggers, Les Brown and his Band of Renown, the scrumptious Raquel Welch and Jill St. John clad in mini-skirts, the pop singers like

Connie Francis or Nancy Sinatra, the whole showbiz Christmas scene, Hollywood on the bulldozed fields of Cu Chi. Beneath the wire and mines and detection devices, under those fields of fire, in the Viet Cong tunnel complex, the flip side of the war, cadres and guerrillas attended their own show at the same time Hope joked. The VC listened to Pham Sang, director of a truly underground theater, present songs and dances, short plays, and exhortation as poetry. Pham Sang sang "Cui Chi, the Heroic Land"—"Our country is a fortress standing against Americans, Cu Chi is a heroic land."[7] His tiny audience was crammed underground into a dank, dark, buried room while above and beyond the American GIs, waving their beer cans, watched Miss World, a world away. They were engaged in a different war, innocent of the Dragonworld beneath their feet, all around them.

Notes

1. After the Bay of Pigs, the great blunder, came a variety of improbable special operations directed at Castro—thirty-three different plans involving the Mafia, biological weapons, and disinformation campaigns. All these seemed to be based on the assumption that covert operations will always remain covert and that romantic scenarios and clever weapons were the way to go. Then and later, the involved could only frighten the orthodox and sensible in and out of the Pentagon—Lyndon Johnson felt the Kennedys had been running some sort of Murder Inc. in the Caribbean—and the Mafia was in reality an incompetent collection of those attempting to avoid stooped labor by recourse to violence—and from time to time turning a reasonable income. A world of poison pens, faked incidents, and sophomoric plots was ample indication that the military should focus on military matters.
2. Douglas S. Blaufarb , *The Counter-insurgency Era : United States Doctrine and Performance, 1950 to the Present* (New York: Free Press, 1977), pp. 50–55.
3. *United States—Vietnam Relations, The Senator Gravel Editions*, Vol. 1. Boston: Beacon Press, 1971–1972, p. 361.
4. Henry A. Kissinger, *Nuclear War and Foreign Policy* (New York: Harper's Brothers, 1957, p. 155.
5. Tom Mangold and John Penyeate, *The Tunnels of Cu Chi* (New York: Random House, 1986), p. 54.
6. Ibid., p. 54.
7. *Ibid.*, p. 54.

9

Vietnam and the American Way of War, 1967–1972

Americans lived out their time in Vietnam above ground, innocent of the nature of the armed struggle, victims of the power of a faith that few understood. The faith, the commitment of the Vietnamese, was the single great enemy asset, an asset that proved beyond American reach, beyond erosion, beyond strategic bombing or firepower, an asset maintained at enormous cost. And in the end the faith proved sufficient just as General Giap had always assumed. The American faith was absolutely in tangibles as the ultimate counter: force that could be organized, seen, deployed, with quantifiable results, force that the reasonable and rational could not forever resist. At the beginning of 1968, the consensus was simply that such power would be telling. The enemy would accept the inevitable. And they did not.

The Tet Offensive and Vietnam: Perception as Reality

By the beginning of 1968, those Americans who sought indicators of progress on their charts still found ample evidence that the Vietnam war ran to assumption. The military was everywhere visible, present, engaged in large-scale operations. Programs worked. The Civil Operation and Rural Development Support (CORDS), the umbrella pacification organization set up by President Lyndon Johnson in April 1967, was effective. Directed by William Colby of the CIA, the Phoenix program out of CORDS had seriously eroded the Viet Cong infrastructure—often by recourse to assassination. Other civic programs were not as effective as Phoenix. Even land reform, so long proposed, was inevitably postponed while the corruption continued. Hearts were not being won. Minds remained closed. The military search and destroy operations proved ineffective, swept and killed and yet seldom kept the country free of infiltrators. The more force deployed the more that seemed to be required. The American presence did not produce pacified zones or democratic mores or a contented populace. The means looked impressive, the statistics looked impressive, but Vietnam looked much the same. And so despite the new methods, despite special operations and the commitment of more American regular units, despite the new bureaus and new programs, CORDS and Phoenix, despite the ingenuity and sacrifice of many dedicated and determined men, the country was not pacified.

The old ways, like the old villages, crumbled. The cities swarmed with self-interest. The government was a facade for the narrow ambitions of limited men. The free market was black and the free world a long way from villages destroyed in order to be saved. Westmoreland and the Pentagon remained convinced that power applied would coerce the enemy, impose cost that could not be met. Then, on January 30, 1968, the war took a new direction for all concerned. Giap initiated a massive offensive on the eve of the Tet holiday. The nationwide attack was as bold, unexpected, and devastating as it was risky. Ho and Giap had jumped to a new stage in the people's war—all the Viet Cong was committed at once, in the open. All the regulars and cadres sent south and many secret assets and even those in Laos were risked along with the Viet Cong. The Tet offensive was intended not only to smash the Americans and ARVN but also to spark a general rising. The concept was as bold as had been that of the communists a generation before in Greece to move into a conventional offensive to support a provisional government in the north. And it proved as fatal as it had for the Greek communists.

The planning had been meticulous and surprise was complete. Saigon and Washington had expected tomorrow to be like yesterday, had assumed that the Viet Cong could not imagine assaulting entrenched, conventional power. Evidence to the contrary was ignored and so suddenly the Viet Cong were everywhere, in the cities, down the street, inside some bases, at the American embassy. All was confusion, fire fights, explosions, rumors, and panic. The Americans at last had conventional war.

The Viet Cong hit twenty-eight of the forty-eight provincial cities and towns of South Vietnam as well as Saigon. Most were attacked on Tet eve or Tet night although one, Bac Lieu, was not assaulted until twelve nights later. An armored column even invaded the South from Laos, collapsing the defense system at Lang Vei. In Saigon the Viet Cong penetrated the American embassy compound while to the north Hue was captured. All this was filmed and dispatched for prime time in America: the Cong at the Embassy. Hue lost and no end in sight. On February 1, for the American public and the many reporters, the situation appeared dangerous. It was not. The Americans and ARVN simply had too many assets to be overwhelmed, even to be anywhere long challenged. Surprise was a momentary matter that assured local successes but no real change in the power balance. Two days after Tet, the initial attacks faltered, the pressure eased, the momentum passed to Saigon and the Americans.

Hanoi had made a grave strategic error in opting for a relatively conventional attack by light-infantry on the core areas of ARVN and American strength. ARVN did not collapse. The Vietnamese people did not rise. And the Americans concentrated their firepower, defended their bases, slaughtered the attackers day after day. In three days momentum everywhere had been lost, in a week all but a few gains had gone, another week and the Tet Offensive was almost over. Only in Hue where the Viet Cong held the Citadel, in some of the suburbs of Saigon, and during a few late attacks did the resistance continue. The television cameras moved to Hue, watched the suburbs burn, reported the chaos live and personal and close-up. During the next week the Americans and ARVN rooted out the residue of the great offensive, closed in on the Citadel in Hue, and moved into the communist-held suburbs of Saigon. On February 23, Viet Cong and North Vietnamese forces withdrew from Hue. Tet was over.

Yet, everywhere it had seemingly been a near-run thing. Within Vietnam many of the promising pacification programs collapsed, many of the supposedly loyal lost their enthusiasm, many of the observers if not already cynical, no longer believed there was light at the end of the tunnel. What mattered most was that until January 1968 the American people had been encouraged to believe that the war was being pursued successfully. In America the Tet television images hardly showed a war won or lent authority to the optimism of the nation's leaders. Many felt betrayed and others vindicated in their criticism of the wrong war at the wrong time. Hawks and doves, isolationists and interventionists, those involved and those vulnerable to the draft, those who hated war and those who did not—all felt betrayed. Increasingly the war was to lose favor, ruin the careers of the involved, run up a butcher's bill that none had imagined.

It had never been a pretty war on the evening news and Tet was bad news. Television brought into American homes the worst images of an unconventional war waged by unappetizing allies for unclear purpose. There was footage of the Saigon Police Chief executing a captured enemy soldier with a pistol: a small crowd, a gun lifted by a disheveled middle-aged man, shot, a young man in a plaid shirt tumbling to the ground. The footage was shown over and over—murder on television, close up in the American living room. Why were prisoners shot by allied generals for television? Why were Marines dying in close combat in Hue in a war that was won? Why were there battles in the compound of the American embassy? Westmoreland would be brought home and promoted. McNamara would resign. On March 31, 1968, President Lyndon Johnson spoke to the nation and to the surprise of all, including his closest advisors, announced, "I shall not seek, and I will not accept, the nomination of my party for another term as your President." Tet, even televised Tet, alone did not drive the war managers from power but their inability to manage Vietnam into victory did so.

Vietnam instead of presenting America with an opportunity to fulfill a responsibility ever so gradually had become a burden increasingly beyond bearing. The country wanted resolution, Americans always want if possible closure. And no one knew how to end the war short of ignominious retreat that might prove to all that America was a paper tiger. And no one knew how to stay to any purpose. The Saigon establishment was not worth saving and there was no other. Replacing the Democrats in 1969, Richard Nixon and Henry Kissinger wanted peace talks, an end of no-war and no-peace. In the meantime they, like the Democrats, allowed the war to run on, eating lives and careers, paying no dividends dividing the country. The whole sacrifice cost too much, produced too many casualties, too many lies, too much corruption of purpose, and too many unanswered questions. During the years after Kennedy's assassination, America lost an innocence about government, about power, about the world. The war was the most visible evidence that idealism had gone with the flowers and the parades. The generals had been given more than enough, enough time, enough men—

in April 1968 the peak had been 543,000 in-country and enough power so that there should not have been a Tet. So Tet was not a Hanoi military victory but a Hanoi demonstration that over the years American power had not prevailed, could not prevail. And this was a victory in perception, a victory of will over power. On November 15, 1969, 250,000 protesters marched in Washington against the war. Johnson was gone and the protesters wanted to make sure that now Nixon would end the agony.

Hanoi had not foreseen such a response as a direct return from Tet. The reality of failure was too much to accept. Officially Hanoi claimed Tet had been a splendid *military* success:

> The General Offensive and Uprising in the early spring of 1968 dealt a lightning blow to the Americans and puppets, not only destroying a significant part of the enemy troop strength and destroying a colossal amount of the enemy's war material but also overturning the strategic position of the enemy and forcing the enemy to suddenly abandon its "search and destroy and pacify" plan and to pursue the defensive and passive strategy of "clear and hold". The enemy had to pull back in order to defend the cities. The South Vietnamese revolution not only had a firm battle posture in the jungle and mountain area and in the countryside but also acquired a new front in the urban areas. The South Vietnamese Army and people brought the revolutionary war right to the hideouts of the Americans and puppets. The key agencies and vital installations of the enemy were dealt painful blows and were paralyzed and disrupted. Our people have become increasingly stronger the more they have fought and increasingly energized.[1]

Reality was not quite as pleasant nor possible to admit even privately for years. American headquarters in Saigon realized just how devastating their defense had been. "The first phase of his offensive has failed." The Viet Cong has lost between 30,000 and 40,000 killed and captured, lost crucial cadres and vast quantities of material, lost the guerrilla infrastructure and the capacity to act for the foreseeable future. In unconventional wars dreams may be murdered. The Americans had killed the 40,000 but not the dream. There were too many Vietnamese to kill. The Viet Cong had been destroyed, mostly in the first week, never to be as effectively recreated. The burden of the war would shift to northern regulars recruited and sent south. The people might have become "increasingly stronger" but the armed units were disastrously fewer. And there had been no rising. The war would become more conventional in that the northern troops would be irregulars, light infantry not guerrillas. And the Americans would bring air war to the North as Nixon sought to force Hanoi to allow an

American withdrawal with honor. Another stage had been reached, a step toward orthodoxy made out of weakness not strength, the reverse of Mao's dictum. Still grand strategy was unchanged, the communists had been promised the future, operated out of a revealed truth shaped by Marx, Lenin, and Mao.

Ho and Giap had always believed that the Americans would eventually tire and withdraw, would not pay the cost forever. The nature of history assured this. And so the communists, the communist armies, the Vietnamese masses would, must pay the cost for an inevitable victory. And the victory was assured by scientific socialism and logic, by the power of the nation and the dream. Tet was suppose to achieve that victory and did not. Hanoi had not foreseen the television impact of the intense fighting on American assumptions and perceptions—not simply on the public but also on the agenda of the political establishment. The Pentagon wanted to persist—Tet had been a military success, destroyed the Viet Cong—but few others in Washington or the country kept the faith. And Hanoi recognized this very quickly: Tet was a victory after all.

Vietnam : Perception as Reality

There was not sufficient American power to put all the Vietnamese dreamers against a graveyard wall. By the time the end of the tunnel was reached, the light only revealed that others had replaced those killed by the ferrets. The Vietnamese could not be destroyed to save Vietnam even if this had been the American war plan. This was the ultimate limit not the restraints on military assets. The Pentagon lacked the power and—as Giap and Ho knew—the will. And even American power was often counterproductive. The United States army drove the Vietnamese from their homes, added many to the body count, ruined the fields, burned the villages, battered the dikes. The refugees fled to Saigon, a place alien to tradition, where their children daily were corrupted and the old ways shamed. No explanations, no formulas read by public information officers could transform the reality of experience. The American presence brought not security but disaster.

Many Vietnamese were not touched by the dream but served nevertheless; underground armies are not composed of paragons and idealists but all sorts. Others, many others, simply went about their way, tolerated the struggle in their name, gave no loyalty to Saigon and so

Vietnam and the American Way of War, 1967-1972 317

provided the water needed by the guerrilla fish. As Giap would note, although the Americans were strong and their weapons strong neither would do them any good because the war was more than a military war. More than military strength and military strategy would be necessary not only to win such a struggle but to understand it. For Giap "In war death doesn't count." Unreasonable to the end, Hanoi was certain of the direction of history and America was a nation without history. Rich, powerful, various, America had choices but Hanoi none, only destiny.

In Vietnam for the militant, the true believers, the nationalists and the communists, those with grievance or no other choice, the war was all, conviction absolute—they could not go home again, would not conceded, would persist. When an American general saw through his binoculars his power deployed, helicopter gunships and armored personnel carriers, great sweeps by heavily armed Marines, heard over the radio the gunners calling targets, the pilots reporting strikes, tasted the cordite on the wind from thundering explosions creeping across the front, he could not help but assume such power irresistible. The Americans could not see beyond visible power. To the targets the display was dangerous but not intimidating. The very size of the American effort bestowed worth—importance on the hidden rebel. For the terrorist on his spindly bicycle, the barber spy inside the base camp, the guerrilla in his spider hole, America resplendent was a paper tiger—its weapons, its power, its array of possession were a curse not a blessing. For those in the grip of a dream, shelled and bombed, exhausted, hungry, sick, vulnerable, less was more. All that power could not change the direction of history or tarnish revelation.

Military power even when deployed without imagination according to faulty doctrine can, however, deny the dream. The American military hoped that by denying the communists long enough, inflicting pain, those in Hanoi, ultimately reasonable men, would compromise. After Tet the American generals were sure that they were winning but increasingly Washington did not care. An American consensus for withdrawal was forming. The American will to war required tangible returns and there was only protraction. The problem seemed to be to manage a withdrawal. What was wanted was an exit without humiliation.

President Richard Nixon proclaimed a secret plan to end the fighting. The sensible would have assumed that America could find a means to close down Vietnam quickly, at least, and at worst declare victory and withdraw. Nixon and Kissinger, however, like Lyndon Johnson, could not easily imagine defeat. America had never lost a war. Losing would as well have dreadful strategic consequences. The necessity for Vietnam was to maintain American strategic parity, contain communism, assure the security of the West. And so withdrawal had to be seen as a result of responsibility discharged not evacuation imposed. The process would take nearly four years and withdrawal would still be humbling.

In the meantime America simultaneously pursued a political end to withdrawal and maintained military pressure to impose such an end on the still reluctant men in Hanoi. The American commanders in Vietnam were often sure, still, that conventional power would impose intolerable costs and so achieve a desired end. And they would fight the war that they came prepared to fight, a conventional war. What else were they to do with North Vietnamese regulars in the field? Certainly the military campaign added up, returned appropriate numbers, battles won, bodies counted, targets destroyed. The second round of Tet in May had failed, no great surprise, damaging some pacification programs, eroding ARVN morale perhaps, but wasting increasingly scanty communist resources. The next year the siege of Khe San was lifted in February 1969, another American battlefield triumph. It was one of many and yet the battlefield triumphs contained more ashes than fire. On the Vietnam battlefield, where so few Americans actually fought, there was an erosion of will.

Almost from the beginning of the American escalation in 1965, the Pentagon had imposed an orthodox structure and conventional priorities upon the military presence. The Lightning Division was moved intact—because it could be, because it was policy; so the computers, air conditioning and miniature golf course, the assets of the home base, arrived on the battlefield. And with the home-base, sooner rather than later, came all the tacky appendages of America, seedy nightclubs and real estate agents, hourly radio weather reports from home, clerks in uniforms working to a clerk's schedule, cold beer, fast food, and bar-girls. The sleeze and froth of an occupation army clogged what was supposed to be a war. Saigon collapsed under consumer goods, recent immigrants, and the American aliens. The war for most of the Americans ran to rote, could be scheduled, included danger only as a random element, a transient rocket round, a freak mine. There had been too long an American sacrifice, one too unevenly spread and

Vietnam and the American Way of War, 1967-1972 319

rationalized by juggled numbers and fixed books and career enthusiasms.

Having done everything required, won every battle, counted up the points scored and the victims slaughtered, filled out the lists and forms and surveys, smashed every visible Cong and communist, the commanders still did not have victory in hand, had not even found a way for America to leave. An elusive enemy, difficult to find, impossible to eradicate, ruthless, brutal, cunning, and no matter how often crushed and disbursed, was still there, still persistent, undefeated. Those with vision or those still innocent would continue to believe that escalation might prove effective even if in the past it had not—an invasion of the North, the bombing of Hanoi, interdiction of the China supply lines, an expedition into Laos or Cambodia, more aid to ARVN, more reinforcements. Nixon would shop from this menu. Something would have an effect. Increasingly the men wanted only to finish the tour, the 365 days. The commanders often performed as requested, increasingly for career purpose and not always to the advantage of their troops.

To the traditional corruption of camp followers and non-combatants was added that of senior commanders on safe heights. Generals had special aides, personal valets, chosen drivers and messengers, slick personal guards and private clubs and special privilege. Only four generals were killed. The rest like most officers came back replete with decorations—half the generals who served received decorations for bravery. Everyone knew there were officers at the front running up statistics for career purpose. The rot within the American military system in Vietnam, not admitted even by the dedicated and honest, grew. Beyond the clubs and duty rosters, beyond even the artillery battalions, the majors and captains who directed combat increasingly risked retaliation from their men if they proved too enthusiastic about building a record, risking lives for statistics. Fragging, tossing a grenade under an unpopular officer was hardly rare. The soldiers in country wanted only to finish their time and return undamaged. And so they went through the motions and sought to survive.

The Department of Defense could move whole armies to Asia, move millions of tons of earth and pour acres of concrete, construct harbors and airstrips and office buildings, dredge rivers and even pursue the war but could not fashion a good war, an effective strategy or even a means of exit. Incremental escalation allowed the Pentagon to deploy the same system and the same formulas and ignore the reality on the ground. All the while the great machine, grown vast, rank, almost luxurious, without a vision beyond the check-lists would not adjust to the reality of Vietnam, the limits of power, the limits of the tangible.

Nixon and Kissinger wanted to move on, to shape a new world order that would include normal relations with communist China and a detente with the Soviet Union, wanted to reduce the risk of mutually assured destruction, wanted a bipolar world that would be stable, in part safe for democracy. The last years of the American involvement through the Paris Peace Conference and the withdrawal in 1972 revealed that little had been learned except the unexpected cost of wars of attrition. America would take away a brutally mixed legacy but one untouched by the reality of unconventional. And left behind was the last illusion: a government in power with an army in the field responsible for a nation at risk.

In 1972 Vietnamese reality was shaped in America by the same assumptions. What looked real was real: a recognized, established regime, a modern army, a viable consumer economy, and a people transformed. America had insisted not only on making the Vietnamese war American but also on making the Vietnamese into Americans. The process of Vietnamization meant recreating the same system that had failed to win the war. And, still, there was hope that Saigon might still evade defeat. This might best be done by transferring more tangible assets, more systems, more weapons to the government. Well before formal American withdrawal in 1972, Saigon had received vast amounts of equipment: 7,000,000 automatic rifles, 12,000 machine-guns, 50,000 wheeled vehicles, and 1,200 tanks, 900 artillery pieces, as well as all the necessary infrastructure of modern, conventional war. There were roads and port facilities, hospitals, command centers, forward bases, housing and clubs, mine-fields, artillery bunkers, filled depots and dumps, the whole enormous American-built, American-equipped military establishment firmly in place. And there was the new model army, ARVN, fashioned in the American image, same uniforms, same doctrine, on parade impressive, on the battlefield often adequate. Once again any visitor could number the assets, sense the power—if not actually find the will to power.

There were also all sorts of tangible problems. Despite all the construction, all the poured concrete, all the new schools and roads and housing, the rural programs and urban projects, much of the country was a wasteland and much of the population urban refugees. The economy had been created to service the Americans. And despite all the military assets, despite the flashy uniforms of the officer corps, the elite units, the guns and tanks, ARVN had neither the seemingly unlimited fire support that had been available to the Americans nor the confidence and capacity of their ally. ARVN was at best an adjunct to the American effort and at times a burden. Alone, ARVN was an illusion. And so too the regime, the institutions of governance and the economy. This was a problem without a solution.

By the summer of 1973, General William Westmoreland felt that "it was virtually impossible for South Vietnam to survive." Westmoreland had retired from the army in the summer of 1972 and like many of the involved still felt vital American interests and honorable commitments were at stake. He was "sick about what was happening":

> And there was nothing I could do about it. Congress had tied our hands. I was disconcerted and very deeply depressed about the situation....Militarily....we won every engagement we were involved in out there.[2]

In 1973 and more so in 1974, there was every indication that "winning" every battle out there had not won the war but rather had exhausted America's patience without creating in the South a viable alternative to the communists of Ho and Giap. As early as 1964, Westmoreland recalled, he had told Robert McNamara that "the thing that is worrying me...is the staying power of the American public."[3] America had not stayed and those left behind at Saigon were assumed by pessimists to be without staying power. They needed more help, more things, more tangibles. Americans had not distinguished between power and will.

Still, even if there were to be no huge transfusion of American aid and comfort, even if ARVN had problems of morale and competence, even if the government in Saigon was flawed, the assets available were impressive, impressive even to those in Hanoi who, while certain of ultimate victory, foresaw more years of struggle. Their assessment proved pessimistic. On March 30, 1972, the North had launched an Easter Offensive that proved a devastating blow to ARVN. In miserable weather, deploying unexpectedly heavy artillery formation, the attack of northern regulars revealed a lack of resolve on the part of ARVN commanders, indecisiveness bordering on incompetence.

The era of the guerrilla had passed. Hanoi pursued a regular war between the northern veteran regiments and the uncertain ARVN units. Some Western observers still felt that what ARVN needed was more, more American aid, more firepower, more kept promises, more of the same. Hanoi knew better—and so too most in Saigon. There was no will. Once the North Vietnamese Army drove south in 1975 in full strength, years ahead of schedule, the last conventional illusions were shattered. Washington, however, was no longer the ally of first or even last resort, no longer in 1975 even very interested in Southeast Asia.

To all the other Vietnamese images was added the footage of the evacuation of the Americans from Saigon, the helicopters at the embassy, the ultimate nightmare in color on the evening news, coda to a ten-thousand-day war. Nothing would be forgotten and each learned what was congenial. The opponents of war were reassured that war did not pay. This war had been wrong, immoral. The military and the hawks remained convinced that if America had laid waste to the entire country, truly used all the power of the nation, then the communists would have reconsidered their commitment to aggression. Nothing changed their minds: more would be more and so enough. The power existed to impose a final solution, kill the revolutionaries, the communists and the nationalists, and so all opposition. Power deployed without hesitation and largely without restraint could be made to tell, could win a conventional victory in an unconventional war. And their fervent opponents denied the validity of that war, the necessity of power, the decency of the system. The Vietnamese war, all war, any war was evil and so too the military-industrial complex, the returning veterans, Nixon and Kissinger, the system, all evil. In the American arena the debate was bitter, uncompromising, long-lasting, and at cross-purposes: the orthodox military defending the conventions of war and their opponents the necessity of war.

The military did not want to understand that military power always has limits—may even unchecked run out of enemies. There are ethical and political limits as well as tangible ones, systemic limits and arena limits and psychological limits: even unconditional surrender is apt to have conditions, implicit or actual. American war aims assured that Vietnam could not be destroyed by unlimited American military power—the battle won but the war lost. The escalation of means would have imposed escalating costs to all but could not reach the essential Hanoi asset without destroying what was to be saved: Vietnam. The means to erode and corrupt the dream of the communist-nationalists had been discarded, unrecognized long before. In fact the astute from the first had suspected that there was no effective counter to Ho's dream, no way to fashion opposition to such an appealing conviction. In any case the American generals were left with power that could not be effectively deployed. The military analysis was that more power could have destroyed the enemy will whether or not Saigon had a pure ideal as banner. In post-war America their opponents believed not only that this power should be denied the military but also than all military power was illicit, inappropriate at best and more likely illicit, immoral. It was, of course, a very American argument.

In a sense everyone was quite willing to avoid the lessons of Vietnam that did not teach what each wanted to learn. The moralists did not want to learn of greed and aggression as a component of the international system: wanted evil taken out of the equation. No one in the Pentagon wanted to learn the axioms of the unconventional, the craft of irregular war and the dynamics of the armed struggle. None were interested in the asymmetrical lethal dialogue between will and power. Vindication or better closure were the goal by 1975.

With Nixon gone, an era discredited, with a truculent isolationist Congress, Kissinger and the new President, Gerald Ford, feared that America would be taken as a pitiful giant. A symbolic display of determination was needed that America was not a paper tiger. Thus in Asian waters a Cambodian communist provocation was met with swift, if ineffectual violence. This was meant to be a show of determination and instead it was indication that the military was still ill-prepared for special operations, very small wars, expeditions and excursions, conflict at the fringe.

The Cambodian *Khmer rouge* had seized the crew of the freighter *Mayaguez* and a rescue mission was dispatched to free the forty men by then already released. Forty-one Americans were killed attempting to free the forty. Intelligence was scanty. The *Khmer Rouge* had by chance been well placed and were by nature determined. The efforts at micro-managing with President Ford in voice contact with naval pilots over the scene suggested that, no matter how small and special, combat need not be pursued in isolation by those involved. Just as the military response had to be cobbled together so too did command and control. Any number of lessons might be drawn from the *Mayaguez* debacle from the continued determination of the President to the inability of America to fight any but big wars with effect.

In fact unlike Vietnam the incident had little impact. There was not going to be another serious unconventional commitment. On that everyone agreed. *Mayaguez* as a symbol of American determination did not sell. America did not seem to care very much except at the needless waste of lives. The event was a few days television wonder, a bungled grace note to cause forgotten. After Vietnam, after Watergate, the country sought peace and quiet not lessons.

Vietnam was over. Rather Vietnam should have been over. The bad news kept arriving. The communists won not a nationalist victory as the American radicals assumed or a bastion for further expansion as the hawks assumed but rather a battered and devastated nation immune to their political and economic formulas. A generation of sacrifice had left the country destitute, and ruled ineffectually by authoritarian old men. The nice little war in Laos fought on the sly, on the cheap, in secret was closed down and the local allies largely abandoned, a disgrace that neither pragmatism nor the rites of the practical could hide. Laos was best forgotten. And for Cambodia the doves found that some wars may have validity, some causes lost initiate real horror. The *Khmer rouge* came to power in Cambodia, declared Year Zero and began to slaughter all those not pure in heart, the literate and the urban, the unlucky. Genocide could not be denied. The horror of Asian communism was not an imaginary nightmare of hawks. Cambodia could not easily be forgotten for even in isolation evil was present at the killing field. It seemed as if all Western, all American efforts had come not simply to grief but also had assured misery, pain, and violence in Vietnam, in Laos, in Cambodia.

Congress had taken, and would continue to take steps to see that such adventures were not repeated. In 1974 American training of foreign police was prohibited. In 1976 Congress would not tolerated involvement in the Angolan civil war in central Africa even after Cuban troops arrived in support of the Marxist regime. Cold War or no Cold War, the nation wanted no more irregular excursions, distant responsibilities, small wars. The Pentagon, too, had no interest in small wars even for vital purposes or in the opportunities of anti-insurgency. If there were to be war, it must be real war. The military establishment had chosen to learn the only acceptable war should be one with a clear mission, with vital national interests at stake, with a general consensus on commitment, with a set timetable, ample resources, full public appreciation, and certain victory assured. This was what was wanted.

This was what the military had learned from Vietnam: America's wars must be real, viable, popular, short victorious, not small, nasty, protracted unorthodox.

The military had learned congenial lessons from Vietnam—everyone, the generals and admirals, those who would be generals and admirals, had made of history what they chose. They military largely felt betrayed. The system had not been allowed to work as intended. The military had been opposed by the media and the radicals, by politicians and intellectuals and those who understood neither war nor risk. And whatever else, the Pentagon became reservoir of a bitter lesson learned: put not thy trust in any but your own. The military must be allowed to pursue war in the American way, deploy power, organize, rely on the elegance of technology, a balance of conventional force, great systems, and main battle units. War in the bush was not for Americans. Terrorists, gunmen, carbombers, guerrillas, and partisans could be left to special operators and special programs for none was a serious military threat to vital American security interests.

In point of fact most armed struggles did fail, were marginal. There had not one, two, three Vietnams. The *focos* of Che Guevara failed and Che had been killed in Bolivia. The Middle Eastern wars were real wars that generated Arab terrorists—not a military matter. The gunmen who appeared in Belfast and Rome and Berlin were a distraction not a military matter. Much of the Third World was peripheral to real power, to the nuclear balance, to the future of NATO. If vital interests were at stake then full commitment would follow, if not no commitment should follow. The country in general no longer felt a need to police the world, "to pay any price, bear any burden, meet any hardship, support any friend, oppose any foe." The survival and success of liberty could not be assured by more Vietnams, had not been assured by a war sold as vital for both moral and strategic purposes. Sufficient duty had been done. The country would move on. In so doing the military wanted to move not so much on as back to the future, back to the comfort of the conventions of the past. The real threat was the Soviet threat, the real enemy armed with missiles and nuclear submarines and tank armies. The real response was to focus on real war, pursue America's interests by engaging in conventional war—or at worse and in rare cases allowing those at the margins to pursue incidental interests. Never again should the nation engaged in a peripheral war unless fully committed.

As for Vietnam everything was forgotten: no lessons in craft, no tactical doctrine, no rules for snipers, or roles for firebases. In the waging of conventional war the Americans did not need Vietnamese lessons. What was learned was one great lesson: Never Again. The radicals and liberals were as one with those of flag rank: Never Again. The Vietnam experience was aberration never to be repeated. What had actually occurred in Vietnam was not as important as the lesson that such experience was to be avoided. Within the military any colonel, every admiral, all those who had been touched by the war took that lesson to heart, obvious, vital, crucial. As always the generals prepared for the future on the basis of past experience: the American example was special in that every effort was gong to be made to avoid what was assumed to have happened in Southeast Asia: the corruptor of American purpose by limited war. For a generation "Vietnam" would cast a cold shadow over innate American optimism. And so all enthusiasm for risk and adventure, novelty and innovation and hence many strategic and tactical future options, political options, military options, policy options, were avoided because never again must Vietnam be risked, Vietnam as remembered, Vietnam as perceived.

Notes

1. An extensive treatment of the Tet Offensive from Hanoi can be found in Chien Binh Commentary, "Heavy Blows Having Lasting Effects, " published in Quan Doi Nhan Dan (Hanoi), February 11, 1968, and broadcast by Radio Hanoi's Domestic Service, 1115 GMT, February 11, 1968. Patrick J. McGarvey, ed., *Visions of Victory, Selected Vietnamese Communist Military Writings, 1964–1968* (Stanford: Hoover Institution on War, Revolution, and Peace, 1969), pp. 267–276. There were those then and later who assumed that Giap had sacrificed at Tet for a psychological victory—casualties did not matter to Hanoi, but it appears that while willing to sacrifice, Giap and Ho anticipated a greater return on their risk.
2. *Washington Post Magazine*, February 9, 1986, p. 10.
3. *Ibid.*, p. 10.

10

Conflict on the Margins of the Cold War

Whether effectively understood or not, the trauma of the Vietnamese experience, a bad war, a special and general defeat, had profound consequences for America. Everyone took away their lessons and the military establishment's new axiom was not unreasonable. Intervention for muddled war aims pursued without sufficient support for a protracted period was undesirable. To be sent on such a mission without sufficient resources was recipe for disaster—and this was true. What complicated matters was that none could tell if resources were ample or effective until deployed. Then it was usually too late. If one never deployed, then the military would not need to fear frustration and futility.

Sill, the military would obviously have to be deployed for war—the establishment was filled with the bright, the perceptive, and the sophisticated and so recognized that the best prospect was to oppose commitments that summoned up too closely the parameters of Vietnam. Intervention in great force would be a congenial assignment as would conventional war. What was not wanted was Vietnam where America's tangible assets for whatever reasons could not be brought to bear. It was assumed that military power in such a war could in some matter be brought to bear against the enemy. Everyone recognized that the war was asymmetrical but the assumption was and remained for a long, long time that this meant the United States had conventional military power and the enemy unconventional military power, covert, cunning, zealous, and lethal.

What was not recognized was that the asymmetry meant that military power brought to bear against military power could not effect essential asymmetry between real power and the will of the enemy. The will was not beyond reach but beyond American understanding. Psyops were little more than bribes and civic lessons. The great asset of the enemy was their dream, the assurance in the reality of the future as read, not even in the real grievances, the blunders of their enemies or the skills of the unconventional. And there were too many committed to the dream for Saigon and the Americans to defeat, too many to kill. The dragon was out of the egg. And so to kill the dragon the Americans had to kill the will and instead sought out one dragon after another to slay with power and elegance and technologically elegant means. They fought a war without ever addressing the citadel of the enemy, the dream. To win a war where perception matters force must be deploy on behalf of ideas and ideals against the dragon's will. This was the asymmetry never addressed, never recognized and so never the foundation for an understanding of the dynamics of any armed struggle. The Americans were, just as they suspected, fighting the wrong war.

None in the Pentagon wanted to reflect too long on the proposition that will had defeated power, that there was never from the first sufficient power and the greater power committed the more will in opposition generated. The Pentagon and the security establishment choose insisted that power must be deployed only when it would be effective—not unreasonable. Few believed that less could be more and that if it were not then more would not work either. The establishment insisted that, military force could work properly employed—and when this was for whatever reason uncertain then involvement should be avoided. After all in some circumstances it was possible to kill the enemy, even in a people's war one could kill all the people necessary. And when this was not possible or too costly or unethical, then there would be trouble. And in a troubled world American should avoid such entanglements—just as the shrewd had always insisted on avoiding a land war in Asia. And what had Vietnam been in the end but just that sort of war: too many targets and no will at home to continue. All this was very sensible, often applicable to strategic matters.

The first military immediate response, underway even as the withdrawal began, was to bury the past as swiftly as possible. No one wanted to know. No one wanted to read Vietnam war novels or to hear old war stories or to publish war poets. And no one did, read or listen or publish. The nation, not just the Pentagon, was eager to turn elsewhere, inward. Ford was an interim president, decent but too much touched by the recent past, by Watergate. Instead Governor Jimmy Carter of Georgia, an outsider, mostly unknown, was brought to Washington in the November election. It was assumed he would oversee the return of the country to normal.

There would be no repetition, no more adventures, no provocative interventions in distant wars for uncertain purpose. America was to return to high moral ground, focus on human rights and decency not on power and pragmatism. Congress would seek to limit the deployment of American power just as did the administration. Anyone who walked the corridors of the Pentagon in the years immediately after Vietnam, talked to veterans now senior commanders on the move up, sat in classes at the war colleges, or discussed military matters would have assumed that there were no lessons to be learned from Vietnam. That war was over. No introspection was needed. There was no analytical examination of American assumptions about the asymmetrical war just finished. All attention focused on real war, on repairing, if possible, the distortions imposed by a decade largely wasted. The fault in Vietnam lay not within American assumptions, military doctrine, and evolving perceptions but in the arena, the uncertain purpose and the failure to pursue the avowed commitment with appropriate resources. Appropriate meant more. And even then with more the enemy could have been defeated but at too great a cost—even the Pentagon accepted that American power had limits—so that power should be shaped for real strategic purpose not even in part for irregular tasking that might lead again to the margins dominating the center.

Year by year the anti-insurgency systems were run down from staff school lecture time to procurement orders for special operations. The budgets sent to Congress indicated the shift—"We didn't want Congress or the public to think we were looking for another Vietnam."[1] In the Air Force the various aircraft—firing platforms—developed for special operations/low-intensity conflict support were phased out. The A-1A, the A-37, and the AC-47 and AC-119 gunships all became legend and transport-delivery capacity was curtailed with only a few MC-130s and HH-53s still available to meet any infiltration requirements. The Navy eventually decommissioned the two diesel submarines modified for SEAL, Marine, and Special Forces reconnaissance teams. The Army's Green Berets were fewer and made do with older weapons, inappropriate communications systems, curtailed training.

Even the word "anti-insurgency" disappeared, to be replaced with "low-intensity warfare."

When any irregular mission was suggested from outside the system, there was no enthusiasm and only the required, *pro forma* discussion until the need faded. Thus the new visibility of the international terrorists, in particular at the hostage-holding shoot-out at the Munich Olympics in the summer of 1972, resulted in demands for a contingency military response. The Pentagon felt in 1972 and continued to feel that there was no effective military response to terror. Papers were written, research contracts given, conferences held, simulations played—until outside interest in a military response to terror died. And when the next spectacular incident indicated that the Pentagon was ill-prepared to respond to the terrorism, there would be a short spurt of concern before the military returned to more serious and congenial matters. As for insurgency, unlike terrorism which was spectacular and analytically trendy, anti-insurgency had no advocates. Little wars elsewhere, Northern Ireland or the Sahara, Eritrea, Chad, stirred little interest. Insurgency as subject had no takers.

The Pentagon in public and in private seemed to have adopted the philosophy that it will not rain if you have no umbrella. The American military did not like and often did not understand special operations, special missions, psychological warfare, deception, black intelligence, political warfare, or socioeconomic programs—all the facets of the unconventional. Elite units were troublesome and covert missions or peacekeeping was unmilitary.

The 1970s had seen a new, largely unpleasant world evolve, more complex, more alien to American interests and historical postures. The continuing wars of the periphery, revolts, coups, rural insurgencies still did not seriously disturb the Pentagon. The Soviet involvement in the Third World could best be handled by other agencies, other means than military. The Middle Eastern wars were contained. The American client and ally Israel was successful. The sprinkle of Third World insurgencies led to no major confrontations. There was no need for intervention. This fit the priorities of the Carter administration. The old responsibilities, the old alliances and alignments had first claim and imposed priorities. This fit the predilections of the Pentagon. The Cold War was treated as a real if unfought war. Force was to be a last and conventional resort even if the President was not always sympathetic to the military's shopping list or the needs of regular war. Thus in a world filled with small wars and irregular means, America prepared for a really big war.

Carter's choice for Secretary of State, Cyrus Vance, was deeply committed to diplomatic means, adamantly opposed to the use of military force as an appropriate policy means, opposed to almost all covert action and any intervention in the affairs of others, a post-Vietnam player. His rival, National Security Council advisor, Professor Zbigniew Brzezinski, an activist, had his militancy curbed by the constraints of the times and the caution of the President. Conciliation in Panama or the Middle East was the first priority. Congress through the War Powers Act and other legislative restrictions curbed potential foreign intervention. There was to be no imperial presidency. The end of conscription and the establishment of an all-volunteer force was an outward sign of the public suspicion of military pretensions. There were restrictions on covert operations. The CIA went through both traumatic investigations and severe personnel cuts that for a time meant that the new director Admiral Stansfield Turner faced more internal problems than operational opportunities. Congress and the public saw that there was no meddling in Angola, no intervention in the Ethiopian crisis, no involvement in the violence in Zaire in 1976 or 1978 nor in Iran's unfolding revolution beyond speeches and public concern for the besieged Shah.

There were, at least occasionally, signs that a reluctant America might suffer from shifts on the margins. And the margins often held real resources: the oil that would when withheld indicate the value of energy—and the cost. The Western position in the Gulf decayed, for example, when the Shah fled Teheran in March 1979 and Iran slipped into revolution. Saddam Hussein in Baghdad ordered an Iraqi invasion of Iran in September 1980. America watched. The most patent evidence of America's new restraint had begun on November 4, 1979, when Iranian militants seized the American embassy and staff. There was no terrible, swift vengeance, only futile diplomatic initiatives and formal protests. The Russians, on the other hand, saw their opportunity and intervened in Afghanistan. All Carter did was boycotted the 1980 Olympics, a gesture of a President seemingly unable or unwilling to act. So in Iran the ayatollahs ruled. At least in Afghanistan the Russians began to discover the real costs of intervention when the faithful opted for a guerrilla war—a means historically valid, ideological congenial, and long practiced that impose real cost.

Carter's restrain was not always sufficient for those Americans still traumatized by the recent past. The radicals and many liberals feared a return of intervention, usually in favor of regimes and causes they disliked. They noted the agitation for a rapid deployment force arising from the events in the Gulf as a danger to the new America. The Institute for Policy Studies in Washington, the bellwether of the Left, began to worry about "interventionism in the 1980s." Their ideological opponents felt that Carter was unwilling to pursue American interests with sufficient vigor. So Paul Nitze's Committee on the Present Danger appear to attack the new isolationism, the focus on human rights and democratic values seemingly at the expense of vital interests and military capacity.

For the conservatives, increasingly coalescing around the drive to put former California governor Ronald Reagan in the White House, the Carter administration's impotence was patent as the Iranian hostage crisis unfolded. After the months of dithering, the military option was initiated in April 1980 collapsed in flames and recriminations. The Pentagon was unprepared for such a mission. And many liberals were not supportive. Secretary of State Vance resigned because the attempt had been made at all. For Vance, for many suspicious of military power, the rescue raid was a resort to force before every diplomatic resource had been exhausted, To the activist critics Vance seemed an apologist for appeasement, innocent of the real world, a clear and present danger as were many of his former colleagues. After the debacle at Desert One in Iran, the Carter administration lived on borrowed time. Risking another Vietnam was one thing, an undesirable prospect not only to the Institute for Policy Study but also for the Pentagon: but the hostage humiliation was another matter. Why could not America do something? Why had the raid failed—another unconventional failure? Why could not the ayatollahs of Iran be punished?

In point of fact, other than a rescue raid there was little that the Carter administration could have done but exercise restraint and seek to apply diplomatic pressure if the lives of the hostages were to be protected. It was a frustrating course of action as in time Reagan would discover. And when the rescue mission collapsed, it appeared that in the case of very small wars very few lessons had been learned. The Iranian revolution had come as a surprise to Washington and the new Islamic Republic had proved beyond understanding or policy response. The ayatollahs and mullahs were not reasonable men, not like us. How could America penalize a people seemingly determined to become martyrs, as eager to die at the hands of the Great Satan as they had died in front of the conventional Iraqi army? What Washington did not understand, what Americans did no understand was the dream that drove the ayatollah. In Iran power could not be deployed and so the nature of the faithful in Teheran was irrelevant—until Reagan later sought to negotiate, to reason with the unreasonable.

Iran for America was an emotional issue unlike Angola or Ethiopia or Yemen. In Iran there were hostages, a human element with regular television coverage, with the President isolated in the White House. There were yellow ribbons around the trees and relatives in tears. Iran affronted American dignity. And even rescue seemed beyond the military. Why? Why could not the Pentagon emulated the Israelis at Entebbe when commandos at great distance had rescued the hostages?

In fact some small effort had been made to cope with the necessity of special operations. By 1979 there was a larger budget, a bit more capacity. The Pentagon established Delta Force—a small, elite antiterrorist force related to the British Special Air Service and the German GSC-9. Israeli experience with the Palestinians was thought relevant to America's new needs. There was some institutional interest in insurgency, a few meetings, some papers. Terror as mission could not be completely avoided. The small opening toward the unconventional marginally improved the presidential spectrum of options but had in Iran not led to competence, much less a rescue.

In time there were answers to what had gone wrong with the rescue mission, detailed analysis, Pentagon reports, and leaks in the media. There had been no adequate special operation force available. The rescue mission had been cobbled together from the various services and bureaucracies, all with different procedures, priorities, capacities, even language. Despite long practice unity had not been possible, even in some cases, trust. In an effort to deploy a lean, hard force, perhaps emulating the perceived Israeli mix, there was insufficient redundancy to cope with helicopter failure and accidents. With the crash and chaos at the first landing strip, Desert One, the mission aborted even before the more serious obstacles had been reached. Recriminations indicated that there had been disagreement from the first about the mission. Everyone felt it was a military mission if covert and special and after that had come confusion. There had been faulty planning, uncertain intelligence, ineffectual training, the almost traditional incompatibility of equipment. There had been all the usual interservice tensions and clashing operational methods. The system could not cope with the unconventional. And the system malfunctioned.

The Pentagon's systemic faults and shortcomings had been present with none of the virtues of the big, brutal, and direct. The American public had felt the administration was to blame. Responsibility ended in the Oval Office. Despite a variety of impressive foreign policy achievements, despite a real increase in Pentagon budgets and capacity, President Carter was perceived as ineffectual. Elected in reaction against the imperial presidency, he had not been, as President, sufficiently imperial. He was enmeshed in details and the common touch. Absorption of briefing books, control of line items, micro-managing did not produce an image of control, vision, or even—after Iran—competent management. In any case, in November 1980, Carter lost his bid for a second term to the former Governor of California Ronald Reagan who in a power suit with cue cards could act like a President. On the day the new President took office in January 1981, the American hostages in Iran were released. The ayatollahs' power revealed.

For the Pentagon whatever the political impact of Iran, the exercise had been a sideshow. The Defense Department spokesmen, the great generals and interchangeable information officers, had increasingly expressed concern not for an unconventional capacity but for massive intrusions of funds to allow a real military build-up. In this they were one with many Reagan advisors who feared a dangerous shift against the country in the nuclear balance of terror. Most, unlike the Democratic advisors, were sympathetic to the orthodox Pentagon shopping lists. The Pentagon made only the most marginal shift in priorities, in doctrine and in strategy: orthodoxy dominated all responses to any challenge. Even the controversial Rapid Deployment Force, proposed to allow American intervention in the Persian Gulf, was conceived as a conventional expeditionary force to be poised to respond to a conventional challenge. If the Iranians deployed unconventional means, infiltration, subversion, terror, as was their wont, the Rapid Deployment Force, still theory not fact, would have no capacity to act effectively. Between a hostage rescue team and a Marine expedition, there was no Pentagon capacity for Reagan to deploy as there had not been for Carter. Those around Reagan, however, were not adverse to deploying any tool against the Soviet empire—the Evil Empire.

The Pentagon like the radical critics of interventionism had been relieved that during the Carter administration there had been no small wars, no expeditionary forces required—only the one Iranian rescue mission. Whatever the reasons, there had been few adventures or provocations. The military commanders chose to assume that any irregular provocation during the next administration could be countered by the regulars or by a small elite force. Some budgetary crumbs could be given to special operations but the orthodox were in command. The Pentagon was thus far, far more focused on avowed Republican intentions to make the American military strategically strong vis-a-vis Soviet Russia than on the other tool, special operations, small wars and subversion. And so the Pentagon misjudged the aspirations of the new Republicans.

The radical Republicans who came to power in Washington in January 1981 were dedicated ideologues, more interested in restricting big government than in the attractions of running the new empire. Only in defense matters was big government a virtue. They wanted, the President wanted, the Republican party and the right thinking wanted the nation to be strong. Defense was to have the funds and hence, it was assumed, the means to assure respect. Details could be left to the professionals, mostly to be found within the defense establishment and its suppliers.

Although a new generation of radical, conservative analysts and academics was on the way, the hard Right at the top in 1981 was composed of those largely innocent of foreign affairs: self-made millionaires, unfashionable economists, libertarians, political operators, Sun Belt anti-establishmentarians. They intended to impose new priorities, cut taxes and spend on defense, encourage the economy, and make American powerful, respected.

As Secretary of State, Reagan in part as a nod to the Republican middle, appointed General Alexander Haig who as Kissinger's protege without a notable previous background in international affairs had come fast and far from handling appointments to NATO commander. Elsewhere in the national security establishment, talent seemed thin on the ground, experience rare. At the National Security Council Kissinger and Brzezinski, Ivy League intellectuals, pragmatic theorists, major players, were followed by Richard Allen and then a series of mediocre administrators expected to learn on the job, Judge William Clark, Colonel Robert McFarlane, and Admiral John Poindexter, a California lawyer, a mid-career Marine, and a naval engineer. Real decisions would be made from time to time by the President's close advisors and old friends. Caspar Weinberger as Secretary of Defense became a conduit for the commanders' conventional agenda orthodox and requirements. On the other hand, William Casey, new Director of the CIA, was, indeed, a covert action buff who would adjust agency analysis to his political convictions and urge special operations.

The new and the old both tended to share an abiding distrust of the Soviets and hence a predilection to see a bipolar, threatening world, a perception that bled out all the subtleties and various shades of difference found in the real world. They wanted hard-edge analysis and action agendas. They were aggressive, articulate, confident, and for a long time, innocent. The immediate result of the Reagan revolution on the defense establishment was patent. Vast sums went to the Pentagon. A six hundred ship navy was on the way. The B-1 was rescued and the M-1 missile to be built. Strategic systems blossomed, were elaborated and duplicated and refined. Strategic redundancy was the order of the day.

Some of the money trickled down to special operations but not much. In the Pentagon the military's capacity to cope with the unconventional remained a minor priority. Casey of the CIA was involved in the unconventional. The Pentagon was involved in systems analysis, administration, procurement, technological research, and weapons development—the sinews of real war.

Almost at once Secretary of State Alexander Haig announced that terrorism was a major threat to America, must be countered, a line must be drawn. Counter-terror was not a mission sought by the military. Haig pointed out that the insurgents within the mountains of El Salvador, a country known to a few Americans, threatened not only our allies in the capital but also all of Central America, all our vital interests to the South. The Pentagon, just like the Institute for Policy Studies, could imagine all sorts of disquieting scenarios, escalation that led from vocal support and a few instructors to a rerun of the great Vietnam trauma. The greatest challenge to the Pentagon was not the subversives or guerrillas but the mission which violated a proper military role. These could best be, should be, assigned to the intelligence community.

The record of such intelligence operations was mixed: the Shah had been put back on the throne but the rationale for the Bay of Pigs had been not only faulty but also a signpost for disaster. And how could the military have deployed proper assets in Guatemala or the Congo?

And behind all of this there was the accumulating evidence, much still classified, concerning the pitfalls of the unconventional: targeting Castro had generated the most ludicrous proposals. The Cold War was replete with examples of American unconventional incompetence, allies lost, mission revealed or aborted or even adjuncts to Soviet policy. The intelligence community seemed to assume any romantic scheme could be implemented, that anything was possible and as a result were apt to reveal vapid assumption, faulty argument, disaster in the field, and no learning curve. To imagine American interests served by poisoning Castro required a real suspension of common sense. Destabilizing radical regimes, eliminating foreign leaders, and underwriting despots often required a feeble grasp on reality. The White House might want change cheap and covert but those who were so charged should have known that special operations were never really cheap nor stayed covert. Yet, the romantics went on validated by Cold War priorities and assumptions, admired by their own, symbols of Amerika gone wrong. Many unconventional tasks were, of course, valid and effectively managed—some, like Laos, were managed so that the commitment did not escalate, some like the funding of democratic European institutions made possible opposition to communist power, much intelligence was valid. Special operations failed when those involved assumed that the means did not determine the end, that cover lasted forever, and that all could be excused under a combined rubric of national security and practicality. So for the intelligence community, unconventional operations were welcomed if not always understood. And time rarely corrected innocence: even the exposed and practical Americans when involved in the unconventional were apt to blunder. Those who did not, who found a vocation in the unorthodox were rare—most of the intelligence community revealed, as well they might, the dynamics of the American ethos, carried their national imperatives into the underground. The generals and the admirals wanted no part of this. The exposure to American misconceptions about armed struggles, alien agendas, unconventional arenas, and the price of secrecy was not what mattered in the Pentagon. What mattered was that such missions fell outside control and doctrine.

If such tasks were required, then the military wanted the intelligence community to do them—as did the intelligence community. The operators assumed their own competence in their secret world, were as innocent, arrogant, and shaped to careers as that of the military. After the Bay of Pigs nothing really change but the personnel. The special operators continued to operate if with more bureaucratic caution. Without introspection or empathy for the alien, the intelligence community still eager for appropriate missions, would insist on existing competence. What was different with the military was a reluctance to fashion any competence that might lead to unconventional tasking, for unlike the intelligence community the generals and the admirals assumed the unorthodox also un-American, the irregular corruptive, and such assignments penalty not promise.

For a generation those most often charged with responsibility for special operations found the assignment congenial and promising and assumed their own competence. The operatives and station chiefs assumed that they were pragmatic when they were merely romantic, assumed that they knew the arena when in reality they merely were in residence, assumed that secrecy permitted rules to be ignored when the greatest asset of the state is legitimacy. There was a fondness of techniques and technology. And most of all there was an inclination to find identical macho pragmatism and mission perimeters: what worked worked. Those who did not approve did not understand reality. In point of fact, the Americans failed to understand the covert world, the power of morality, of legitimacy, and the difference in agendas of those not paid with government checks. To be unconventional does not mean to be covert, cunning, and ruthless but to be forced into irregular means to maintain the ideal. The Americans were apt to discard the ideal in the name of pragmatism, embrace the covert without realizing the enormous costs exacted by secrecy, and assumed that American power can easily adapt to unorthodox missions.

The military sought to avoid responsibility for missions that were peripheral to interest and capacity and risked deeper involvement. Those who had suffered in Vietnam from mission creep were rising to flag rank with their lessons learned, their minds set. So there was delay, inaction, a few bureaucratic maneuvers, the odd report or commissioned study, and the hope that the administration concern with the irregular might be a passing fancy. It was not that the Pentagon felt that insurgency was a mystery or that the military if so ordered could not cope but rather that such missions were secondary, distractions and best undertaken by others. Yet, for a variety of reasons the Republican administration was ideologically sympathetic, in what unconventional missions might offer in countering communist subversion and the revolutionary aggression of new Marxist-Leninist regimes. Many wanted not to respond but to take direct action—unconventional action.

While Secretary Haig did prove to be a passing player, his concern with Soviet provocations remained. Worse, for Pentagon purpose, there was within the Reagan vision a new posture, a new direction that opened the unconventional door from the Right. The new radical Republicans felt the Nixon-Kissinger detente had permitted Moscow to scatter new centers of communist subversions. Crazy states like Libya and transnational terrorists like the Palestinians had violated law and world order. And not all radicals were covert or hidden. The invasion of Afghanistan created a communist puppet in Kabul. The Cubans had propped up new Leninist regimes in Angola and Ethiopia. Mozambique was Marxist. Nicaragua was going Marxist. And the new militant governments in South Yemen or Syria or Libya were avowed enemies of the West. To add to the old guerrillas were these new Leninist outposts. These regimes might be vulnerable. Communist expansion might be reversed—rebels, as in Afghanistan, could be allies. In Washington there were those who wanted to support insurgency and subversion, wanted to deploy guerrillas.

The Reagan ideologues brought a singular and unexpected shift in strategic thinking about unconventional war. Except for the activities of wartime partisans, Americans in the twentieth century—and long before that—had assumed the rebel as opponent. The American army fought irregular campaigns against the Indians in the West, against guerrillas in Mexico and insurgents in the Philippines. The Marines had been involved rooting our insurgents in the banana wars of Central American and in regular Caribbean expeditions. The Navy was intimate with gunboat diplomacy. Everywhere from the Chinese Boxer rebellions to the collapse of our wartime ally Chiang Kai-Shek in China, the rebels had been an enemy.

America was formally anti-imperialist but in Latin America and Asia Washington was apt to support order and continuity at the expense of change. And often colonial rebellions benefited the Soviet empire. The Malayan and Chinese and Indonesian rebels were communists and the Arab and Africa nationalists often sounded so. Those who had seen Mao as an agrarian reformer, Nasser as a simple Arab nationalist, Castro and Che as democratic idealists had been betrayed. The American Right had proven right. Rebels wanted radical change, benefited by chaos and violence, were if not communist agents, then, still a clear and present danger. The messianic terrorists in the 1970s used the language of the far Left, bombed and killed in the West not the East, sought sanctuary behind the Iron Curtain. Arabs or Armenians, Italian sociologists, German intellectuals, even South Moluccans dreaming of never-seen Spice Islands, all damaged Western democratic order. Rebels were as always a disruption and disaster—except when they conspired against the Soviets. The Marxist-Leninists, however, had been brutally effective in responding to conspiracy and insurrection—Hungary, Berlin, Poland, the undergrounds of Eastern Europe or China. So America had found few subversives to foster, few guerrillas to support.

In Central America, where Secretary Haig wanted to draw the first line in El Salvador, traditional anti-insurgency could be abetted in traditional American ways: but, on the other hand, in Nicaragua, anti-government insurgency could be aided in line with a new aggressive policy. In both cases the insurgency game was to be played for ideological reasons, for reasons of national security and not simply as a defensive reaction to perceived danger. Reagan wanted the nation strong, the Soviet empire weak and his associates saw no reason to worry about the military lessons of Vietnam. That was then and this was now. In Central America all the new Reagan initiatives could be deployed. A sense of urgency existed. El Salvador was under threat from native rebels, from Nicaraguan Marxists, from Cuba, ultimately from Russia. And the Nicaraguans in turn were still vulnerable, not yet firmly set in place. And there were the regional considerations, Panama and the canal, Mexico to the north, the other Central American states, wretched, disorderly, often guerrilla-infested. There were real dangers of communist escalation. Unconventional action was needed.

In El Salvador, for most Americans as mysterious and unknown as Vietnam had once been, insurgency was seen to threaten crucial American interests. The arguments for aiding the government in San Salvador were too reminiscent of Vietnam for many. The regime was unsavory, a political, economic, and social elite who exploited the national assets and was protected by killer squads and an undisciplined army. Their fate roused little sympathy. Even when the moderate Jose Napoleon Durate, a Notre Dame graduate and a dedicated democrat, became president, El Salvador appeared filled with unrepentant reactionaries bent on killing a peasant quota.

Congress did not approve. Only fifty-five trainers, not "advisors" as in Vietnam, were authorized and only limited aid. When television news cameras picked up American trainers carrying M-16 rifles, there was an outcry—the first step into the swamp as seen on the evening news. There was repeated concern when Americans, not unexpectedly, were attacked. In 1983 an off-duty naval officer was shot and killed, and in 1985 four off-duty Marine guards from the American embassy were machine-gunned to death while sitting at a sidewalk cafe. In 1987 a sergeant was killed during a surprise attack. Four advisors and two medics died in a helicopter crash the same year. There was grave public concern about unnecessary losses in a far away place. The administration persisted, if on a limited budget, in supplying military equipment, in training the army in and out of the country, in dispatching funds. Durate was welcome in America, his reforms praised, his political opponents on the far Right given no encouragement. And the guerrilla war was not lost, if not won. The American concern had mixed results, no one won and much later no one lost, an accommodation was found and in Washington was El Salvador forgotten.

In the case of Nicaragua, in 1979, a radical collection of rebels overthrew the corrupt and discredited family regime of Anastasio Somoza. This was the first triumph for an armed struggle since Cuba, a victory made possible by the decay of the center, by special Nicaraguan historical conditions, by the usual image of the rebels as idealistic, nationalistic, bold and daring, and by reliance on the traditions and experience of rural guerrillas, urban conspiracy, and a little outside help. The new Sandinista government was dominated by Marxists, sympathetic to Castro, and devoutly anti-American but sufficiently fearful of Washington to permit some opposition and thus hide the long march to the Left from the innocent—conservatives felt that the Carter administration had been filled with innocents. Western radicals were charmed with Daniel Ortega Saavedra and the new radical government in Nicaragua. The Reagan people knew better, wanted to find a better way, find freedom fighters. They found the Contras, a mix of exiled National Guard members, clients of the dispossessed elite, and in time a few disenchanted moderates. In 1981, when the Sandinistas began firming up ties with Cuba and the Soviet Union, the new Reagan administration approved CIA aid to the main rebel opposition on the ground, the Nicaraguan Democratic Force—the core of what became known as the Contras. Over the next decade, the old National Guard, local peasants recruited with pay, Miskitos Indians, the Somoza elite, and a few disenchanted Sandinistas and middle-class reformers made up the shifting Contra organizations. It was a mix that engendered little enthusiasm outside administration circles and the European hard Right.

Despite all sorts of misgivings, including not unexpectedly those of the Pentagon, the aid continued. The fighting escalated. Casey had moved Duane "Dewey" Clarridge from Rome because his dashing style pleased the Director. As the CIA operative on the spot, Clarridge was given an opportunity to deploy irregular means, run a small war without oversight, ignore the practical. And then in 1984 came the revelation that Clarridge and CIA had mined Nicaraguan harbors. The CIA had been caught playing foolish games—again, had been caught without adequate cover, had confirmed America's notions about the unconventional, anything goes and no one will know, the fix will work. Congress cut off of Contra aid.

The Congressional ban on military aid to the Contras in 1984 and 1985 meant that the CIA, Poindexter and Colonel Oliver North of the National Security Council, General Paul F. Gorman of Southern Command, and Thomas O. Enders of the State Department had to play the system to assure the Contra support that President Reagan wanted. The continuing reluctance of Congress to underwrite the Contras and the end of direct military aid meant that the Contra armed struggle could not be protracted unless payments were kept up. The administration in Washington never saw this as an insight into the motivations and priorities of their proteges—only as a technical obstacle to protraction.

So the Restricted Interagency Group had to circumvent the spirit of the Congressional restrictions but not necessarily the law to fuel rebellion until Congress relented and permitted further funding: insurrection on the installment plan. In time this policy, run out of the NSC offices in the White House by North, became entangled in an equally covert effort to free American hostages in Lebanon by arrangements with "moderates" in Teheran in violation of avowed anti-terrorist policy. And so the Contras were privately funded by arms transfers to the "moderate" ayatollahs in expectation that American hostages would be released: foreign policy made real only because the involved were innocent and romantic and entranced with the covert. The Iran-Contra exercise was more special than most American covert operations in that the adventure incorporated so many misperceptions. Essentially even before the Iran-Contra adventure became scandal, Congress wanted all special operations closed down. When the Iran-Contra scandal finally revealed the extent of the NSC activities involving the negotiations with Teheran and the transfer of Iranian money to the Contras, the administration was all but ruined. All the general and special doubts about American involvement in the unconventional emerged.

By then, however, Reagan, one of the most popular of American presidents, had more than any other chief executive deployed special operators and programs aggressively—not simply as response to provocation real or imagined but as means to change events, effect history. The Marines were sent into Lebanon as a highly visible gesture supported by the Western alliance and the Lebanese government, popular with the public for a time, even acceptable to many in Lebanon for a time. And when there was violence arising from the uncertain mission, the President was popular and sufficiently powerful to withdraw if not admit failure. Elsewhere withdrawal was not necessary because as in El Salvador the commitment was narrow, visible, and seemingly adequate to the challenge.

The administration seeking means to punish the Soviet empire found in Afghanistan a means and a method that was acceptable to the political consensus. The Islamic rebels, no friends of the West, could be supported simply by secretly distributing the vital ground-to-air missiles that canceled the one great Soviet advantage of close air support by helicopter gunships. That the mujahedeen were enemies of our enemies was sufficient to underwrite the program. America had sufficient missiles and a covert delivery system and no need—or opportunity—to be further involved. The loss of missiles by sale and theft, the hoarding of missiles to be used later was considered small price to pay for the enormous military and psychological costs the mujahedeen inflicted upon the Soviet empire. In Africa covert aid to the nationalists leader Jonas Savimbi in Angola frustrated the Cuban-backed government's efforts to impose control—Marxist-Leninist control—over the country. If Lebanon were a well-meant failure, Afghanistan and Angola were pragmatic successes in the short term given the administration's assumptions.

In Nicaragua the American intervention, although costly for Managua, did not induce change—the Contras were no closer to victory. The American illusion was that an armed struggle could be summoned up for policy interests, by the distribution of funds and weapons. In Afghanistan and Angola what was done was aid existing struggles, one driven by Islamic xenophobia and one by tribal loyalties. No amount of money or equipment will engender faith. Operations can be bought. And if such operations do not pay the going wage or offer prospect of advantage and this was the case in Nicaragua, those involved do not sacrifice. Many do not even persist.

The lessons learned in Central America were adjusted for Washington reality: do not get caught. Most of the involved then and later remained convinced that Western freedom fighters could be encouraged, armed, organized, and deployed to mutual advantage in Nicaragua or in Africa. And most recognized that the process might be unpopular. A few Americans killed caused grave worries. A few special operations engendered great political turmoil. A few such proposals certainly upset the Defense Department. So it was the domestic political implications not the reality on the ground that most concerned the administration over the Reagan years.

A few American soldiers in El Salvador carrying M-16s had caused grave public concern so most unconventional initiatives must be covert. Yet, if exposed, the operation must be unfortunate but politically acceptable, as was the Bay of Pigs, as was Laos, as had been involvement in Iran and Guatemala—not popular, not everywhere acceptable but not a scandal. The covert maneuvers in aid of the Contras were scandalous, violated law and propriety, revealed the seamy, simple-minded side of the administration, revealed an innocent arrogance and demonstrable incompetence. The whole affair was later disguised in public with patriotic rationalizations. Despite their public assurance those involved did not know what they were doing, Rear Admiral Poindexter, Bud McFarlane, Oliver North at the National Security Council, Casey at the CIA, and sympathizers in the Pentagon or State showed not the slightest capacity to adjust to reality. North made it into a slide show and the ideologues into another test case for the proper world view. And the ayatollahs were moderate only for the National Security Council. None of the foreign players acted to the American script, acted as Americans were wont and as the Americans involved assumed that they would. There was no guidance from the top, no strategic vision, no specialist knowledge, but no shortage of assurance. As always the political advocates of special operations offered a quick fix to complicated problems: sponsored revolution on the installment plan, get the hostages back.

The entire Iran-Contra exercise gave all special operations a bad name, limited prospects in the future, reinforced American liberal opinion, and confirmed the Pentagon's suspicions. Increasingly there were those who wanted America out of all special operations—dirty tricks. The orthodox did not like them. The Pentagon did not like them nor State nor most Americans. In any case what should have become clear was that nothing applicable had been learned from Vietnam and nothing inappropriate forgotten. For the Pentagon what was never forgotten was the danger of the irregular. Although trapped in Lebanon, the military had avoided the worst of other adventures. The best irregular mission was no mission at all. The Pentagon was still loathe to learn even technical and tactical lessons from Vietnam so even irregular war still lay outside the margins of conventional competence. In any case Americans still assumed that covert craft was the driving force of an armed struggle, of rebellion and revolution, rather than the imperatives of the will. Money could buy craft. Skills could be learned. A fix could be funded. Anyone could be a gunman or a rebel—and so Americans believed that the beret and the AK-47 made the guerrilla, not the ideal, the sense of history, the factors that shaped the will. Thus if tasked the Department of Defense first sought to shape a conventional force and if this were not possible then conventionally to arm, train, and deploy irregulars regularly.

In Salvador the Pentagon shipped in the usual American weapons and recycled tactics from Vietnam. Why wouldn't the locals use what Americans used and use them the same way? The "trainers" sought to fashion out of poor, untrained farm boys clones of Westmoreland's men who had won every battle and lost the war, ARVN on the cheap. The Salvadorian army received on-the-shelf American items, M-16s or M-60s. The *compensinos* did the best they could. For lack of alternative, for example, they carried the belts of the M-60 machine gun crisscrossed over their chests, photogenic, *macho*, but assuring sweat-caked dirty cartridges and subsequent jams. This was a difficulty that had never bothered the Americans who at the front in Vietnam received their ammunition in closed boxes delivered by the trucks and helicopters—systems not available to the Salvadorians. No one had planned ahead for local reality. The American system was apparently universally applicable: all war is alike.

American vital interests were, it appeared, not so vital that the Pentagon could furnish appropriate weapons or relevant tactics or even useful advice. The internal requirements of the Pentagon determined who would fill the fifty-five "trainer" slots in El Salvador, each action bureaucracy wanted a share but often sent the next name on their list. The same eagerness to participate despite a lack of capacity had troubled the Iran hostage rescue mission: too many systems and services, too little understanding of the arena. And Iran had been an all-American disaster. Not only was there no effort to relate American procedure to Salvadorian reality and capacity but also no recognition that there was a problem. The problem was in perception: if there was to be Defense involvement then it would be on Defense terms not those of the mission—a poorly defined mission at best. In the core of the Pentagon, the generals feared a replay of Vietnam but on the ground the Americans were allowed to repeat the past. Between 1983 and 1985, the Salvadorian army tripled in size and journalists soon reported heavily bombed towns and free-fire "zones of persistence" but no lack of guerrillas.

In Nicaragua the Contras did no better as rebels than the government in El Salvador did as the army—training, weapons, and advice was at best simplistic and at worse useless and therefore in the field lethal. In Honduras an American build-up to intimidate the Sandinistas and encourage the government of Salvador revealed, again, the grand designs of the Pentagon: 7,400 acres were needed to train Salvadorian and Honduran troops. The Americans always seemed to need more, more acres, more slots for career enhancement, more layers of command and control, more weapons systems, more air-strips—the C-7 Caribou could land materiel and troops on any of nine hundred carefully plotted Central American strips. In 1983 General Gorman, both shrewd and reluctant, arrived in Panama as head of Southern Command to "counter Soviet and Cuban militarization and other destabilization undertakings." In two years he accomplished much in preparation for any potential conventional war, lots of plans, lots of exercises, lots of assets in place; but the rebels in Salvador persisted, and the rebels in Nicaragua, our rebels, apparently totally dependent on American aid, were, if a drain on Managua, not a threat.

The Reagan adventures in Central America would in time fade off into textbooks, if not doctrine, forgotten except as scandal; and they were, indeed, textbook cases of the American response to the armed struggle. The Reagan Republicans were confident in the American capacity to cope with the unconventional despite limited evidence to that effect. All presidents had been attracted by what could be seen as a quick, low-cost, secret fix to a troublesome problem: get rid of the communists in Guatemala or Castro in Cuba, help the democratic forces in Italy or Germany, keep Russia out of the Congo, and overturn the Sandinistas in Nicaragua. And do so quietly, cheaply, efficiently. Thus it was always assumed a problem could be managed off center stage, out of sight, for low cost, and if need be the exercise could be denied. This is how American fought the Indians, pursued vital interests in Latin America, countered pirates, or dispatched troops to Russia after the communist revolution: legally but quietly, irregular missions. All this every, every time had great charm for most chief executives until the Bay of Pigs invasion collapsed or the Iran-Contra exercise displayed the inept: those who claimed they could cope were innocents abroad and perhaps criminals at home. No matter, special operations have always had special charm.

Closed wars, limited to the participants, waged in secret, were possible but even then not fully secure, certainly not at the end of the day. Journalists appeared with mini-cams for Central American massacres, could be found on tour with African guerrillas, and arguing pool coverage with Shi'ite hijackers in Beirut. For American purpose this meant that unconventional conflicts could not always or often be hidden away. Sooner rather than later, the American denials would be implausible. And the operations might well have then and later undesirable, unforeseen fallout—Afghanistan mujahedeen moved on to become an Islamic foreign legion opposed to American interests. Unconventional initiatives seldom were as sold—cheap, effective, quiet, and plausible—but remained popular.

When unconventional operations were the responsibility of the military as in Greece, as in Vietnam, the military sought to transform the covert and illicit, the underground into more congenial form. The Americans did not want to accept the asymmetrical nature of conflict only of the tangible assets involved. What was accepted as unconventional were the irregular means—those techniques employed underground when no others were available. Actually such means were the only way for the weak to deploy the power of the dream. The Americans assumed such means were identical with the dynamics of the underground, were first choice not last, were all that need be learned to pursue the unconventional. The American focus remained on techniques and capacities—and tangible assets. Special operators were trained to endure, to cope with limitations in far places, to survive and strike but those so deployed would always do so as soldiers not as guerrillas without prospects or pension, those on tour for life driven by history not capacity. Such enemies were not congenial and so, as in Greece, the military preferred to counter the irregular with regulars, the dream with brigades when the dream was recognized at all.

In El Salvador the army chose to create a clone as in Saigon. Still while the government did not win, the rebels did not win either. In Nicaragua the rebels, the new Reagan freedom fighters, were dependent not on faith but funds and advisors, not at all like the mujahedeen in Afghanistan who went their own way. Allah was the answer not missiles. The asymmetry was not between helicopter gunships and rifles but Allah and helicopter gunships. In both Central American cases the direction of a small and distant war fell outside the conventional military province. Washington knew what it knew but knew little of the reality of the armed struggle. What did it matter? The Contras were bought and paid for and bled the Marxists. And in Salvador a new army was established to pursue the obverse of the same struggle. No one needed to understand the nature of Central American reality or the dynamics of an armed struggle. In Afghanistan this did not matter: the mujahedeen knew what they knew would be ample and so it proved, costly but ample. In Central America the result was mixed, a stalemate and an embarrassment but the embarrassment had not been a military one but one for all special operations.

In all cases there was the irresistible urge of Americans to act as Americans, the persistent reluctance to adjust to local conditions or to learn from past disaster. The foreign friend or enemy were assumed explicable, driven by the same agenda and assumptions as Americans. The *campensinos* of Nicaragua and Salvador were country boys and could not integrate sophisticated weapons. No matter, they were integrated into an American anti-insurgency campaign. There is little doubt that in Afghanistan, as in Salvador or Nicaragua, Americans of all sorts and persuasions would have liked to impose priorities of efficiency, would prefer a united resistance instead of tribal feuds, internecine bickering, wanted decent allies. Other nations were apt simply to make do with the locals, accept that Berbers would be Berbers and Arabs be Arabs; but Americans wanted their associates to be as Americans, to be as well good and decent, to advocate modernization and democratic practice. The bad guys made the American public uneasy and only the pragmatic romantics could believe that the Contras were without blemish or even that Savimbi's UNITA in Angola was untouched by tribal greed. The hard-edged operatives insisted this did not matter; but, of course, it did matter to most Americans, to most in Washington, and most of all to those attracted to the American ideal. Magsaysay was thus a genuine American hero, an idealist for an ally instead of the usual despots and generals. Mostly even before a campaign began, the Americans were apt to have discarded the power of an ideal for convenience sake—given up to the others the dream in return for perceived pragmatic assets. The hard-edged special operators were as innocent. Ideals have enormous power, killing power, every time, and often offer the only power to an underground or to a flawed government. And greed, evil, and corruption cannot be hidden forever as acceptable by-products of the irregular or as an acceptable aspect of governance. Forever can be a very long time, however, and many of the practical take short views, tolerate corruption or despotism, find rationalization for deploying the unsavory.

No matter, no matter the risks and costs, the embarrassment and scandal, what was wanted by the Republican administration was to confound the Evil Empire. And this was done. The transfer of funds, the deployment of sophisticated weapons, the public support and the private aid, all caused the Soviets and their government allies in the Third World problems. In Afghanistan patronage had been a net gain, a good match with the American way held at bay. In Africa American protegees, if they did not prosper, persisted. In Lebanon the entire mission collapsed and was abandoned as ineffectual in imposing peace on those not so inclined. In Central America the line had been held in El Salvador, even if the Nicaraguan game had not gone as well. The ultimate failure of the Sandinistas under economic pressure and inept policies came almost as unexpectedly as had their victory. They would lose in an election few orthodox communists would have held and fewer accepted the negative results. In any case once the Marxist were out of power Nicaragua could again be neglected by Washington, no more lines need be drawn.

During Reagan's tenure America may not have played the new revolutionary game very well, but the game was not over. Yet, the years of the Republicans brought revelation about the nature of unconventional no nearer. Many American understood such wars but few were found in position of authority, with flag rank or easy access to the Oval Office. The doctrinaire felt no need for enlightenment for they were confident of orthodox skills, intense training, high-tech systems and their own priorities. Americans would persist in being Americans.

Americans have had little recent experience in revolution. The American dream supposedly for export is apt to be considered neo-imperialism. Despite this the American way has enormous attractions but not as a lever for swift, radical change. American offers a system that works not a dream to change history. The American way offers opportunity not a sense of urgency. Once democracy was radical and dramatic but no longer—desirable, pragmatic, but without the incandescent appeal of revealed truth, deep faith, the compulsion of resurgent nationalism, the urgency of Islam, or even the imperatives of the clan or tribe, the attractions of vengeance. America often had little to display but skills and techniques, easy money and elegant arms, and things that could be counted, American assets rather than the American dream. That dream was shaped to practical accomplishment rather than an appeal to grievance, to pride and vengeance, to the addictive absolutes of history. Those denied thus perceived a world quite different from the tangible and quite alien to what American imagined wanted—especially since so many obviously wanted what America had to offer.

For the powerful perception does not count for much, deception is unnecessary, and subversion a lesser option. Thus few Americans are familiar with the dynamics of scarcity where all that can be deployed is the energy of the faith. Energy requires not skill, but commitment. Bandits can be hired and irregulars, small forces can be recruited, deployed, a professional army paid and trained, can begin and end with tangibles as did the Contras; but for an armed struggle there must be a dream to supply the energy and persistence. And there are those without a dream, killers in the ghettos, bandits beyond the law, village defenders and tribal warlords. Some seek redress for grievances Americans cannot imagine but all are apt to use unconventional means. Terror and gunmen are un-American. Armies without banners are not part of America's recent past. Rebels do not invest, allocate resources or buy on margin or on time, do not wait on certification or the monthly net returns, do not operate to budget or to rule. Militant rebels find no charm in decency, compromise, conciliation, fair shares and a modest pension, no charm in a Raspberry Reich or Aldo Moro's *compromesso historico* in Italy or the Free World of Ronald Reagan. For an IRA history had to be changed. Irish republican volunteers will starve to death for the Irish Republic as imagined—a method not amenable to easy funding—do not wait on foreign funds or the correlation of forces. They will even try peaceful means to achieve the vision. The welfare state does not address the welfare of the visionary. The American way is not their way. America was apt to offer not ideals but tangible systems. And systems are more valid than mere dreams except underground when without the faith there is no deed, no armed struggle, no prospects at all.

The rebels live with enormous intensity and at great risk on the dangerous edge. They can not go home again for the underground is their only sure environment and for most a final, lethal environment. The orthodox in general and the Americans in particular seem to offer friendly rebels dangerous entanglements, well-meant instructions, and the doctrines of regular military institutions. The orthodox often impart an ineffectual copy of the conventional—as true for the Soviet Union as for the Untied States. Even those courses and schools that graduate assassins, agents, and skilled terrorists often produce soldiers not idealists with guns. One can learn how to kill but not how to perceive history on the move.

So the Americans chose instead to create a bad little army of irregular regulars, a small copy of the American way, an inexpensive export model. This worked in Greece because the communist ineffectually chose the same strategy. It was unnecessary in the Philippines because Magsaysay made any army unnecessary, addressing the grievances that generated the Huk underground faith. In Vietnam the effort engendered catastrophe, avoided in Laos because isolation and cunning kept the irregular as system, the war small, running to seasons, protracted. When conventional power could be deployed, focused, then intervention worked as in the Dominican Republic, when power was limited but so were the aims, then simply not losing, protraction was a real return was the case in Laos and El Salvador. When power could not be deployed against the perceptions of those beyond reach of the tangibles then, as Vietnam indicated, the Americans were assured of frustration.

The professionals army always wants to encourage others to their forms and agenda. And professional aid is often inept. Soviet cold-weather equipment has been dispatched for warm weather wars. The PLO received not only tanks and artillery from patrons but also delusions of grandeur. Often without empathy for the rebel, the conventional offer only their familiar assets. This is true for each conventional power but has long been more so for the Americans because the establishment was so isolated from the factors that generated an underground. Thus in opposition to the unconventional Americans deployed the conventional and in alliance with the unconventional Americans offered conventional aid and comfort.

In the years after Vietnam, unconventional conflict even within the context of the Cold War became even less conventional. The era of the communist insurgency, the years of the guerrilla could all but engender nostalgia once the terrorists were perceived as threat. Yet, however irregular the operations of the *Brigate rosse* or the PLO appeared in Washington or for that matter Moscow, all were classical armed struggles, each followed the rules, revealed the same dynamics. Each was organized, controlled, and historically valid if often tactically unsavory. What was less controlled and more unsavory were true irregulars who were driven by greed and ethnic malice. Increasingly in the free fire zones of the Third World, the irregulars were joined by warlords, tribal levies, criminal gangs with banners and tribes without flags. Children became soldiers rather than starve, killed out of curiosity, brought fear and death and anarchy to the countryside. A failure of the state meant an opportunity for those with guns. And those who took such a course were not traditional revolutionaries, did not fit easily into Cold War models, offered little advantage to anyone, and so were apt to be filed and forgotten in the outback, not relevant.

For Americans the world was becoming more foreign, less explicable. The motives of the Iranians who dispatched teenage children across mine fields to attack Iraqi positions were beyond American experience and comprehension. The warlords of Southeast Asia, the guerrillas for hire in Latin America, terrorist hijackers, the cunning agents of international crime posed novel problems. American wanted to find moderates, find the reasonable—even in Teheran—and instead there were terrorists boarding the flight. As long as the Cold War imposed models and priorities, these terrorists and their guns could be defined, assigned significance, encouraged or denied not only by Washington but also by Moscow. Such a definition of gunmen as players in the greater game was apt to hide the nature of anarchy, the persistence of ethnic and religious grievance, the random violence of the greedy or undisciplined. The vile and malicious, the narrow hatreds of the parochial had always been there but amid the great bipolar confrontation after 1945 were ignored. All during the decade of the 1980s, the key to low-intensity violence was Moscow for Washington and Washington for Moscow.

The Reagan ideologues did not want their mind cluttered with exceptions, historical explanations, local data or special cases anymore than Kissinger and Nixon were interested in the small picture, the reports of consuls, and the tempering conclusions of CIA or State analysis. Each knew what they wanted to know. Few knew that they knew little of the underground, terror, or the dynamics of the unconventional. And if so, these flaws were minor. The power of American would make good any failure of analysis. Power compensates for much. And so the Reagan administration like the Democrats before them, shaped unconventional war to conventional American image and priorities. When the military was dispatched on special missions, tradecraft and anti-insurgency doctrine were deployed: hearts and minds did not matter much nor the dynamics of the underground, only targets and skills and weapons. This was the American way, the military's way. And the American military had evolved into an institution singularity inappropriate to respond to small, irregular missions.

More can almost always be deployed but not less. Less imposes choices and denial. So, sensibly, the bureaucracy wanted more. There was an instinctive bias for high-tech whether deployed in the tunnels of Vietnam or sent to a client underground. Sometimes, as with the missiles in Afghanistan, the system proved effective but often the military solutions, just as the CIA initiatives, seemed worse than the problem. In October 1985, the Marine base at the Lebanese international airport had been destroyed. No one at the Pentagon was very sure why the Marines were there or to what purpose. What was intended? What were the Marines to do? And having been target, the mission was soon canceled, American forces withdrawn, terror triumphant and Lebanon left to others. Then and later the Pentagon felt badly used, not at all responsible for the disaster—that had been fate, the anarchy of the times, technical errors in command but not the system. The system stayed the same. Even the techniques of deployment in a terror zone stayed the same. Mission flaws had been at fault not really the Marines. And, of course, the Marines should not have been dispatched on such an unconventional mission at all. If it were not a real war, the system was not prepared for a discrete or elegant response—the Marines could be dispatched.

Later in Panama and Grenada the military with administrative authority would dispatch huge conventional forces to solve a small local problem. The Pentagon was apt to find no middle ground between too much and nothing. The muddled middle ground the generals felt should be left to others within the national security establishment. Yet, some low-intensity capacity had to be shaped to quiet congressional critics if nothing more, to respond to narrow mission parameters—rescue hostages or secure an airport. If there were unconventional operations, the task should be amenable to a small elite force. Such small and anticipated threats could be managed. Proper missions, even small wars could be managed. When dispatched on a such mission, the military felt that it was possible to cope just not desirable to do so. Because many unconventional challenges were neither predictable nor congenial to doctrine nor fitted easily past experience, the whole low-intensity requirement thwarted the military. Failure, however, was un-American and so failure in Vietnam or Lebanon was not a military failure, a systemic flaw but the inevitable result of inappropriate missions. The flaw lay outside the system. So the system was never called into question, nor doctrine nor experience nor the logic of denial. Instead of organizing a response, the establishment denied the need to be prepared. If requested to prepare, offices were opened, papers written, seminars held.

The American military even under the activists Reagan circle was determined to deploy as always, treat every challenge as conventional or amenable to the accepted conventions of the unconventional. No more telling example of this mind set could be found that the expedition dispatched in October 1983 to Grenada in the Caribbean just after the Lebanon disaster. The administration wanted no more Cubas, wanted to close down the radical New Jewel regime in control of the tiny former British colony. The horror of the Beirut disaster was still tangible in Washington: the Marines slaughtered, American prestige in tatters—and why? In Grenada there was a clear mission for the military, a necessary response to another unfolding challenge to American prestige. Reagan acquired token aid from six small Caribbean countries and asked the Pentagon to restore order and democracy and incidentally rescue the American medical students on the island.

The Grenada expedition appeared to present the Pentagon with an orthodox challenge: deploy the appropriate military means in three days to occupy a Caribbean island thereby removing a regime perceived as illicit, communist, and potentially dangerous to American interests. The opposition on the island was anticipated as real, if small, buttressed by Cuban advisors and supported by some civilians. The Pentagon system went into gear fashioning at speed an expeditionary force that would include a selection from the available units, representatives from all those who wanted or could wrest part of the action. There was no real thought to an indirect approach or reliance on special forces or on irregular means. The powerful as first choice use power. In three days the Americans landed with overpowering force—no entanglements, no protraction, no irregularities, no problems. In the fighting, more extended than anticipated, against a smaller defense force than foreseen, forty-five Grenadans were killed and twenty-four Cubans with an additional 600 other Cubans captured. The Americans lost eighteen killed and 115 wounded. The communists were rounded up and their regime closed down.

The diplomats took over. The administration was satisfied. The Pentagon was apparently delighted. The army issued 8,612 decorations, including 170 medals for valor, even though there were never more than 7,000 men on the island. And then reports of bungling began to appear: Special Forces misused and endangered, inept planning, interservice rivalry, flawed communications and control, a clumsy campaign, stalled by a handful of amateurs and Cuban construction workers, not a triumph at all. The Pentagon did not agree: America had won and won without complication on short notice without resort to the irregular or the unnecessary. This time, unlike the Teheran mission, there had been enough helicopters, enough of everything. In such matters one cannot have too much.

The Pentagon, in particular the army, was not unduly discouraged by the record in post-Vietnam conflicts. Such low-intensity conflicts were all obscure, marginal, forays and displays, more noted by cameramen and commentators than strategists. During the years of the Reagan administration, despite the enthusiasm for an aggressive response to Soviet provocation, America had avoided most of the real wars and nearly all of the unconventional conflicts. Most generals would have preferred to have avoided every one. The Marines should not have been in Lebanon. Little could be done in Angola, should be done in Ethiopia or the Sudan or Yemen. A military response to terror was an undesirable mission. The Delta Force that had emerged as the answer to hijacking would serve. The CIA could cope with small missions, dirty tricks, special effects, and commitments that did not require a military component. If more were wanted, then everything was on call. The expectation is that there would be few calls. The public and the politicians might want something done about terror but the military knew better. More assets deployed would not help, nothing within military control would help. The problem with terror was that there was no solution, no prevention, and no assurance of effective response. So entanglement had been avoided to democracy's advantage, avoided in the Philippines, in Haiti, in South Korea. The line drawn in Central America had great political and regional cost but very little for the professional military: the CIA and the National Security Council and the President were blamed for excess not the Pentagon.

Elsewhere, a terrorist attack on Americans in Berlin had been traced to Libya. Americans in general and the administration in particular were inordinately frustrated by the inability to trace terror to its lair, discover the responsible. The common wisdom was that there must be order in random terror, a conspiracy, control and command, surely communist subversion but at the least as culprit the crazy states that rationalized atrocity. State terror was more explicable than random gunmen and stray hijackers. Lebanon had indicated the difficult of bringing power to bear on irregulars—and the cost of deploying elegant combat aircraft over a free fire zone. What had been remembered as lesson, however, was that the more regular the force then the more vulnerable to American power.

So Libya discovered as culprit meant a tangible target. An air strike against Colonel el-Gadaffi's Libya was ordered and carried out as a conventional military exercise. The targets showed up as blips on a target radar, were not snipers hiding in hills beyond reach of a battleship's guns. Even if the strike were not surgical, few air strikes can be, it was assumed effective. Washington observers noted a decline in terror statistics. There was as well the seizure in mid-air over the Mediterranean of a few terrorists from the hijacked *Achille Lauro*. These were minor matters but ones that indicated the new aggressive anti-terror policy using orthodox means. What was wanted was assurance that such exercises would stay minor: Vietnam had begun with a few advisors, no great commitment.

Coalescing even before Saigon collapsed in 1975 the military's reservations and aspirations were to remain almost unchanged for a generation: if American was to go to war that war should be an American war. There were to be no more un-American wars. In this they were one with much of Congress and many of the people and all of the radicals who were fearful of American interventionism. America should not send soldiers to do a policeman's job nor intervene militarily in troubled zones nor oppose every provocation with expeditionary force. While there might be a disagreement over what was the American way between radicals and the Pentagon, there was none on American military expeditions. For both such unconventional missions were unacceptable and if for the radical critics any such mission was unacceptable, the military establishment merely wanted others responsible for the unconventional.

Few at the Pentagon chose to consider that the way in which the military waged the war eroded public confidence or that political criticism might reflect genuine distaste with a war without aims or prospect of a good end. Many in the Pentagon felt in 1975, in 1985, in 1995, that the Vietnam war could have been won, if the public had maintained faith, if the limitation imposed had been lifted, if the appropriate strategy had been deployed. Next time, if there need be a next time, the people must give the military *carte blanche*, the war must not be protracted, limited, or unpopular. From the secretary of defense down to the war college instructors, the military wanted a general understanding, no surprises, no Vietnam, no unconventional war at all really. So too their critics.

The Pentagon made no secret of what kind of war would appeal, regardless of any political considerations or the desires of the commander-in-chief in the Oval Office. First, any unconventional entanglement must engender general popular support. No more anti-war movement, treason preached at universities, civilians spitting on veterans. The Pentagon wanted a guarantee that the American people supported the war. No support, no war. This next and future war would have victory parades. It would be short, violent, orthodox, and successful, systemic error forgiven by triumph: Grenada on a grander scale. General John W. Vessey, Jr., Chairman of the Joint Chiefs of Staff in September 1985, on the eve of his retirement, mused on the war problem:

> Americans are not patient with long drawn-out protracted wars, wars that dribble away our public support and strength ... would support military venture if it were sensible ... (generating support) was the job of the political leadership of the country ... Don't go it you don't have to ... but if you have to go, go in a fashion that's going to get public support for what you're going to do. Do it quickly. [2]

Time would erode only some of the military angst that Vietnam had generated but the assumptions were institutionalized. Only certain kinds of war were appropriate—and the more unconventional the prospects the more likely to be inappropriate to the military.

If the venture were popular, precise, militarily viable, if adequate forces and freedom of action were assured, no carping over means, no need to explain why the village must be destroyed in order to be saved, then a quick war could be won. In 1982 and 1983 the Pentagon had urged against any Lebanese commitment. In March 1983 General E. C. "Shy" Meyers had warned against putting troops in Central America. The Pentagon wanted to be left alone to design, budget, procure, organize, and deploy the means to wage real future wars, to be left alone to its own devices and desires. In October 1985, at a speech to the National Press Club in Washington, Defense Secretary Weinberger summed it all up—again. American forces should only be used in limited situations when the troops had public support and stood a good chance of success.[3]

The public, like the Pentagon, liked swift, technologically sound solutions, believed such solutions could be organized, did not want to protract matters or to depend on stratagems, deceit, secrecy, or the irregular. The new doctrine of real wars was articulated repeatedly by General Colin Powell as chief of staff: no more flawed crusades. So the generals wanted generals' wars, the sound of trumpets and the promise of popular esteem after swift victory won by technically elegant, massively deployed force. And why not?

Ultimately at the end of all sorts of excursions and alarms, irregular missions and new roles, after expeditions dispatched in haste and entangling alliance, after the Soviet empire imploded and China adopted the capitalist way, the Pentagon would get the war wanted a decade later in the Gulf. Yet the Gulf War was anomaly in the contemporary American experience. The small wars, unwanted, often unproductive, dominated the military menu. Even traditional hawks, in some cases especially traditional hawks, felt that in an unconventional world, the Pentagon should have unconventional assets. Under pressure, ever mindful of political power and the next budget, the Pentagon had already slowly reversed the post-Vietnam rundown of anti-insurgency assets and systems.

Beginning in 1979, there were a few increased line items in the

Conflict on the Margins of the Cold War 359

military budget, no great enthusiasm but still indicator. The debacle of the Teheran rescue operation in April 1980 focused some minds both in Congress and hence in the Pentagon on the dangers of a nearly total lack of special capacities. Money was shifted if quite tiny sums in Defense terms. The small Reagan budget for special operations went up in small annual jumps from 1981 to 1986. Congressional support appeared with Senators Barry Goldwater and Sam Nunn critical at the slow response to potential unconventional challenges. New equipment arrived. The Special Forces often found the infusions a mixed blessing. The Pentagon proved no more effective in procurement of special equipment than in deployment of special forces. After seventeen years the army's Special Forces received a reliable, multi-capacity radio, the AN/PRC-70, that unfortunately weighed forty-five pounds and was too large to be effectively portable and required additional batteries that added still more to the weight package. Inept procurement policies, giantism in design, and over management unrelated to combat were hardly problems for Special Forces alone. The unconventional continued to suffer not only from the conventional flaws of the Pentagon system but also from a continuing lack of command enthusiasm. Senators Goldwater and Nunn could make waves but they could not greatly erode the bureaucracy's concern with other matters.

There were some gains. The numbers grew so that by December 1984 there were some 14,081 personnel in various special operations forces, Army 8,331, Navy 1,550, Air Force 4,200. Such numbers were marginal in a military with 2,200,000 members. No one in command was interested much in the cost benefits of the special, the enormous impact of a few, the rewards of elegant, indirect means. In fact in their efforts to deflect external concern with special operations, the establishment was apt to respond by assigning more, more administrators, more organizers, more to give an impression of concern. More bureaucracy meant less field efficiency. The commanders did not recognize nor care that in special operations numbers could be counter-productive, generating not capacity but administrators and formal system. The "elite" units either made do with leftovers and came at the end of lists or were confused with numbers under the assumption that absorbing the special into the general would end demands for an unconventional capacity.

Mostly special forces had obsolete psywar materiel, no air or naval delivery systems, elderly communications equipment at base and in the field—the new AN/PRC-70 radio that did not work was at least new. There were few mortars, limited specialized weaponry except for the tiny anti-terrorist units, no real or simulated missiles or tracking equipment, only limited foreign weaponry. There were lots of spectacular training and in-group enthusiasm, no place in doctrine, no friend in the White House and, most important, no support in the Defense hierarchy. In January 1984, Goldwater and Nunn reported to Secretary Weinberger that the Pentagon had made "few significant improvements." They pointed out that special forces still had the same number of specially equipped planes that had been available for Teheran in April 1980, four years before, and two fewer long-range helicopters vital for deep penetration raids. On October 3, 1984, Deputy Secretary of Defense Paul Thayer issued a directive which "covered" a series of specified force upgrade "urgings." In effect it had become necessary to put the administration's priorities on record within the Pentagon so that the system would respond:

> Resource decisions for current and programmed SOF (Special Operations Forces), once made at the Secretary of Defense level, will not be changed or reduced by OSD or the Service Staffs unless coordinated by the Principal Deputy Assistant Secretary of Defense (International Security Affairs) and the Assistant Secretary of Defense (Comptroller) and approved by the Secretary of Defense.[4]

At last there was supposed to be movement, an end to directives ending in out-baskets without action. A Joint Special Operations Agency under the Joint Chiefs of Staff was designed as an advocate for revitalizing special forces.

Still more money was spent. More troops were moved to elite units. And yet little changed. The air force attempted to turn over responsibility for special operations helicopter missions to the army and thus concentrate on their tactical fighter wings—the ace's image at work. The army would have readily accepted such helicopter missions, a bureaucratic gain, even though the service did not have any of the long-range helicopters with night-vision gear needed to deliver special forces on deep raids. Thus the air force would have discarded the dull helicopter missions to keep a "thoroughbred" role and the army would have enhanced its role and bureaucratic responsibility. And special forces would have been left on the ground beneficiary of a bureaucratic deal. As Representative Dan Daneil said in January 1986:

Conflict on the Margins of the Cold War 361

> The problem is, they're trying to conduct unconventional warfare with conventional plans and commanders... They don't want elite forces. And they don't want to give up any turf... We need an advocate for special services.[5]

They did not find one in the Pentagon where advocates of unconventional war were dismissed as snake eaters, combat-focused specialists.

Despite an activist administration enthusiastic about unconventional missions across the Potomac, the Pentagon and the system repelled directives to change. The few military people charmed by the lure of the unconventional served elsewhere, were like Oliver North seduced by power, by covert affairs, by conspiracy, and the illicit. In the Pentagon power grid more responsible commanders were not so innocent. They were far more wary of presidential priorities or Congressional strategists and reformers. They had no great hopes for the Contras even as they accepted the limited benefits of maneuvers in Honduras or an expanded Southern Command. Sent to Lebanon in a limited role prone to mission slippage, they had gone—and look at the results. Sent to Grenada they had gone but in force—and look at the results. A mission had been flown against Libya—one mission and out. This was what they continued to want: out.

In 1986 General "Shy" Meyer, the former Army Chief of Staff, in statements widely published indicated sound Pentagon thinking on the Nicaraguan application of the new Reagan anticommunist guerrilla doctrine: "If the objective is to try to cause an internal overthrow (of the Sandinista government) it is not something that can be done through military means. You do it through people power, as was the case in the Philippines." General Bruce Palmer, Jr., too, in equally widely published comments, felt that the Sandinistas could not be removed by subcontracting the job as was done in the Bay of Pigs but only by a formal United States invasion—a project that would entail five divisions and 100,000 men and if successful would then cause the Nicaraguan population to rally. "Once they see who's going to win, then they rally." And obviously Palmer and Meyer and the others could not imagine America tolerating a formal invasion. Since irregular means would not work and formal invasion was non-starter, the Pentagon need not be involved. When the Iran-Contra affair broke and the cost had to be paid in Washington, paid by the administration, not the Pentagon, paid by Poindexter and North, not by Generals Gorman and Meyer, the Pentagon was reconfirmed in its priorities, prejudices made fact. Why any longer go through the motions of concern?

The military had never wanted to deal with terror, with terrorists, with urban guerrillas, criminals or assassins. Terror should not be considered a war matter. Terror was best treated as a criminal matter. During the time of terror, regular hijacking, gunmen in the streets, atrocity as revolutionary means, endless futile and fearful spectaculars, none in the Department of Defense wanted responsibility. Since an armed struggle in America was unlikely, the Pentagon had only to evade becoming involved in those abroad. Beginning with the Munich massacre in 1972, anyone concerned could follow the lack of military response of the military through one report after another, decade after decade, each alarm and terror spectacular indicated first that the United States had no effective strategy to cope with an increasing threat and second that the military did not want to be part of the solution—assuming there was one.

An army-air force joint team pointed out the bureaucratic infighting and an American inability to "comprehend the nature of this kind of conflict." Their solution was an unspecific recommendation that a response should involve "all the national resources at our disposal, military and nonmilitary, lethal and nonlethal."[6] The Pentagon's purpose of the exercise was to be able to point out that with the recommendation something had been done so that in effect nothing need be done. A 1986 report had called for a policy revolution, an unlikely eventuality, while other reports often had merely urge the Department of Defense to consider the terrorist problem, unlikely. There was no policy revolution, no allotment of all lethal or non-lethal forces, no serious agenda adjustment, no resources allocated for anti-terrorist tasks.

Six months before another task-force, led by Vice President George Bush with the former Chief of Naval Operations Admiral James L. Holloway III as executive director, found existing policies as sound and proposed a National Security Council staff position to coordinate an anti-terrorist program. Rather than the call for a broad integrated strategic reappraisal this report was more traditional. In matters not pressing to the bureaucracy, whatever their urgency in the outside world, the traditional American response is to appoint a commission that recommends another bureaucratic layer or office or coordinator under the banner of reform, a new entity where the issue can be buried under meetings, reports, and staff discussion. On "terrorism" it was a ritual quite satisfactory to the Pentagon interrupted during the Reagan years only by the real raid on Libya. Many, including most in the

Department of Defense, simply did not see how the military could come to battle with a few elusive, secret gunmen emerging and disappearing into the shadows. It was a matter for intelligence agencies, for the police in various foreign countries, a matter that only rarely could military force have an effect. In those areas where military force might be deployed—in rescue operations, for example—there was very limited capacity.

Even the Congressional efforts to assure a greater special force capacity had been futile. In October 1986, Congress passed a bill ordering all special forces to be brought together under one command, an often-postponed project. Somehow the Pentagon failed to comply with any of the bill's requirements for good bureaucratic reasons. No commander was appointed. The new organization went largely unfunded and understaffed. A headquarters site had been allocated in Tampa, Florida, out of sight, out of mind. Congress sent a strong letter, then a stronger letter, and finally a "totally nasty" letter, all without effect. In March 1987, still another letter was sent to Defense Secretary Weinberger indicating that Congress was "disheartened and dismayed." The Pentagon was neither disheartened nor dismayed, only uninterested in dangerous marginal missions. Even when at last moved to action, the result, like AN/PRC-70 radio, was apt to be inept, bureaucratically sound but combat faulty. The Department of Defense is an enormous, bureaucracy that was at the best of times difficult to shift. Doing nothing is a finely honed bureaucratic skill that year after year frustrated the external advocates, even the Reagan ideologues.

After all the excitements, raids, and excursions of the Reagan years, the consensus on unconventional matters of the Department of Defense was untouched. What had to be done to quiet Congress and the administration had been done but limited and hedged by the commitment to the orthodox. There was an assistance secretary of defense who had a desk devoted solely to low-intensity conflict and special operations: SOLIC. Not only did few pay him great heed but also his area of purview now was open to bureaucratic priorities. There were more people commanding the Navy special operation Seals than there had been Seals when real missions were on offer. And there were increasingly no missions on offer.

In 1987, world wide, there were only 610 American soldiers stationed abroad as trainers, all prohibited from combat involvement—even as observers. The CIA would take some time to recover any enthusiasm for activist measures. In any case the Soviet Union seemed to be having serious problems within the empire even without a militant American policy. There was a new stability in some old combat zones, no novel challenges in sight. And few felt the elections in 1988 would initiate great change: the Democrats were opposed to adventures and no one would consider Reagan's chosen successor George Bush an adventurist. As everyone expected, Bush won handily over Michael Dukakais. Everyone anticipated less of the same: fewer adventures, more maturity of advice, some caution, Republican decency, and the priority of domestic matters.

Few new president's have appeared on election as conventional as President Bush. His record was easily read. His service especially as advocate of the Reagan administration was known. No one expected surprises, not the intelligence community nor the average citizen. Unlike Ronald Reagan, Bush had served in combat, fought in the great war, the last really good war, and gave no indication that he truly shared personally or as a matter of policy the enthusiasm of the ideologues for anticommunist adventures. When Bush moved into the Oval Office in January 1989, he very quickly was presented with an emerging foreign crisis that might require a response. There was trouble in Panama—again. The local dictator, General Manuel Antonio Noriega, was most unsavory, suspected of murder and complicity in drug dealing. Crude, tasteless, tawdry, and violent, he depended upon the Panama Defense Forces and his Dignity Battalions militia and increasingly as policy virulent anti-Americanism. And Panama could soon control the canal. America was not, however, in the business of removing the unsavory from power even in a client state like Panama.

And Noriega had been a client. In fact at one time, he had reputedly be in the pay of the CIA, congenial to assisting the Contras, a key asset in Reagan Washington, and an unattractive if loyal American ally. By 1989 he was even more provocative and more obdurate. For two years the campaign to remove him run by the proper people, middle class friends of American with long established ties to Washington, had faltered before crude force. Noriega had no compunctions about the use of force. He was a media villain. He was also brutal and efficient. The decent were neither and so they needed help to reestablish democracy, to remove a tyrant.

Increasingly Washington had become involved in aiding and encouraging Noriega's internal enemies, the middle-class, the disgruntled officers, any rebels. This aid went on despite rumors that Noriega would reveal Bush's connections with drug smuggling, despite risks of stirring anti-American demon throughout Latin America. No doubt if asked the Panamanian would have been delighted to see him dead or gone. Unfortunately neither the establishment nor his regime colleagues had the luck or daring or skill to achieve this end. And Washington could cripple the economy, isolate the dictator, encourage in word if not deed conspiracies but not rid Panama of Noriega. Many commentators somehow felt that this was President Bush's fault. And yet America's vital interest did not seem threatened—the canal remained open and Noriega was not much worse than many who had been America's Latin American allies.

What evolved as America's vital interest was the administrations perception of danger of displaying a lack of will. Challenged by a dictator, American did nothing—and this might well erode other more vital alliances and alignments. Doing nothing was actually doing harm to American interests. What was wanted was Noriega confounded. The target was one man, the cunning and ruthless Noriega, not even a regime, certainly not a system. Almost all of Panama and all of Washington wanted an end to Noriega: he had no real constituency, no policy but greed, no legitimacy, and perhaps control of the canal. And yet there he was and there he remained transforming America into a pitiful giant unable to act. A local coup collapsed. Pragmatically the most convenient expedient was for Washington was to place a price on Noriega's head. This, however, violated American law and practice. A legal excuse for intervention, however, could always be found, even found for special operations. The real problem was to organize an intervention that would not lead to escalation or to American entanglement against an underground resistance. If it were to be done, it would best be quick and with assurance. There was no unconventional solution that satisfied the responsible. A special non-military operation evoked anguish over old criticism of the CIA or new concerns about state-sponsored terrorism. A military operation by the very few might fail leaving no immediate second option and an enhanced Noriega. Besides in their heart of hearts, many of the responsible lacked conviction about America's capacities in special operations. Psychological warfare, the power of money, the use of unpleasant resources all came without guarantee and the same old luggage that lumbered other special operations.

In any case none of these were the American way. America had no special office for the unorthodox often imagined by thriller writers, no faith in the operators in the black world, no toleration for the covert. Even Special Forces, a regular, uniformed military unit, overt and eager, could not guarantee no American losses and Noriega to hand. The Pentagon and many in the Bush administration did not trust special operations to succeed. The favorite options in the Pentagon and elsewhere was a replay of Grenada but improved and up-dated. You could count on the 82nd Airborne to succeed without need of adjusting to an unorthodox mission.

At one a.m. on December 20, with *Operation Just Cause*, exploiting Noriega's statement that Panama was at war, the United States intervened. Bush ordered a invasion of 27,000 troops. *Apache* helicopters and Air Force AC-130 *Spectre* gunships fired into the *commandancisa* with devastating effect. *Cobra* helicopters riddled the guards barracks. Noriega had declared war and war he received, real war, the American way. The radio station was knocked out—"When we got around to it, we knocked down five antennas until we got the right one." A bombardment was carried out by two ultra-sophisticated F-117 stealth attack aircraft. The F-117 was an enormously expensive and elegant machine crafted for far more serious matters but instead were assigned to drop 2,000–pound bombs for use as "giant stun grenades" to distract attention from a Special Forces landing. Not really a proper mission but an opportunity to deploy the latest technology in a combat situation. On the ground the troops used field artillery against apartment blocks and fired machine guns at any signs of resistance.

Making use of the secure American Canal Zone, the limited Panamanian military capacity, surprise if breached at the last minute the operation blew aside formal resistance. That had been the plan and that was the over-all result. There were lots of tactical and technical errors in execution—for example, too many of the 3,900 troops of the 82nd Airborne Division still wearing 100–pound rucksacks jumping onto concrete runways from 500 instead of 800 feet suffered broken ankles. In the confusion of real war such misjudgments were to be expected. Noriega disappeared and surfaced later secure in the Vatican Embassy. Most of the paramilitary forces disappeared into the slums and the outback. The American military moved out in stage two and secured the entire country, isolated Noriega, who was wanted in the Florida on a drug charge, and flushed out nearly all of his supporters.

The intervention was hardly a surgical strike, not with over 30,000 Americans involved, not with regular troops on the ground. And it was not a very serious war either. The American military had lost twenty-three killed and 324 wounded while Panamanian military losses were 314 killed and 124 wounded. Civilian casualties were considerable and long in dispute—but United States estimates were 202 killed. The fear that Panamanian nationalists would go underground and wage guerrilla war proved unfounded. *Operation Just Cause* was by most criteria a success. It was also one of the largest and most expensive manhunts ever organized.

Early polls indicated that ninety percent of the American people approved of *Just Cause*. There was no early withdrawal. Law and order depended on the American military, rebuilding both the smashed cities and the ruined economy was a long-term and expensive undertaking. Recreating a decent police force and a competent and honest army of some sort was no easy matter. In the months after the invasion, murder was up by six times to thirty a month. Robbery was rife and the drug business booming. These, however, were all problems with conventional solutions, amenable to money, to organization, and to American skills and Panamanian involvement. There was very little organized violent resistance to the American presence. On March 3, 1990, two unknown men, shouting "Long live Noriega!" tossed a grenade into a discotheque wounding sixteen American service men and eleven others. As one member of the Panamanian officialdom explained, "It was against the *gringos*." It was also an isolated instance.

By the spring of 1990, with Noriega in a Florida prison awaiting trial and most of the American military back in barracks, *Operation Just Cause* faded from the news. Panama had once again underlined all the American virtues as vices in small wars and armed struggles. The idea of sending the 82nd Airborne backed up by naval and air units—including the new F-117 Stealth bombers—when a Special Forces team might well have brought out Noriega clearly indicated Pentagon assumptions about force, power, and the American way. To send too much was better than not using enough. Criticism did not faze those who had such power to deploy. America did not need deception or elegance or an indirect approach once there was war. Trying to be lean and mean meant not enough helicopters dispatched to Iran. No need to contemplate that at Teheran the Pentagon had to stitch together an unconventional force because none had been previously prepared and relied on very limited CIA intelligence because no one knew the mullahs. The entire Iran exercise indicated as always that the special or the discrete use of force was not a Pentagon priority and intelligence about the unconventional not readily available. Panama was not placed in the hands of the special or the marginal. Real and conventional force was deployed including in proper subsidiary roles the special forces as adjunct not alternative.

By the time Noriega was in prison and the Panama operation in the history text, the world had changed utterly. Always before, in the deployment of American military power, there had been a Cold War context but not in Panama. Grenada had been threatened by Cuba, Angola by Marxist-Leninists and Cubans, Southeast Asia by communists and so too the Philippines and Greece and Central America. There were no communists concerned with Panama. In fact by the time American troops were dispatched to Panama, the communists were seized on their own collapse. Marxist-Leninist history was coming to an end. The familiar world of iron curtains and ideological alliances, a world of subversion, maneuver and espionage, where even the most isolated Kachin gunman or Irish carbombers could be assigned a wider role, was in rapid melt down.

The Soviet Union had spontaneously imploded on the installment plan. The Cold War had been won. A new world order had arrived if not a kinder, gentler order. Bush faced not the traditional strategies of containment and deterrence, a defense budget shaped by nuclear weapons as much as conventional force, but rather a postmodern world with new missions and new roles and no real text. Except for a very few, the collapse of the Soviet Union came as entirely unexpected. The Soviet Union and the Eastern European satellites had not worked very well and by the 1980s the failures began to accumulate. The Soviet military-strategic component required an enormous portion of the national output and even then as weapons systems grew more complex, more dependent on high-tech, computer driven engineering, even that was insufficient. The existing system was too closed, too rigid, too oppressive to allow rapid growth, flexible economic reforms, the free play of ideas or the transparency necessary to an advanced post-industrial society.

Rather than hold fast to the past Mikhail Gorbachev attended to reform the system but maintain the control of the communist party. There was to be *glastnot*, a new open society, and the restructuring of *perestroika*. The result was unexpected and rapid expansion of aspiration of those who wanted national reform, further freedoms, real democracy, evidence of rising living standards—everything—and simultaneously anxiety on the part of those within the communist parties who saw the system in Russia and Eastern Europe under terminal threat. The genii, however, was out of the bag: the reforms continued, the ambitions of all were unappeased, and on August 19, 1991, there was a confused and clumsy coup against Gorbachev led by conservative communists who did not control sufficient coercive force or general enthusiasm. Order was restored in large part because of the actions of Boris Yeltsin, who would replace Gorbachev by the end of the year. And by the end of the year, everything had utterly changed. The communist party was gone. The Soviet Union dissolved into the component republics—the Baltic and Caucasian states independent, the Central Asia republics independent and so too Moldavia, the Ukraine, and Bylorussia. Russia re-emerged as republic overseeing a reversal of a millennium of expansion. One by one the communist regimes of Eastern Europe were transformed, a violent coup in Rumania, Czechoslovakia divided, Bulgaria holding to much of the system, Hungary and Poland moving toward Western example, and then Yugoslavia collapsing into anarchy as the Serbs, Croats, and Moslems sought historic ambitions—and the Slovens and Macedonians withdrew into their own miniature states. Even Albania, the most isolated communist state of all, was transformed.

There was no Soviet Union, no Warsaw Pact, no Cold War, and no consensus on what came next. The West for the foreseeable future did not have to fear a militant Russia. America did not need to fear a Russian nuclear strike—the danger was from decaying weapons and the threat of nuclear theft by terrorists. The West did not need the same systems or doctrine. And what was the future of advanced delivery systems, further deployment of strategic defenses? The Russian military was visibly decaying, morale gone, equipment aging, salaries stinted, brought home from Afghanistan, brought home from Eastern Europe. And what need for NATO? The world was no longer bipolar, the far places were not longer counters and arenas. For the innocent and optimistic it seemed as if history was over: and America had won. Times had changed so swiftly that in Washington none knew what to do, what to keep and what to dismantle. What would be the new missions? What sort of role should the military have? What must be kept and what could be cut? And what would the morrow bring?

Even before the final visible ruin of the Soviet empire with the coup in 1991, America discovered that not only had history not ended but also neither had war, international responsibilities, or the nation's vital interests. In August 1990, eschewing cunning or subversion, the Iraqi dictator Saddam Hussein, one of the Middle East's more unsavory characters, had invaded, seized, and annexed Kuwait. Baghdad assumed that America who had tilted away from Iran and toward Iraq would acquiesce, that Kuwait had few friends, and that once his troops were in place the deed would be done. The problem was that states were sacrosanct. Destroying a member of the United Nations by fiat broke the international rules. As a dominant and unsated regional power Iraq threatened major interests of the West and much of the Middle East. Kuwait appeared a domino not only to Washington but also to Damascus and London and Cairo.

Iraq, untrustworthy, ambition unstated, ruled by a very nasty dictator would dominate the Middle East oil reserves, threaten Middle East stability, such as it was, and successfully defy the international community. Law, justice, and the American vital interests and access to oil were threatened. Only a few Arabs detested the West more than they feared Baghdad. It was clear from the very first that something would have to be done. Many hoped for negotiation, a few to find a means to accept aggression, but others foresaw that only force would prove effective. Saddam Hussein was divorced from any other kind of reality. So that even as the Soviet empire collapsed, the United States was presented with a major problem, one that might well only be solved by recourse to force.

The world may have been in the process of being turned upside down but Saddam had offered the conventional a righteous cause, a grand mission. President Bush was presented not with a small irregular challenge, another Panama, but with the prospect of real war, the war for which the Pentagon had long prepared. It would by necessity be a big war, a war dependent on technology and organization, a war that would be huge, costly, and require great power to be deployed at a distance.

Notes

1. Neil C. Livingstone, "Mastering the Low Frontier of Combat," *Defense & Foreign Affairs*, December 1984, p. 9.

2. *New York Times*, September 3, 1985.
3. *New York Times*, October 10, 1985.
4. Ross, S. Kelly, "U.S. Special Operations II : Issues, Challenges, Threats," *Defense and Foreign Affairs*, October 1984, p. 27.
5. *New York Times*, January 6, 1986.
6. *Washington Post*, April 13, 1986.

11

The Postmodern World and Dragonwars: Into the Next Century

As the Bush administration watched Saddam Hussein, the world continued to turn. The direction of events in Eastern Europe, as one Warsaw pact government after another gave way, Rumania violently, to elections and new democratic groupings, without always an end to old habits and old alignments. Communism was gone, the party if not the habits. The Soviet Union, the Warsaw pact, the evil empire was gone. Democratic norms, capitalist economies, and open societies might not be assured but seemingly were the goal: even in the Soviet Union, especially in the Soviet Union soon dismantled into a new Russian republic. No one in Washington had imagined the rapidity of change and few could imagine the extent: history had not ended but was certainly being played out on the evening news from Warsaw and Budapest and Moscow. For Washington, the news was good—and not limited to the transformation of the Soviet bloc.

Violeta Barrios de Chamorro's election in Nicaragua on February 25, 1990, was a minor miracle: a communist government imposed by the gun was removed by the voters. American pressure had helped the anti-Sandinistas, but the Nicaraguan people had decided on their own future—and the communists had accepted the result. And El Salvador was in process of accommodation, the guerrillas were going to join the system that would no longer function merely as adjunct for the interests of the rich. Central America was no longer on the front lines of concern, no longer of concern. Latin America seemed more stable, more democratic even if in Cuba Castro remained unrepentant, an affront.

Elsewhere the post-modern world seemed a more hopeful place. The release of Nelson Mandela and the opening of a dialogue within South Africa offered a way out of what for a century had appeared an impossible and intractable conflict, one beyond winning or compromise. There were even moves toward democracy elsewhere on the African continent, welcomed with vast enthusiasm by the optimistic if less so by the experienced. Yet naive or not, some unsavory regimes were replaced by new elected alternatives. Some African regimes could not cope, could not rule, remained tiny fearful elites isolated in killing zones. Some regimes, notably in Nigeria, could cope, closed down democracy and ruled by force displayed, if not well then beyond serious challenge. The failure of the armed opposition left and right, to upset the uncertain Aquino government in Manila indicated that in Asia too the world was changing. The new Asian economic tigers prospered and the real China appeared tempted by the capitalist road. Indian and Pakistan remained seized on the old issues and new ethnic violence. The real novelty appeared to be the Islamic revival. This proved not so good news for the West.

The shaping of Islam as a crusade against the West had first generated international concern when the ayatollahs of Iran marshaled their variant of fundamentalist Shi'ite Islam against the Great Satan in Washington. The Marines in Beirut had been victims of the new jihad. The advent of fundamental Islam, a regular phenomena in history's long cycles, was an unpleasant surprise to those without a sense of history or great interest in Islam. Many in America understood Islamic fundamentalism only as a new and novel assault on the West, on the modern, on the secular and sensible. Civilizations clashed. Certainly within the Islamic world many were drawn to the fundamental, to the new jihad, as well as to the religious comforts of the old creed in transitional times. The jihad was diverse, often parochial, had different enemies, took special forms in Algeria or Egypt or for that matter in the Bronx and Brooklyn. The guerrillas of Afghanistan become the basis of a transnational crusade against the West: zealots deploying dollars and missiles and years of practice. In the new decade the revitalized faith was a reality, often an unpleasant one for Americans, for the West.

There were other advocates of violence, armed struggles, recourse to terror, and other victims. There always seemed to be terrorists, the IRA intermittently pursued the long war and ETA guerrillas could still be found in Spain. Yet, the bipolar world that made sense of the parochial was almost gone. And so there were no longer easy missions for America. There was no need to oppose the government in Nicaragua or support the one in the Philippines. All the tides ran to Washington's avowed desires: deterrence no longer needed, missile defense obsolete, the Warsaw Pact in shreds and the Soviet Union going, China intrigued by capitalism, the worst despots replaced, Mandela free, the Palestinian fedayeen frustrated with a terror strategy—only a few, seemingly irrelevant undergrounds persisting.

Before the West could fully grasp the dramatic change in world order and the implications for old assumptions and ambitions, there was Saddam Hussein's challenge to world order, international law, small countries, and to vital Western interests. It was just the sort of challenge that the Department of Defense had long imagined: orthodox, regional, and conventional. Saddam Hussein was a brutal and crude despot, cunning, sly, parochial and ambitious, deploying last year's military systems. He appeared formidable, had weapons of mass destruction, had sought a nuclear option, claimed his elite Republican Guard the equal of any. What could anyone do? The oil states had money but neither will nor effective armies. Syria would not attack and Iran could not. The West was far away and recently congenial to Iraqi interests. And so America would have to lead and could do so with confidence in overwhelming coercive power. The Pentagon with time had no doubts of the outcome: Iraq, unlike the terrorists and bandits, was vulnerable to American capacity.

President Bush almost from the first, despite qualms and hesitation in other quarters, was determined on taking action—he might have had difficulty in explaining why but none in giving the military what they needed to pursue war in the Middle East. They asked for much, more was better than too little, and amazingly were given all. Unless Kuwait was evacuated—and Saddam Hussein could not do so, could not be seen to be less than all powerful, in command—then there would be war.

The Pentagon did not want war, soldiers never do for it puts at risk all that they have cherished and organized. Having fought wars, the generals and admirals are still aware of the human cost, the pain, misery, horror, and corruption of combat. In 1990, however, the system, the Pentagon, the military was tasked with war. And the Pentagon, if reluctant, felt confident of capacity and prospect. The war against Iraq would be a proper war fought from strength—fought in the American tradition of rushing to make good lack of preparations and coping by reliance on organizational skills, military training, technological capacity, and patriotic determination. The Gulf War's Operation Desert Storm was the fulfillment of the Pentagon dream: a war to be won not only the American way but also won without need of change, reappraisal, and especially recrimination. All of America's assets could be brought into play, all assumptions validated. There would even be time to make good surprise, to mobilize while Iraq remained in place.

In Vietnam the gradual escalation had meant that the Pentagon system could gradually adjust without basic change in assumption and control that the cruelty of real war would have imposed. A major war swiftly removes the incompetent, those useful only in preparation and peace, forces a change in the habits of bureaucratic stability and attitudes, doctrines, assumptions and systems not before tested in battle. In Vietnam the Pentagon could endure a reality creep making only minor incremental changes. The result had been a growing gap between the Pentagon war and the real war, a gap hidden over time from those in higher command. In the Gulf the system did not have to adjust but could merely display long prepared competence at half speed: an entire war pursued at the Pentagon pace. There could be months to prepare for an overwhelming attack instead of a rush to defend the helpless Arab regimes had Iraqi attacked the oil states. What was planned was a lengthy and deliberate air assault against almost no opposition, an attack that would assure a subsequent swift and successful ground attack. If all went well, the major land attack would be a parade past the ruins of the dispersed Iraqi forces, never very compelling in any case. There was even a shift so that the attack would not go right up the middle.

And this is what happened. All the requested troops and their equipment—everything, high-tech arms systems, barracks and beans in cans and can openers, forks and plates, soap and pails, towels and main battle tanks—were flown into Arabia in the proper order, dispersed, deployed, and prepared. The media came and was housed and briefed. Troops were housed, fed, entertained, and maintained. The fleet stood off the coast and the advanced aircraft flew into special bases to special care. And Iraqi refused to withdraw. On January 16, 1991, the airstrikes opened the real war and continued until the Iraqi infrastructure was pulverized and the Iraqi military shattered. Then came the ground assault across prepared ground against a ruined defense. And the Americans won, won big, won in alliance, won legitimately the American way.

The flaws and failures of high-tech systems, the inability to prevent the missile attack on Israel, and the disputes and flaws in command and control were simply very minor matters once the war was won, quibbles concerning a vast, effective and resplendent campaign the equal of any in history. There had been almost no Allied casualties and complete battlefield control. The limits of victory were decided in Washington solely on assumed American interests—the field was cleared, cleared by the American way of war in alliance with the world, with the support of the nation, in pursuit of vital national interests and the new world order. President Bush had stopped the advance on Baghdad and avoided the entanglements of occupation—and so inadvertently spared Saddam Hussein and his regime. No-fly zones were announced and enforced limiting that control. The Kurds were encouraged, protected, and supplied limiting Hussein's Iraq. A boycott was put in place limiting Hussein's options and beggaring all not in power.

The war itself was largely isolated, short, brutal, and impersonal, made photogenic rather than dreadful by the media, made grand by victory. The American way of war was not so much vindicated—the Pentagon had always been confident and assured about conventional capacity of the Desert Storm operation—as it was displayed. Nearly everything went as planned. Afterward came the problems: what sort of peace to impose, how to cope with irregular allies and enemies, how to deploy for an uneasy peace.

In Iraq the American security establishment almost from the moment of the cease-fire discovered that normality meant unconventional tasking. To protect the Kurds from Hussein's vengeance something had to be done. The Shi'ites in the swamps of Southern Iraq were ignored and so decimated by Baghdad; but the Kurds for the moment had Americans as friends, had the American air force protective cover, and soon had American weapons and money and advice. And the Kurds persisted with internal wrangles, private quarrels, and special agendas. Protecting Kurds was unconventional. So, too, was aiding the Iraqi dissidents in exile eager to replace the odious Hussein but at no risk and at other's expense. The Israelis and Syrians and the others returned to their old postures. Not only had a Middle Eastern triumph brought complexities, exposure to intractable habits and rivalries, the limitations of the involved, the distance between perceived truth and the evidence on the ground, but also terror.

In Saudi Arabia few of the devout had enthusiastically welcomed the Western presence—huge, alien, necessary and unwanted, disruptive, often impious. Necessity generated shame and anxiety and anger. The result was a car bomb outside an American military barracks, ill-protected and so vulnerable as had been the embassy in Beirut and the Marine barracks at the airport. Seemingly no lesson learned and another butcher's bill to pay. All of this, the feuds and quarrels and lies and terror, were the ripples that spread out from real war, evidence, if evidence be needed, that in the Middle East—in much of the world—the unconventional was the convention, terror a constant, turmoil the usual. And the Holy Jihad was not limited to Lebanon or Saudi Arabia, the Hezbollah, the devout and fanatical, could move easily in the transnational medium, bring the horror of the margins into the parking garage of the World Trade Center, bomb in the midst of New York City as easily as in the wilds of the Arabian peninsular. The congenial war of *Desert Storm* was aberrant, not the gunman and the bomber.

The new world order had changed so swiftly that the orthodox could not adjust. And there was much adjustment necessary, a great many regional conflicts, small wars, chaos, just as they were a great many novel opportunities and triumphs for open societies and democracy. Yet the era of great armies did not seem over to the planners in Washington who sought resources to pursue future regional threats, to keep carrier task forces, to replace the old satellites and fund the next generation of weapons. Tomorrow could not be ignored in the Pentagon even if there were no immediate prospect of a great war. As for the immediate prospects—peacekeeping or nation building or crime fighting—enthusiasm was expressed but limited.

In a sense George Bush, the most successful commander-in-chief imaginable, was the last symbol of the old order: the bipolar world of international conspiracy, great power rivalry, massive risks, client states, and mutually assured destruction. That world was gone. And most Americans assumed that the concomitant responsibilities and necessary capacities had gone as well: alliances had to be rewritten, military budgets reconstructed, the Pentagon downsized, and most of all domestic priorities had to have first priority. There were only going to be small wars and most important very little domestic interest in war. The foreign policy system, the military, and the national security apparatus was ignored for lack of popular demand—was seemingly in danger of being closed down for lack of popular demand. In November 1992, with the economy sputtering and *Desert Storm* forgotten, William Clinton won a surprise victory for the Democrats over the victor in the Gulf War and a sitting president. Everything had really changed and not to the sound of trumpets but amid nostalgia for normalcy.

Yet the world was not transformed into normal simply because the American public was no longer focused on foreign affairs. In fact the world, although no longer made coherent by the diagram of the Cold War, was often all too interesting, filled with horror, pogroms, plagues, failed states, warlords, and narcoterrorists. None of these newly visible threats could destroy the West overnight as the Soviet Union once could. None would require a vast conventional military establishment to defend the West. None were amenable to American assets. How could divisions deploy against a madman with anthrax in a bottle?

The American public, like the Pentagon, preferred to watch the spectaculars rather than participate, preferred to think about biological warfare or terrorist nuclear devices as scenarios for films. Certainly for the national security establishment, for the Pentagon, this was logical. The system was not shaped to the unconventional, courtesy of a generation of neglect and an American public rarely felt threatened. And some threats were beyond defense. What had caused fear was the real threat, the symmetrical challenge of the Soviet empire. These old fears were going. America could still trace every Soviet submarine, threaten any enemy missile silo, move carrier task forces on global missions, read the license plates on the limousines parked outside the Kremlin, but the Soviet submarines were rusting in docks and the missiles no longer functioned. Carrier forces and satellites had no compelling mission. Despite all this conventional power, American could not police the world, impose order on the wilds, or intervene to prevent starvation or pogroms.

The post-modern world presented the Pentagon with the most unpleasant of all possible vistas: declining resources and rising unconventional responsibilities. The effort to maintain the system so that downsizing would not destroy capacity could not easily be sold to a skeptical Congress or nation. Who threatened America? What was NATO for now that Russia no longer need be kept out of Europe?

What was the defense establishment defending against? What could be cut in the Pentagon's budget and how soon? What rationales were compelling for more missile systems, new intelligence satellites, more advanced military technology? What did the military establishment have to offer the postmodern world presided over by President Clinton?

In 1996 Congress created a commission to consider what the military should look like in two decades and it reported in December 1997: smarter, faster, and possibly smaller. The nation needed first a transformation strategy to move the establishment away from the old two-war strategy and Cold War architecture into the next century. The panel recommended a variety of directions to slim down and harmonize the military and noted that although the defense of the nation from all enemies, domestic and foreign, was the prime purpose there were new missions. And the new missions were now the familiar agenda: terrorist cell with nuclear, biological or chemical weapons, drug smugglers—unconventional threats. And what American defense needed to respond was swift, small, stealthy special operations soldiers capable of infiltrating cities not slow, massive movements of troops from overseas bases. The analysis hardly came as a surprise: the common strategic wisdom. And the military insisted that the new direction was a proper and acceptable one. The new century would bring new missions and a new military readiness. Yet, if they were committed to the new missions of a post-modern world, why did the establishment insist on the old systems, more elegant weapons, not quick small units but more of the same, Marine divisions and huge nuclear stock piles and retention of most bases? Why did those responsible not want more human intelligence but a new generation of satellites? The establishment might be concerned about cyperweapons and transnational terrorists but hardly to the point of adjusting priorities or agenda.

The military establishment had a maze of schools, universities, think-tanks and friendly academics to find appropriate missions, to defend old capacities, to offer doctrine for the new age; but what was really wanted was more conventional power or as much as could be salvaged. An army exists to pursue war as an army. In the Pentagon's case as an American army pursuing the American way of war. The missions visible were irrelevant to that central purpose: peacekeeping, interdiction of narcotics traffic, anti-terrorism, hostage rescue teams, nation-building exercises, or expeditions to disorderly arenas for uncertain purpose. Colonels might write papers about these missions but colonels wanted to be generals in real armies. The great purposes remained: American must deploy conventional power, prepare for the next generation of weapons, assume the emergence of serious enemies rather than wait until too late. Such power was expensive—nuclear carrier task forces, fleets that cost billions, or the RAH-66 Comanche scout helicopters at $35,000,000 each—and the cost was already unpopular and apt to be more so under a Democratic administration. The system should not be run down. And a focus on the peripheral, the small missions for small units did not change basic priorities: terror was trendy but main battle tanks were always in fashion. Clinton, however, was focused on domestic matters, eager to reduce military expenditures and more concerned with human rights grievances than new satellites that cost more than the CIA spy budget. The next generation of Stealth Raptor F-22 fighters were to cost $160,000,000. Each Raptor would buy a great deal of health care, a great many computers in the schools. No one was sure what a Stealth F-22 would buy in security or capacity.

Nothing made the transformed international climate more apparent than in the rise of the narcotics trade as international villain. The era of the guerrilla had seemingly passed. The small wars in Eritrea or the Sudan, the armed struggles of Sri Lanka or Spain, the heirs of the Euroterrorists in Italy or German barely infringed on American interests. And even these began to flicker out—Mozambique at peace, more or less, Angola on the way, Eritrea nearly free if the Sudan as usual in chaos, and even the IRA was in the midst of negotiating a ceasefire. The terrorists, however, were still loose and made more fearful in alliance with the great drug cartels. Countering narcoterrorism had become a priority. Drugs had domestic impact. The drug traffic could be understood by the public, by the media.

In particular the mix in Peru and Columbia of criminals and guerrillas and their reach out through the supply routes toward the United States made transnational crime an American domestic industry. American agencies became involved abroad imperceptibly. The State Department found helicopter pilots flying combat in Peru on their payroll and George Bush had denied that American troops had been involved in the slaying of Narcotics kingpin Jose Gonzalo Rodriguez Gacha, his 17–year old son, and several bodyguards in Colombia in May 1990. "U.S. troops in Colombia? No. That's the answer," said the President for attribution. Of course, later it was admitted that the United States had trained the Colombian forces involved as part of an expanding covert role in interdicting the flow of drugs. The arrival of Clinton or the shift in the patterns of the drug trade, in the nature of the cartels was not going to mean an end to concern. Narcoterror was on the mission list if near the bottom; countering the guerrillas of the Shinning Path of Peru and their drug trade friends in the Andes highlands could be left to State Department or Drug Enforcement Agency contract pilots, to others more intimate with the covert.

Clinton found that the problem of narcoterrorism and transnational crime would not go away. Relations with Mexico were complicated by drug crimes. The illicit narcotic industry corrupted governments, imposed penalties on American society and assumptions—and could not be easily countered. Noreiga in prison in Miami as a result of military action had no effect on the production of the Colombian cartel, did not prevent the corruption of the Mexican political systems, or the rise of Nigeria as entrepôt for criminal organizations. There were those who felt that the military should address the narcotic system that generated billions of dollars of illicit gain, ruined political establishments, and corrupted Americans. There were always those who felt the military should be brought in to solve complex and intractable problems: built dams, evacuate victims of natural disasters, stamp out yellow fever, or integrate the schools.

It was not so much that such a mission played a considerable role in Pentagon thinking only that it played any at all. And this was in part not that the times had changed but that the Pentagon was often seen as a resource of first resort. Americans, including and especially the President, turned to the military for solutions, to rescue hostages or to dispose of despots. Perhaps dirty tricks or special operations might provide a quick, cheap fix; but for real repairs the Marines or the gunboats or the engineers were needed. Why should not the Pentagon take on the drug trade? What was obvious to most specialists was that the military was singularly ill-prepared to interdict drug shipments. The Coast Guard could deploy, a few helicopters could be dispatched; but the key remained various cooperating law enforcement agents and the cunning deployment of development money. So military aid and comfort was to the relief of the Pentagon limited, specific, and minor.

Clinton was the first new-model President, the first President since Hoover who did not need, however reluctantly, to direct a nation under potential terminal threat. He led a nation in a complex lethal world that posed lots of little problems but no great one. The White House could move foreign affairs down on the list of urgencies and from all reports the American people did so at once. In a transnational world with porous boundaries and borders, great capital firms, weak states, and shifting population, isolation was not possible but still had enormous appeal to the voting populace. Clinton was lumbered with responsibilities few Americans wanted to admit were necessary and fewer want to fund. In a sense the Pentagon reflected this dilemma, unwanted missions difficult to fund and necessary to undertake.

Every new problem in the post-modern world was not amenable to a military solution—what was worrisome was that many unconventional problems were. Clinton very swiftly had to devise responses to chaos in Haiti and Somalia, the escalating conflict arising from the collapse of Yugoslavia, to the reality of international terror, and to the rise of a new generation of African famine, plague, and tribal war. The Kurds did not go away nor the Holy Jihad nor the old intractable conflicts in Ireland and the Middle East, in India, in the *altiplano* of Latin America—but some at least fell outside of direct American concern and others were not beyond concern but beyond military tasking. Still Americas believed in solutions. Despite the horrors on color television, the protracted small wars, dirty, nasty, massacres in distant places with strange names—Grozny, Mogidishu, Colombo, Americans still assumed that much of the world was like America if underdeveloped, badly organized, querulous, and quarrelsome. The world as perceived in Kansas or inside the Beltway among the policy people was different but not fundamentally strange. When the strange and horrible intruded, Americans declined to accept that others could be so different, so bitter, and so intractable. Few wanted to hear about ethnic cleansing in Bosnia and only the intrusion of film clips of starving children in Africa engendered interest: and even then few wanted to accept that warlords deployed starvation as a weapon, that hunger was not a result of bad management but of cunning. That was too unconventional to accept. As for the rest of the world, they were not so different.

Everyone wanted peace, stability, a chance to flourish. Many from the Caucasus, Africa, or Asia came to America and did so, ran the newspaper stand, invested in fishing boats or laundries, went to universities, dropped old habits, and became new Americans. So why could not any people, any place organize, reason together, fashion a civil open society that offered opportunity? When confronted with other values, other assumptions, with those who ran to a different time, would not give up old grievances or sought to inflict pain instead of seeking profit, Americas were not simply taken aback but appalled.

An armed struggle is at least a highly structured process, follows in technical and often tactical matters a viable dynamic. An armed struggle is reasonable, explicable without the need to understand the motivation, the faith. Once the structure of the faith, the ideology of revolution, is diluted or absent, once grievance and venom dominate cause and effect, Americans are truly lost. A gunman may be misguided but seems involved in a traditional conflict with real goals, kills for a united Ireland or a Tamil republic. The director of the cocaine cartel is criminal but Americans are familiar with the dynamics of organized crime—and it is organized. On the margins of the new world order appear assassins without agendas, criminals without structure. Who in Kansas know what to make of warlords and village killers: those who slaughter for imagined grievances, for a lesser god or for some unsavory cause, those who kill children for recreation or torture for pleasure. All these are beyond understanding and so beyond manipulation.

Americans failed to grasp the power of hate in the Balkans, the general joy of an atrocity performed in Liberia, the pleasures of genocide, the affirmation of treachery or the exhalation of the false. These are not armed struggles but anarchy. Somehow the reality of Stalin and Hitler, the experience of generations of war and horror, the endless example of human sin at work in history is not incorporated into the American psyche either in Kansas or the Pentagon. Confronted with horror the American response jitters between accommodation and annihilation, between withdrawal and engagement. The Pentagon's desire to avoid the unconventional is simply most American reinforced by the nature of any professional military establishment.

None but the mad kill for political cause in America, lynching is unfashionable, the good Indians run casinos and cults, atrocities are not wars but incidents. History is not so much over as diluted. Americans thus expect the best from others, reason, compromise, enterprise, moderation, decency, and find all the alternatives not simply disquieting, appalling but inexplicable. Americans want answers that make sense. Who killed Kennedy since one man is inexplicable? The politics of paranoia is merely the American way to find explanation that eases fears and anxieties generated by the strange. And the post-modern world found on CNN was for most absolutely alien, The wandering killer bands of Liberia togged in stolen bridal gowns, the Hutu of Rwanda slaughtering their neighbors by hand, machete slash by machete slash, the old cankering hatreds remain inexplicable. Americans prefer tidy violence, explicable mysteries, an O. J. Simpson murder trial or Princess Di dead in a car crash. These other the low-intensity horror are too horrible. The Americans in general and the military establishment in particular is in effect disarmed. Not simply taken aback but quite unprepared to cope, often unwilling to cope.

This denial of the dreadful has been an American constant along with an inability to empathize with absolute commitment that energizes the armed struggle or with the weight of history and grievance on contemporary assumption. American optimism, so admirable, so useful, so patent, so innocent, had been immune to such reality. Rather than seek means to manipulate those so involved, corrupt the armed struggle, or deceive the liars, Americans fall back on the congenial and comfortable. The nation is repulsed.

Urged to action Americans and the Pentagon would opt for surgical airstrikes and so stay beyond contamination. Protraction may bring contamination. No one remembered centuries of entanglement along the western frontier. No one recalled the whole long, quiet history of American military intervention in Latin American, intervention that was largely ignored even at the time. Americans all remember that Black Jack Pershing never caught Pancho Villa not that the Marines occupied Haiti for nearly twenty years. What the military want, what the American people and political establishment want is swift justice, swiftly forgotten, cheap and distant, kept to the margins. Almost no one wants a blurred mission, conflict short of war that may engulf the nation in an alien arena. So the desire is to send the Marines for a swift victory on the beach, but the acceptable reality has often been leaving them in place but forgotten. The danger is that on some strange beach where reason does not run the Marines will require attention, reinforcement. So the best mission is none at all. This has been the American way.

The Pentagon, the military, is the most American of all institutions, is open to the new and the ambitious, incorporates the nation's ideals, response to challenge and crisis as do the people at large. Doctrine can be adjusted and weapons systems introduced, new missions may be given but the military will remain, as it should, American. The military has a role largely unadjusted over time: to defend the nation, to obey civilian authority, to shape in war and in peace an institution that reflects the values and aspirations of the people into a defense of their interests. And no finer institution than the military, most American of all, kept pure, uniform, coherent, dedicated, and stable if not docile.

There are a great many kinds of Americans, many that fall outside the most inclusive stereotypes. So the defense establishment is various as well as coherent: for some the American frontier tradition is inert, unknown, and unfelt, for others its pursuit led them into the military. There are Americans who love the margins, the dangerous edge, find the underground congenial, empathize with the irregular. And some of these wear America uniforms. Some within the Pentagon do not believe, as Burke insists, "There is nothing which will not yield to perseverance and method"—as do most Americans. The iconoclasts, alien strains, dissenters, and heretics may within the huge establishment find a role, but they do not dilute the essential American nature of the institution. The orthodox do not easily suffer the unorthodox. Billy Mitchell was disciplined not only for being right but also for being novel and visibly so. Lansdale could be novel in the Philippine outback and in the special programs of Vietnam, could even be visible as token but could not really be important, influential.

Unconventional threats pose very serious structural threats to any establishment. Because the threat is unexpected and irregular, there is no very satisfactory doctrinal response or bureaucratic preparation. Special operations are special because they are unforeseen. This means that a national security establishment has all sorts of assets of response but they are not carefully calibrated to the unexpected: drug wars or transnational terrorists do not generate a prepared and particular response. The system must make do and by the time an institutionalized response is in place, in all likelihood the threat has mutated. This is everywhere a problem in deploying power. Spain has the *Garda Civile*, Northern Ireland the Royal Ulster Constabulary, Britain the SAS, some countries have an armed militia, others paramilitary national police, and many special forces and intelligence operators. All organizations do the best they can—and the longer the challenge, the more familiar and so the more conventional the response—and yet even after two hundred years London still finds the Irish gunman elusive, mysterious, and dreadful.

In any case there has been little American attempt to institutionalize action on the margins, deploy innovation. For Washington the margins keep changing and with the end of a bipolar world the organizing element has been lost. When the unconventional imposes on the military agenda it is treated as any other problem: seminars, analysis, a new course, advisors brought in—and what else could be done? This is the nature of great organizations: organize everything. And the American security establishment because it is America as well as an establishment is particularly prone to evade the irregular if possible and to regularize it if not. Most of the CIA does not really like to run spies nor the army encourage the Green Berets. The suspicion of the special extends even to special operations, the use of partisans, the returns of novelty. The military, any military, is innately conservative, conserves past wars, tangible assets, consistency. Change implies risk and even the American military arising from a society faithful to risk accompanied by good works cannot entirely divorce professional habits.

The nature of armies, the nature of large bureaucracies, is not especially American. In a sense each nation has the army it deserves, each nation comes to terms with military necessity. A few armies transform the volunteer: ideological armies, mercenary units. Most, however, reflect the ethos of their soldiers, especially national armies. And any army is confined not only by the nature of the troops but by certain military imperatives. The master sergeant is the same in the Grenadier Guards or Trotsky's armored train. Other ages have other armies but all need sergeants. In modern times the national ethos is not easily denied, easily transformed by Lenin or Islam. Even without a state Serbs stayed Serbs. And if there is no nation, there are often tribes with flags, equally special and persistent. The ethos largely determines the dynamics of armies. And an American army is the most national of all for nationality in America is a special process, result of birth for some but conviction for many. A sergeant or a general can be of any race, religion or background, come from any descent but always nationality will out—reveal the American writ small. For many Americans nationality is recently implanted and for all absorbed from the arena, boiled in from the melting pot even when the special ingredients can still be found. Hyphenated Americas are apt to be more American that any, the conviction of the convert, even with accent and alien habits. Americans share not religion or race or history but habits and identity, assumptions and agenda.

As America goes, as America is, so too the Pentagon. To make a people pessimistic is beyond legislation and often events. To make the Americans different is not really possible. The endless flow of emigrants does not dilute the pool only make it more variegated and the new more American than they imagine. Emigrants can rarely go home again, for they are in process of transformation—and whatever the loyalty to the old, their children become one with their neighbors. So, any effort to reform the defense establishment by imposing attitudes and assumptions not common to America is futile. Even to introduce the obvious that for many is history for some is not a matter of text and examinations but rationale for murder is largely futile. Americans may recall the Irish famine and contribute to a holocaust museum but these are politically correct, charitable causes. They may study their revolution or the civil war but this is the printed word, a school requirement. American history for most Americans is not injustice learned by ancestral experience nor the residue of memory, does not live but is isolated as prose in a book. Americans believe that history should be kept in texts not transformed into grievance and so policy. The sensible sensibly believe that justice can be measured out if not pure and absolute then in ample dosage, a concession here and an apology there. They believe even more so when the car bomb goes off under the World Trade Center. Islamic terrorists are mad dogs, dangerous, not real people. American paranoids who bomb and kill for fantasy are sociopaths, mad and dangerous, and typical of nothing, not real Americans.

Americans are most reasonable, confident of method and perseverance and if war be needed then best it be swift, absolute, and short. Taught terror they cannot imagine the gunman's grievance. Americans are not for protraction. American was industrialized, urbanized, tied together with railways and telegraphs, built in a generation, and all the while the army pursued the indigenous dissent out in the badlands with a few units and limited commitment. If there is an acceptable unconventional mission it must be won swiftly or left to the margins, out of sight, low-cost, a matter of Indian raids. If war is worth pursuing, it is worth winning promptly, worth the cost. Americans do not like to wait on tomorrow but by shaping the day assure the future now.

Thus nearly everything that works to deflect, delay, or defeat an armed struggle, an unconventional challenge, deception, sedition, or subversion requires habit of mind alien to the American psyche. The British after generations of practice have learned to tolerate the revolting Irish—each horror inflicted on the British engenders only momentary indignation and then the nation moves on assured that the defense of the realm will in time punish the guilty if not prevent the next atrocity. Not so Americans: indignation and outrage must have tangible returns. Americans want not only action but also visible action. They can only be so obliged by the tangible: American jets attacking il-Qaddafi's compound, X-ray screens at every airport, a battleship dropping enormous shells into the Lebanese mountains. Patience in adversity, the toleration of affront, acceptance of provocation are not American qualities.

This was essentially the America Clinton had to lead into the uncertainties of the postmodern world, a nation that did not cherish foreign affairs or the alien, wanted prosperity and to be left alone, safe, secure, free of crime and fear of the future, both conservative and liberal, optimistic and narrow. And what Clinton found was that the postmodern world was far more unconventional, far more demanding than hoped. Bush had *Desert Storm* but Clinton had to tinker at the margins, great risk, little profit, no real enthusiasm. Some within the administration were more militant, but there was no overriding purpose, no need to thwart the Evil Empire or even make the world safe for democracy. What was in a sense most unconventional about the spectrum of low-intensity conflicts and irregular threats was that the past was such a limited guide to action. Little could be done in response to classical armed struggles, even when in the case of the transnational terrorist of the new Islamic jihad America was prime target. Little could be done with small wars that did not directly impinge on American interests—and little need be done when the violence did not attract the media. The whole world of plagues, famines, murder, and failed states was not a specific American responsibility or involve vital American interests. Still, the ends of the world were but one CNN broadcast away.

At the far end of the world, brought into the West as usual by the media, another all but inexplicable crisis appeared in the Horn of Africa: Somalia, a nation few in American, few in Washington could find easily on a map. This time there was starvation just like the previous Ethiopian crisis where children died on camera. This time there was starvation amid anarchy. For the humanitarian agencies, it was hard to sell again and again starvation as focus for compassion; but in the case of Somalia, the violence attracted intense, international, media attention—starvation with war visuals, starving children and fire-fights dispatched by satellite links.

Relief aid could not get to the starving and rotted on the dock or disappeared into the black economy. And the starving continued while teenage boys in rags tore through the streets in "technicals"—stolen four-wheel-drive Toyota pickups and land cruisers. Armed with light machine guns, the boys in the "technicals" were eager to shoot as display and sought conflict as a clan right, a clan rite. No one ruled Somali. The old dictator Siad Barre had fled and his faction was isolated, if still capable of violence. The succession was in dispute. Brokering accommodation had failed and the major contenders Mohammed Ali Mahdi and Mohammed Farah Aidid led private armies. There were other factions, other schismatic clans. The North, once British Somaliland, formally seceded, a country recognized by no one and ruled by no regime. The south was divided and the capital Mogidishu was reduced to ruins. War was an ambition for many Somalis not a last resort and war had destroyed the vestigial political order. Without order starvation became part of the arena of contention, food a means to punish and reward.

The world community felt that something had to be done: at least the starving had to be fed. Private initiative could not work because distribution was impossible. Even if the food could be imported and protected, shipments could not be guaranteed to the starving. Americans, too, felt that famine at the end of the century was intolerable. This was again especially so when famine could be filmed—starving in isolation in the Sudan was one thing but starving in Mogidishu on CNN was another. Washington in 1992 was, of course, largely seized on the election campaign. No one wanted to make Somali an issue—misery could be found closer to home as the rafts filled with Haitian refugees were returned to the island by the Coast Guard. Who needed Somalia? So one paid undue attention to trouble in a far away, unknown land until the media focused interest. Round-eyed, rag doll children dying on camera, dying needlessly because food could not get through was not an issue but an affront. Thus, just at the moment of transition, with Bush a defeated president and Clinton an unknown factor, decisions had to be made about Somalia. And such decisions in November and December 1992 seemingly came easily. There was an opportunity to do good at small risk, to great advantage. All that was needed was order on the Mogidishu docks and food distributed. And the Pentagon was not foolish, not without specialist advice. Experts explained the Somalis as martial. Intelligence reported on the wide distribution of arms and on the inclinations of those directing the irregular war. So the Americans, if sent in harm's way to do good, were to be deployed with some care.

Of all the American's entanglements with the unconventional, the Somali crisis epitomized in a relatively small compass the crucial problems and faulty responses, responses shaped not by doctrine as much as by assumption. On paper the Americans felt that they understood the Somalia issue, the challenge, the mission, the prospects. General Colin Powell knew the estimated number of arms within the country, the capacities of the warlords, the dangers of irregular attacks. He also knew America wanted no surprises, no unconventional challenge, certainly no irregular war that violated the parameters of the Pentagon's war of choice. The Somalis would not be accommodating nor reasonable nor dedicated to what were assumed by Americans to be universal values: starving was as much a means of war as poisoning the wells with dead camels—necessary and expected. Everyone agreed that Somalis might be different and Somalia was not Wyoming but most of the Americans proceeded as if this were to be the case.

They knew the risks, that there were risks. In Somalia, nevertheless, they perceived reality as anticipated. Assuming that most people are much like Americans, they had no empathy and so no understanding. The hurriedly assembled experts and quick-study analysts could describe and deploy the mysterious clan names, the Dir, Isaak, Darod and Hawiye, the Digil and Rahanwein, the history of the dictatorship of Mohammed Siad Barre and its collapse in 1991, the economic statistics, the road maps and population estimates. They could not offer a sense of the alien Somali mind—a mind shaped by a millennium of dry seasons and cruel choices, bleak, brutal, and subtle.

The American contingents would be part of a United Nations force dispatched solely "to establish secure supply routes" so that starving people could be fed. The entire mission was humanitarian and the American Marines that would arrive first were "friendly forces" in every way. Who could oppose the mission—to feed the starving? Already 300,000 had died and 1,500,000 were in jeopardy. Who would oppose a limited, temporary, presence with no other purpose but to aid the helpless and desperate? And so the friendly forces came armed but without malice, without expectation of advantage, without arrogance. Of course, as Secretary of Defense Richard B. Cheney told reporters, the American forces would be allowed to take "preemptive action" against anyone posing a threat to their safety or that of the relief workers. Still who, even in Somalia, killed kindness? Especially when kindness came fully armed with weapons beyond the imagination of warlords and teenagers buying AK-47s at the suq for $100. The United Nations, the Americans, the Marines anticipated Somali enthusiasm and the acceptance. Even the warlords could hardly advocate further famine.

The Somali mission sold in Washington was bought by Americans no different than those who were responsible, those who would be in charge of the expedition, those who advised them. For the military, moreover, a congenial mission to aid the starving Africans amid the anarchy of Africa, a popular mission, a necessary mission presented to the public, to the politicians, further evidence of the value offered by the Department of Defense during a period of drawdown. *Operation Restore Hope* was so entitled for any who might miss the point. After all Panama had been *Operation Just Cause* and few assumed otherwise. Names matter and images in unconventional missions. Perceptions, however, may differ: one general's hope may be a warlord's despair, what was just for one client might not be so for another. What was, indeed, assumed was that hope and justice, food and fair play would carry the day—it was an opportunity as well as a risk.

As Chairman of the Joint Chiefs of Staff, General Power delivered a "paid political advertisement" on behalf of the Pentagon: "The nation is blessed that it has a military capacity during a period of historic drawdown that can respond at a moment's notice to this kind of operation.... We've got to be very careful as we manage this drawdown.... we don't want it too fast and to make sure we don't go too far." The expedition under the auspices would be a showcase of military capacity—a capacity that should not be drawn down too far. The initial Marine Corps unit felt that their mission was a public relations bonanza at just the right time. "Here's what looks like a good news story. American service personnel are helping to solve an absolutely horrible situation, and these are things the American people should be aware of." Marine Corps Commander General Carl E. Mundy called the operation a "damned good" showcase for Marine capabilities. "This I think would be the classic case again of a situation in which because of the infrastructure that's available, the primary means you have for entering is from the sea." [1]

So the 1,800 Marines and then the other Americans, 5,000 from the 10th Mountain Division, and more Marines for a total of 15,000, and the United Nations contingents and the United Nations administrators, the outriders of decency from the foundations and relief agencies, the media, all would enter by the sea—and by air, enter in numbers, enter to do good or oversee it being done. General Powell would deploy the Americans so that they were in sufficient numbers to intimidate but not so distributed as to appear threatening—not in a position to secure the country only to feed the starving. The Americans would be in Somalia but visibly ready to leave. Getting out was a quite different matter than getting in and most important to a military that remained haunted by the evacuation of Vietnam.

The Americans might assume that they were decent but not the Somalis. They assumed from ample historical evidence that everyone had other motives, special ambitions, and a hidden agenda. And the faction leaders were obviously in a position to cause trouble, for everyone recognized that the country was awash in weapons even if few in the West could grasp that one of these was starvation. United Nations Secretary General Boutrous Boutrous-Ghali thought it would be desirable to disarm the population before the international forces arrived but proposed no means to do so. The Americans had hopes that such arms could be bought, a prospect that violated all Somali practice—arms made the man not just war. Yet nothing unexpected happened in Somalia. The Marine landing was reminiscent of the long-ago landing on the Beirut beach in 1958. Anyway who would oppose sophisticated forces replete with helicopters, high-tech communications, with a fleet offshore and an airlift in operation, professing only benign aims? The clan claimants for some time maintained a seemingly friendly watching brief. What they were watching for was weakness, opportunity, and vulnerability, deploying in the meantime guile and cunning in apparent acquiescence.

What occurred was the slow and inevitable decay of clan restraint in matters of violence. Foreigners were not welcome. The Somalis had been colonized not only by the Italians but also by the British and French. Others had denied the unity of the country, imposed defeat and the lost of the national territory. All foreigners were suspect, even those who worshipped Allah. All Somalis had national grievances and each clan special ambitions, every clan leader an eye for the main chance, and all the men a desire to use arms.

The obvious Somali target, if there were to be target, was the Great Satan. The Americans were the epitome of the foreign, not just different, but alien in religion, habit, and experience and arrogant in the display of wealth and power. They were a natural enemy to those who by nature sought enemies. And so as soon as practical the clans began to resist—none could imagine easily that the incursion was truly limited in time or disinterested. Feeding one faction harmed another. Paying for labor penalized those not paid.

Besides the play of jealousy and pride, some Somalis simply wanted an opportunity to inflict harm on the alien and arrogant whatever their departure date. There was, as well, all that potential loot, full warehouses, rows of vehicles, barracks and depots and convoys. So the lull after the Marines walked off the beach was not permanent but rather a precautionary pause. Disorder soon involved the United Nations. Thus "securing supply routes" meant defending positions, engaging in erratic and irregular incidents, becoming involved bit by bit. What the United Nations had taken on was to impose order. This meant in effect establishing a protectorate for the good of the general population assumed cowed by the militia gunmen and warlords. The early mission had slipped away into a far broader if not articulated mandate.

What also occurred was that despite extortion, theft, graft, and waste a great deal of food was distributed, further famine averted. As promised, good was being done if at increasingly risk. Stability did come to parts of Somalia—partially through exhaustion and the diminishing returns of looting and theft. Starvation was over in most parts of the country because the crops were being harvested. A few felt the entire expedition had been unnecessary—the famine had peaked. The United Nations, the various national contingents, most of the relief agencies differed, felt intervention had made a difference, was going to be a success. Success, however, was overshadowed by the rising shadow of irregular war, ambush, snipers, armed raids against the United Nations: convoys in danger, patrols required. In May 1993, the basic mission shifted to peacemaking, nationbuilding by force deployed: order was to be restored so that Somalia could emerge now that famine had been averted. Turkish Lieutenant-General Cevic Bir took over the new peacemaking mission. Those most visible as Somalia clan leaders, as warlords, became targets.

The attrition of stability during the spring of 1993 was transformed when on June 5, 1993, the militia gunmen of Aidid in Mogidishu ambushed a Pakistani unit. They shot and killed twenty-five soldiers, United Nation soldiers: an appalling and unexpected loss. It was a bitter blow to the prestige of the United Nations and to the Americans, identified by many with the entire enterprise. Americans had increasingly been targets, a worrisome development that produced a call for a greater commitment, commandos and gunships and equipment. In June the United Nations went on alert. There could be no peacekeeping until the streets were safe. The mission goals began to shift.

The Pakistanis on June 13, 1993, killed twenty Somalis in a fire fight. Americans enraged at their own losses placed the blame on Aidid and his militia. No longer a faction leader or even a warlord, he became a wanted criminal, His picture was on a poster with an offer by the United Nations of $25,000—Alive or Dead. The poster could be found pasted of the walls of Mogidishu—fame not notoriety. The America Major General Thomas Montgomery wanted to crush the Aidid faction. He had asked to redeploy attack helicopters and had received reinforcements: Rangers and more armored vehicles. What was needed was more power on the ground to respond to armed provocation. American helicopter gunships raided Aidid's suspected command and control centers. Informers were paid, patrols dispatched, hostilities assumed. The primacy of mission was to eliminate Aidid, punish the provocative, and thus establish American and United Nations preeminence.

America was engaged, once more, in an unconventional mission in an alien arena to no clear purpose deploying tangible, military power against those driven this time not by a dream but by clan loyalty and an affection for the guerrilla game. Precision bombing could and did smash Aidid's safe houses but Aidid could and did escape pursuit again and again in the maze of Mogidishu. Driven by ancient animosities and great ambition, Aidid proved cunning, ruthless, sly, and at ease in his own country. It was the very situation that the Pentagon had feared: an indeterminable unconventional conflict with mission creep and no exit date, no public acclaim, no way home.

The terminal disaster came when a new Ranger unit responding to provocation was flown in by helicopter to restore order in part of Mogidishu. The Americans were ambushed. Eighteen were killed. Somali outriders rushed through the city and hurried media representatives to the scene: a mob dragging a Ranger's body down the street, howling in glee, "American, American, American." The footage went out worldwide. Photographs made every newspaper—a Pulitzer prize was awarded. On television the people could see an American soldier, desecrated, naked and ruined, a plaything in a dirty street at the end of the earth, murdered by those he had come to help. America had been humiliated once again, this time by a clan lord and his rag tag gunmen. Somali once a showcase had turned into killing zone.

There were those to complain that if more sophisticated gunships had been sent to Somali the Rangers could have been protected: more was needed. The fact was that there was ample American and United Nations power in place for a limited mission but not enough to impose order on the nation or for that matter on Mogidishu. Aidid could not be destroyed at an acceptable cost and no one had factored such a cost in the expedition to "secure supply routes." And no one in the American military had really factored in the risks of confronting Aidid. And so the Rangers had been sent into a city they did not know to confront an enemy that they did not understand for purposes that had not been on the original agenda and could not be achieved in existing conditions.

There would be an intensive review in Washington of all military aspects of the Somali adventure, everything from the difficulty of deploying non-lethal means to allocating blame to the United Nations or the failures of intelligence. Many at the Pentagon felt that United States Central Commander General Joseph Hoar should have remained the military authority. The switch of missions to peacemaking under a Turkish general was not popular. Many also felt that the Americans should have kept more troops in-country, but retrospective analysis is always quite keen. A Senate Armed Service Committee report released in September 1995 detailing error and recommendation read remarkably like that of the Long Commission on the Beirut International Airport incident in October 1983. The system needed to be pulsed and tightened but was essentially sound. More helicopter advanced gunships might have helped: they frightened the Somalis, created a vague sense of menace. In essence nothing much had been learned.

The American system properly deployed would have worked—worked as it did in Panama and in Grenada. The system, however, was not tuned to the unexpected, to the alien, to the unorthodox. Fine tuning the Pentagon system did not change the systemic assumptions. The basic problem had been an inability to grasp the unconventional nature of the arena, an American problem rather than a military or bureaucratic one. Somalia was not Panama, not Grenada. The Americans were unlikely to be successful without paying more than desirable. The expedition could be swift but ineffectual or extended and—perhaps—successful. The commitment could not be short and successful, might not be successful without high costs in any case. To explain this in the autumn of 1992 in Washington would have meant that the responsible would have had to be converted to an un-American perception of reality, Somali reality. Washington assumed starvation a remediable evil and war a final option. Somalis assumed starvation had advantages and war a desirable career choice. The administrations of first Bush and then Clinton were not so much ill-advised but simply assumed that the American world view, their view, was valid. They hardly recognized that there might be other priorities, another agenda amid starvation. In fact no one imagined that there was another perception of reality. Being told so by specialists did not mean that those who made policy could truly grasp the Somali mind and so Somali prospects. And they did not.

Another problem was the assumption that famine could be treated in isolation, that meaning well would immunize the expedition. Somalia was replete with those with a vested interest in both famine and in opposing any alien incursion. Worse Somalia was filled with many who sought only opportunity to do harm. In Lebanon the warlords and ethnic factions, the advocates of Holy Jihad or simple atrocity, had been so awful that all had been labeled as aberrant. The Somali agenda was beyond American imagining. Who would seek war? In America the military was shaped as defense, as last resort, as necessary in a troubled world. And that world was often, usually, amenable to reason and accommodation or should be. Americans believed in solutions, in organizing for the public good, and in "doing good." The Somali assumed war a necessary and desirable rite.

Finally, the Americans who were sent into harm's way were neither trained nor prepared to cope with the unconventional, evade entanglement, corrupt the opposition, avoid confrontation. The 5,000 soldiers of the 10th Mountain Division were proud to go, to do good: "It is going to be nice to help out, to show the world that Americans really care."[2] The Somalis did not see such soldiers as "caring" but as targets. Even the Marines assumed their mission humanitarian. In any case a Marine is a Marine not trained to be a humanitarian ambassador nor in this case trained in Somalia survival. They were soon involved in a "conflict short of war" pursued by the savage, brutal, and elusive, by those without remorse or restraint. An irregular war by clans on their own land deployed against one more invader. And all the gunships could not bring them back.

Some one had to be at fault—this too was the American way since the contingent and unforeseen fell outside the assumed capacity of the nation to overcome obstacles by deploying skill, reason, technologies, and perseverance. So the Secretary of Defense Les Aspin in time resigned and those who had seen a bonanza were not to be found. The United Nations were at fault, the chain of command, the changed mission, political laxity, the lack of more weapons. Someone had to be at fault because the Somali mission was visibly an American disaster in planning, in implementation, and in retrospect. On the other hand, Aidid, revolutionary triumphant, gunman into statesman, gave the keynote address at the Pan African Congress meeting in Kampala in 1993 and lasted until he was shot and killed in Mogidishu in 1996—a Somali original. The United Nations contingent was withdraw in 1995, the famine over, the country divided, the warlords in place, the world warier. And none more wary than the United States Department of Defense.

In fact, as the Somalia crisis had unfolded one disappointment and disaster after the next until final exit, the Pentagon had managed to scamper through another unwanted expedition, unwanted exposure to vague missions and alien culture. Clinton had inherited a curious problem from Bush in the flow of unwanted, illegal Haitian refugees using derelict boats, refugees who arrived destitute and eager and claiming political haven. Bush had established a Coast Guard blockade and returned those of the migration who could be detained. Clinton was outraged at the denial of safe haven when campaigning but once in office and aware of the numbers involved had maintained Bush's policies—potentially 200,000 Haitians might try to escape. His advisors, not to mention the friends of the exiled Haitian President Jean-Bertrand Aristide, were inclined to seek means to displace the regime of Lieutenant-General Raoul Cedras. This would almost surely mean force. Cedras and his associates had no intention of giving up the advantages of control merely to please the President of the United States: greed, arrogance, and isolation dominated the regime's analysis. Cedras assumed that Washington would not want to pay the cost of intervention so nothing need be done on the island but delay and pursue advantage.

The refugees continued to embark insisting that Haiti was in the hands of despots and killers: 500 civilians had been massacred immediately after the coup that had driven President Aristide into exile. Murdering potential dissidents, murdering the vulnerable had long been a means to maintain control in Haiti. The United States could, of course, do nothing, avoid entanglement since no vital interests were at stake and force was always risky. This was generally the view of the Republicans and many others, but activists especially in the new National Security Council urged intervention in the name of democracy to stabilize the island—and to end the stream of refugees. President Clinton tended to wait on events.

Haiti had drifted at the edge of American consciousness for a century, a tourist stop, a source of Creole-speaking emigrants, a primitive island culture of voodoo, and an arena for regular expeditions to restore order. On July 28, 1915, the United States Marines had moved into the Haitian capital Port-au-Prince to make the island safe for democracy and instead initiated nineteen years of guerrilla war that cost between 300 and 400 American and 40,000 Haitian lives. The Americans withdrew on August 21, 1934, without having established Haitian democracy although there was order. No one before or after had ever been able to rule Haiti both effectively and fairly, most who tried had died or fled—and the two most famous recent dictators the Duvaliers, the father Papa Doc and his son Baby Doc, were not fair but brutal and ineffective and greedy. When on February 7, 1986, Jean Claude Duvalier—Baby Doc—fled to ultimate exile on the French Riviera aboard a United States Air Force jet, the island revered to confusion.

There were elections and constitutions and coups and five governments before the radical priest Jean-Bertrand Aristide was elected president in December 1990. The Tonton Macoutes, a private Duvalier bodyguard of criminal killers in Ray-Ban sunglasses and blue jeans, staged a coup that collapsed into riot, more confusion, and a military takeover. Aristide was expelled in September 1991 and Cedras took over, one more despot in Haiti's game of musical chairs. All this replayed on American television had a certain awesome horror. There was as well the footage of thousands of refugee fleeing the island on make-shift boats, fleeing to the United States. Haiti had emerged again as nightmare and an American concern. Those in control on the island were odious, offensive to American tastes and assumptions; moreover, whatever the rights of Aristide, Haiti appeared likely to collapse into horror.

The Pentagon had almost no historical memory and remarkably little symbolic or unit tradition. Each challenge is new even if countered in the congenial, old ways. So Haiti was nearly as mysterious as Grenada had been. Intervention in Latin America or the Caribbean would certainly involve the risk of entanglement and no matter how dreadful the regime sooner or later engender regional distaste. What the President wanted was for Haiti to disappear as crisis, become a small obscure tourist island with an elected president, modest development, and a docile population no longer driven to illegal emigration. And his advisors insisted the only way to manage this was to restore order in a legitimate and authorized expedition that would establish order, restore democracy—albeit a radical president—and encourage stability, a decent police force, and some prospect of hope. What was *not* wanted was entangling island resistance to what might appear as imperialism however benign.

What the military wanted was to be passed over or, if not, to be able to limit involvement, limit island exposure, limit responsibility. The risks were obvious as was the route: international and regional cooperation, diplomatic efforts, an embargo, and the prospect of force. On schedule an agreement was signed on July 3, 1993, that would bring back Aristide. Pressure on Cedras and his allies was maintained. Then, on October 11, the *USS Harlan Country* carrying United States military trainers, dispatched according to a United Nations agreement, was prohibited from landing by gunmen on the docks of Port-au-Prince. The paid mob turned back American power. Anyone could watch: "They were probably the worst television pictures of the administration. The Image was 'American turns tail.'.... Never again, never again."[3] On the other hand what was to be done? On October 30, General Cedras officially reneged on the agreement to allow Aristide to return. America and incidentally the United Nations had been humiliated.

As month followed month through 1994 and into 1995, pressure was increased on Cedras but without effecting the basic dilemma: the general would not go and so force was the last and increasingly the only option. American involvement was legitimized, made international under United Nations auspices with the adoption of a resolution on July 31, 1994, authorizing "all necessary means" and the formation of a multinational coalition to restore democracy in Haiti. Clinton agreed to an invasion on August 19. A month later on September 18, General Cedras accepted the inevitable and with his chief aides agreed to step down a month later. The next day, September 19, the Marines arrived in Haiti.

The intervention despite alarms and excursions was a success: the president returned, elections held, a police force organized, order restored, infrastructure improvements made, the country pacified if not transformed. The military had to weave between military tasks and civilian, between this definition and that. "If the mission bleeds into policing, people say you've gone beyond your mandate. But if you stand back and watch, they say, how can you do that."[4] Such missions are apt to supply not answers but incidents. There were incidents—a firefight when the Marines returned fire and killed ten Haitian security forces. There was confusion about the police, the army, the various armed militia, who was to be tolerated and who was a danger. Cedras and his lieutenants did not resign until October 10 or flee the country until October 13, two days before Aristide returned to power. The military had to cope, impose order, support the effort to train a new police force, and protect the government. And so they did.

The intervention essentially went just as intended. Most important the Americans were withdrawn on schedule. By any reckoning, Haiti was a success, the horrid in exile, the gunmen gone, a revitalized police force, and a government in place. There would be problems, murders, and tension. The island would not really be stabile but the dictators would largely be forgotten, the Toutons Macoute gone and voodoo in films. On December 1, 1997, the last 1,400 Pakistani and Canadian troops would leave behind only a small group still struggling to train an honest and effective police force.

The Pentagon felt not so much proud that the mission had been accomplished but secretly relieved that the exercise had not been afflicted with all the potential ills imagined: mission creep, divided political counsels, armed opposition on the ground, a irreversible decay of public order, loss of international support. Instead, most had been foreseen—including the bits that did not work as planned. The mission might have fallen prey to arena forces, Haitian factors, the climate of chaos even when there was little prospect of a classical armed struggle or a national rising. If there is time spent in-country, there is time for crisis and change, for resistance. Haiti was not a simple mission in and out: the military had to remain, become involved, act a part. Then they could leave. To everyone's relief there were few surprises and few alarms.

The things that went wrong in Haiti over the sixteen months of commitment were small and largely predictable. The military was asked to do what was really not within the capacity of those so charged: determine who was legitimate and who not, who had the right to bear arms and who did not, who was in charge in the countryside, where crime began and politics stopped. The Special Forces or the Rangers were not diplomats and could not be so trained. They could not suddenly be taught Creole and recent history and given a list of suspects if for no other reasons than there was no American consensus on who was suspect and what was the direction of history. The moments of violence, humiliation, and despair were few, disappointments small and subsequent criticism rare. For activists Haiti was example of what might be done: limited mission, specific goals, carefully monitoring, international cooperation, decency encouraged, and no great losses. For the military, still a reluctant participant in any operation that risked assets and especially one open from the first to mission slippage, the contingent and unforeseen, this was not entirely the case.

The basic military assignment had been to control the countryside, the country. "Get on the ground first" and "don't worry about the details," suggested former President Jimmy Carter to the Pentagon when he returned from his Haiti mission.[5] And the Special Forces A-teams sent in did so, got on the ground and then ran into details. Rather swiftly some 95 percent of the country was monitored by Americans soldiers who knew little of Haiti. They found those armed and haltingly organized to oppose the return of President Aristide were a mix of old Tontons Macoute, police without uniforms, soldiers without units, the new *attachés* armed gangs blending into the militia FRAPH (Front for the Advancement and Progress). And the mix had American friends and mentors. The CIA had a built-in disposition to see Aristide as undesirable, a radical if not a communist, who would be opposed to American interests. The American CIA people involved were the same romantic pragmatists who assumed covert operations inevitably pragmatic and effective especially because they were covert and illicit.

In Haiti the CIA people put pressure on the Special Forces command to ignore provocation, crime, or intimidation if FRAPH were involved. Those America soldiers so pressured in the countryside soon found their initial instructions, their avowed mission, and their orders contradicted not only by those in the embassy but also by their own commanders who yielded. In the end the individuals in the embassy concerned were transferred, efforts to damage President Aristide failed, and the Special Forces moved on as originally planned. In the sum total of Haiti it was a small matter but an event not to be found as textbook example, not a matter of doctrine.

What the military saw, once again, was that even within a United Nations peacekeeping operation, especially within an unconventional peacekeeping operation, political priorities, shifting missions, perceptions imposed through irregular channels all warped a straight-forward effort. Shooting ten armed Haitian when they opened fire was an action that meshed with expectations. Having a mission apparently corrupted by the CIA was unorthodox, irregular. All unconventional missions, no matter how straight-forward soon became complex, dangerous, uncertain, and not military at all. Time in-country, time on strange ground inevitably erodes mission edges, introduces not only the unexpected but also the unconventional. Expeditions are rarely simple when extended and what the intelligence community wanted was a clear enemy and the military wanted instead an empty battlefield. What they found was the real world of Haiti. What they would always find would be the contingent, the unexpected, the confusing and strange.

Always despite the advice from the Pentagon both on practical and policy matters, decisions were made, as in the case in Somalia and Haiti, that shaped the unconventional missions largely to other priorities: those of the administration. In turn such factors often evolved from arena conditions, a shifting agenda, the personalities of the moment, alliance responsibilities, and public concern rather than military considerations. The Pentagon was apt to assume that military considerations were now and always pure, sharp-edged, a matter of logistics and tactics and doctrine, that political factors never were weighted by Wellington or Eisenhower, that the unconventional as defined by orthodox is in fact unconventional when almost all missions from the invasion of North Africa in 1943 to the Haitian assignment have unconventional components. Purity of battle where only deployed forces matters is rare: but still the Pentagon in Grenada and Panama as in Greece long before sought to impose battle priorities on a mission.

Even before Clinton's election, the world was seized on the wars and horrors arising from the collapse of communist Yugoslavia. Suddenly old, forgotten national aspirations emerged from time, Balkan ghosts and so too cruelty, vengeance, pogroms as policy, and ethnic cleansing as aspiration. War, turmoil, and the slaughter of the innocent came to the Balkans, to Europe, to the American media. Violence moved up the structural scale from random murder by the village defender, to paramilitary units sweeping whole provinces, new nations and new armies and old wars. American might have no vital strategic interests but many Americans, many Americans within the beltway in Washington, many within the administration were quite interested: a European war unleashing passions assumed long dead.

The prospect of direct American involvement in a Balkan war held no charms for the Pentagon or apparently most Americans despite the call for a response to atrocity. If there were to be involvement in the Balkans, the Pentagon could only hope that practical considerations would determine the scope and agenda of intervention not wishful thinking. Allied with the Pentagon doves were a great many who feared entanglement in contrast to those who placed international responsibility above risk and insisted America, NATO, the West, and the world should not stand idly by while innocents were slaughtered. What was needed was action to protect the innocents from the odious and duplicitous and brutal.

Any such action would require force, military force since there was no reason to assume the factions loose in the Balkans would welcome intervention. The actual military component in such a mission remained simply a potential factor for those discussing future policy. American policy evolved in response to the crisis—meetings, massacres, allied distrust, battles and betrayals, NATO involvement, proposals, missions, pogroms. The military would only become a dominant consideration when committed. Then, of course, it would be too late. Before that diplomacy, assumptions, domestic politics, allied priorities, public opinion, politics, and the evolving arena were what mattered. In the past and in the Balkan conflict, the military was apt to advocate accommodation. What was wanted was a settlement, but the military as always had to wait on events often urging moderation and diplomacy.

The Pentagon rarely imagines war advantageous. Commitment is apt to mean the failure of diplomacy and the reality of danger if even, as in Somalia, this is not apparent at first. The admirals and generals are all too conscious that mission creep, the unexpected—the unconventional—will find Americans in harm's way. And those engaged in murder in the Balkans could engage any international force in alien territory, rely on large popular support, could protract irregular war, inflict casualties, could persist as guerrillas beyond reach of high-tech systems or conventional response. A Balkan war would be a nightmare, costly, futile, foolish.

Thus one of the great, post-Cold War crises where the ancient aspirations of Serbs, Croatians, and Moslem could not be adjusted by negotiation and certainly not by force was viewed with alarm by an American military. Both Croatians and Serbs wanted a greater nation at the expense of the other and certainly of the Moslem of Bosnia. There were mutually exclusive hopes of a greater Serbia disguised as a new Yugoslavia and a Greater Croatia arising from the newly independent state. The Moslems hoped for the best. To the north the Slovens withdrew into their own state albeit with some problems and to the south a new Macedonia, landlocked, impoverished, vulnerable, anathema to the Greeks, was announced. The Albanians in Kosovo were for the time quiet. The crucial arena became Bosnia-Herzegovina where Serb, Croatian, and Moslem sought if not dominance then more. The world, the West, NATO, and America watched massacre, private murder, atrocity, and killer legions—Tigers and Wasps and Green Berets—engaged in slaughter. All, including the Moslems, supported by the arrival of the "Afghanistani," the new mujahedeen, free-floating Islamic guerrillas joined by 400 volunteers from Saudi Arabia, sought to do harm and protect their own. All factions slaughtered the innocent and vulnerable for the folk and the nation and the faith, pursued domination, defended their maximum aspiration as justice sought. None admitted fault. None could be trusted.

The appropriate international response was not found largely because those with the power to intervene would not pay the expected price in lives. Everyone was outraged and indignant, appalled, but not beyond reason. No outside vital interests were at stake beyond decency and civility, not for Europe, not for NATO, nor for the new Russia or America. In the West those who urged intervention did so out of compassion not for advantage. And no conciliatory proposal or mission or threat had effect against those killing their way to further power. Without countervailing, coercive power—or exhaustion, the crisis seemed likely to continue. All the involved were unsavory, the more powerful the more unsavory, so that Serbia became a pariah, blockaded and damned. Alliances shifted, strange unappetizing leaders appeared, lies were convention, wanton violence normal. Ultimately, on March 8, 1992, the United Nations authorized the dispatched of 12,000 troops from thirty nations supported by NATO. The UN force had a minimal mission with limited military forces and doubtful prospects. In fact, soon, the United Nations Protection Force proved hostages to fortune and was riven by divided counsel and evasive leadership.

For Pentagon purposes the United States would act in concordance with NATO, would not have troops on the ground, would not intervene. The United Nations were in Bosnia only to offer humanitarian aid and their "good offices" in a negotiated settlement: evidence of international concern but hardly in an effort to keep a peace that did not exist. Everyone feared escalation, mission slippage, real war that would involve everyone, the United Nations, NATO, even the United States, in an endless and vicious Balkan war to no advantage. Everyone on the ground in the United Nations feared crossing the Mogidishu line, painting the white vehicles of peacekeeping green in order to engage not in peacekeeping but in peaceseeking. Bit by bit, Washington had become involved, involved with the NATO commitment, involved in sending troops under United Nations auspices to protect Macedonian independence in May 1993, involved in a promise to send 25,000 troops if the United Nations forces were endangered. The United States still had no troops within Bosnia, on the ground where an incident could lead to chaos and conflict.

Washington urged airstrikes, surgical, limited, low-risk, and high visibility if there were to be force deployed and used. The two great presidential first options are always special operations, covert and cheap and deniable, and surgical military strikes, overt and elegant and visible. For those with United Nations troops on the ground any such symbolic strikes would not be cheap. The British and French and the others felt their forces would be held hostage after any airstrikes. No one wanted to match the local forces, dedicated, well-armed, and aggressive, much less engage them—fight the Serbs in the wilds of Bosnia. Washington was willing to go for "lift and strike," airlift arms to the besieged Moslems in Bosnia and dispatch airstrikes against the Bosnian Serbs. Those in the UN on the ground continued to oppose airstrikes—making peace from 10,000 feet—because it would not work and would endanger their troops.

Finally, when the Serbs blatantly failed to fulfill their agreements and attacked the United Nations safezone at Gorazde, two NATO airstrikes against Serbian positions were authorized in April 1994. On

April 10, two United States Air Force F-16s dropped three bombs on an artillery command bunker. The next day two Marine Corps F/A-18 Hornets dropped three bombs an a group of Serbian tanks and armored personnel carriers. Military force had been deployed by NATO—for the first time—under the auspices of the United Nations: low risk, high visibility, but, of course, quite ineffectual, no lift, not much strike. For the Bosnian Serbs, in fact, the exercise was symbolic of the West's weakness, the limits of the United Nations. On September 16, the Serbs announced—prematurely—the capture of Gorazde. Another airstrike produced the loss of a British Sea Harrier jet hit by a surface-to-air missile. The mission was canceled. More Serbian promises were broken and there was more trouble at Sarajevo. Those with troops on the ground still feared an extended air war would leave their forces trapped, still insisted that peace could not be made by airstrikes. And it was not.

Even in the midst of the Bosnian Serbs' truculent aggression, the balance within the arena was shifting. The failure of greater Serbia, the international odium, the impact of the economic boycott, and most of all the reality of Bosnian war finally tempered Serbian ambition in Belgrade. What was good for the new Yugoslavia was not necessarily good for the Bosnian Serbs, aggressive, arrogant, and overextended. By 1995 Serbian President Slobodan Milosevic accepted the necessity to compromise the aspirations of the Bosnian Serbs. The crisis moved on, complex, shifting without accommodation and often without prospects but there was no general war, no NATO war, no American military involvement. In time Croatia forces were armed and trained. On May 1, 1995, the Bosnian Croatians launched an offensive against the overextended Serbs largely cut adrift by Belgrade. On May 25, when the Serbs failed to removed their heavy weapons from Sarajevo as promised NATO attacked a ammunition dump. More dumps were hit the following day and the Serbs took United Nations peacekeepers hostage, ultimately 370 were held.

For Americans the drama of events was less the risk of escalated war than the rescue of the pilot of a downed F-16 by a Marine team on June 8—ten days later the last United Nations hostages were released. The Serbs overran the Srebrenica safe area after NATO airstrikes failed to deter their troops. On July 28, Croatia sent troops into Bosnia. The Serb defenses began to collapse and Belgrade did nothing. Nothing was easy in Bosnia but the pieces were in place. On August 28, a mortar round landed in Sarajevo and killed 37 civilians—a round investigation indicated came from Serbian position. The last such round on February 5, 1994, had killed 68 people shopping on a market square and had not been traced. This time the incident set off the drive to compel Bosnian Serb agreement to an accommodation.

On August 31, NATO launched three days of airstrikes. The Serbs were not convinced. The strikes were renewed for two weeks: 3,400 sorties, 750 attacks against 56 military targets. The United States planes operating out of Italy and the Adriatic were fully committed to the NATO-UN strikes. It was the kind of commitment that the Pentagon preferred, high-tech strikes from a distance. The attacks had come simultaneously with heavy Croatian gains on the ground in Bosnia made possible not so much by "lift" to the Moslems but by America ignoring violations of the arms boycott by Croatia. So America got lift and strike, the NATO allies had ample time to protect their vulnerable troops, the Croatians and Moslems redressed the balance of control from fifty percent to seventy, and the Bosnian Serbs discovered no comfort in Belgrade but pressure to compromise. On October 5, Clinton announced that there would be a ceasefire in five days. The airstrikes were over—and perhaps overemphasized for it was Belgrade that had neutered the Bosnian Serbs. And at that an accommodation was still no easy matter. The interested parties met on and on at Dayton, Ohio, in November, before finally resolving the issues. There would be a NATO-led United Nations Implementation Force, IFOR, that would include 20,000 American troops with a headquarters in Tuzla in Sector North. IFOR was amenable to all. There would be no Balkan war or so all hoped, none more so that the Pentagon.

So the new United Nations IFOR arrived. There were no incidents—60,000 troops on the ground, the 20,000 Americans, 13,000 British troops, 8,000 Italian, 2,300, French, 2,000 from Russia, and 2,100 from the Netherlands, a Nordic Brigade, troops from all sorts of places, thirty countries were involved, Malaysia and Morocco and Egypt, Hungary, the Baltic countries and Bangladesh. They were in place for a set period, awaited the election that was held the next year on September 14, 1996. The odds were long on a stable and democratic Bosnia but there was no war. The most odious figures slipped away—war criminals, fanatics, and zealots on pension or retired to exploit advantage, living quietly or in some cases arrogantly in gaudy new houses displaying their sleek families in town, on television. The wars seemed over if not the recriminations. There were no atrocities and little for the Americans to do but wait out the dull days. The early times had been filled with risk, the need to establish bases, build roads, cross rivers with portable bridges, overcome the terrain. Then came waiting.

During Clinton's first administration when all attention had been on domestic matters, cutting the budget, downsizing the military, adjusting foreign affairs as an aside, the military had been presented one unconventional mission after another—and at times missions that did not offer prospect of escalation or entanglement. The decay of order in Africa meant that Americans were repeatedly at risk and a task force was dispatched to take out those citizens who would go, wanted to get out of Liberia or Rwanda. The emergence of failed states and the rise of chaos, free-fire zones, and anarchy encouraged the Pentagon to shape a Marine special forces for such special and unconventional assignments: swiftly in and swiftly out, citizens saved and the imperatives of the arena immaterial only the level of violence, the security of the airport, the numbers, and refugee sites.

In April 1997, 1,388 Marines of the 26th Expeditionary Unit, most abroad the United States Navy carrier *USS Nassau* were positioned off the coast of Zaire in case an evacuation was necessary. The 26th had taken out the American citizens under threat in Albania several weeks before but this was the first time in Africa for most. "I've been drinking a lot of bottle water. . . . The heat slows you down a lot. . . . I thought the rain forest was pretty cool."[6] Africa was a stage set for a carefully prepared exercise not a potential arena of commitment. The army and air force had 670 troops across the Congo river in Brazzaville while other small units were in Liberville in Gabon and in Kinshasa. At Brazzaville the temporary base for 1,800 troops had generators, computers, showers, offices, and a medical unit, everything was flown in, no one drank the local water. The Marines could take out 600 people an hour using helicopters and boats. Who in Kansas could find Brazzaville on a map? How many Americans cared about Zaire? How swiftly would CNN be to film chaos and anarchy if Kinshasa collapsed? By then the Marines could be in and out.

The fact that Americans no longer were engaged in underwriting one regime in Zaire or opposing another in the Congo meant little to those in Kansas in any case. Only when chaos emerged on television were most Americans engaged. And all hoped that those days were gone. Only the unforeseen might involved American forces intent in rescue operations in a Zaire conflict: but the future is especially difficulty to predict. And none of those concerned was apt to consider that the tiny force available for rescuing American citizens was more formidable than that possessed by Zaire. A rescue operation was for the involved not necessarily a mere display of competence, especially when the future was unpredictable. Even for the Pentagon the future is indefinite.

There were peacekeeping, peaceseeking, and nationbuilding responsibilities. There was an sullen Iraq and the decay of the Soviet nuclear stockpiles. There were the old alliances and new alignments to pursue. Reluctantly the Americans, much like the Pentagon, accepted such responsibilities, some directly to advantage but others less so. America could not withdraw from the new world order but could not engender a great deal of enthusiasm for the investment in that order. There had been expeditions and alarms but never a threat of real war except perhaps in Bosnia where the peacekeepers were as much as anything a poison pawn to keep the conciliation game going. And once in place with no conciliation there was to be no easy withdrawal. In December 1997, NATO officials indicated that the force of 34,000, including 8,000 Americans and the force commander, would have to remain after the original evacuation date of June 1998. Then the threat moved to Kosovo.

These new and supposedly necessary tasks were all that were apt to be allotted to the military: the wonders of *Desert Storm* were long past. Nothing had indicated the new world disorder as clearly as the car bomb that detonated outside the military barracks in Saudi Arabia: the replay of Beirut. There was the same lack of physical security, the same assurance that perception was reality and that the American presence was welcome and if not welcome at least acceptable. Most of all, there was the frustration of the irrational at work: fanatics killing Americans in pursuit of an impossible dream. Fanatics suddenly emerging where no conventional force would have dared appear. Everyone had officially worried about Islamic fanatics but the security measures had not changed since Beirut. Every, every time, the unconventional came as shock, beyond doctrine and system, unwanted, lethal, not readily countered, unfair, un-American, and seemingly inevitable. In 1998 there were bombs at the embassies in Kenya and Tanzania—and airstrikes in reatalition in the Sudan and Afghanistan: the past replayed in the post-modern world.

Yet, it was within this tumultuous post-modern world that the American military was offered missions that did not require massive military deployment: Marines or air transport, a few troops on the ground, or ships off the coast. Terrorism, peacekeeping, drug interdiction, special operations, hostage rescues, all the extras were still peripheral to the major Pentagon purpose and most careers. The need for countervailing conventional forces, new high-tech weapons systems, another generation of satellites would remain prime priorities even if the immediate future offered only the irregular.

In each of the Clinton cases, the military had been exposed not to classical armed struggles, much less real war but to old conflicts in the post-modern world: clan war in Somalia, a failed state in Haiti, Islamic grievance, anarchy in Africa, rampant and parochial nationalism in Bosnia—chaos, crime, disorder, and faction fighting within the new world order. In Mogidishu anti-insurgency tactics and techniques might have been more effectively deployed. In Haiti the Special Forces were deployed not for more special military purposes but to fill a vacuum, to establish order, to do good and monitor gunmen. In Bosnia before the Daytona accords, the United States as a component of NATO authorized by the United Nations had an unexpected mission—the first NATO war action ever. What the first mission was to achieve was not as clear but the Pentagon had no problem with airstrikes and so none in performance of the mission especially when the second wave of strikes was real, not symbolic, required a campaign. After Bosnian Serb concession came still another international mission not as congenial. In Bosnia the United Nations-NATO presence after Daytona more than all else was an outward and visible indicator of concern, might at best radiate a sense of menace and at worse be hostage to warring factions. The mission entangled the American troops in a conventional deployment for unconventional purpose, put the forces at risk not readily weighted as had been the air defenses of the Bosnian Serbs. The Pentagon was simply ordered by the Commander-in-Chief to station 20,000 troops in the midst of Bosnia and trust to events.

Some of the responses to unorthodox missions could be foreseen. In effect where once unconventional war had conventions, could be taught if to doubtful purpose in the academies and schools as low-intensity conflict, so too war short of war was amenable to formal doctrine. A military force must have doctrine even for the irregular. One terrorist bomb in Saudi Arabia engendered a budget line for millions to teach terror, to investigate the newly perceived menace. Biological warfare and nuclear devices became the province of the national security establishment as well as the subject of films and thrillers and the focus of agencies concerned with safeguards and security. Rand could be asked by the air force to define a role for the service in the struggle against terror but such a role was not really relevant to career patterns, weapons systems, or actual interests.

Many of the new missions deployed military force not for coercive purpose but as psychological counter, as the authority of last resort, as nation builders, peacekeepers, symbols, and at times hostages. Shifting priorities, sudden tasking, missions withdrawn or altered or refined. Can Special Forces cope or the Marines? Will there be water for the troops, will there be friendly forces, will there be any government at all? Will the mission be aborted, redefined? The only continuity was that there was no conventional war. Flying airstrikes was almost a relief.

In a sense the military did not have to cope with much that made the unconventional frustrating. Many of the missions involved what Americans did best, administer and organize, cooperate and deploy effective technical means: feed the starving, direct traffic, rescue the stranded, intimidate the guilty, encourage the innocent and even order airstrikes. Most of the missions did not find the troops deployed against real enemies only those who did or might pursue their interests at American expense. In Lebanon, in Saudi Arabia, in East Africa there had been those committed to the punishment of the Great Satan. The Moslems of Somalia, however, were only opposed as always to the vulnerable and strange, assumed that alien power, no matter how visibly awesome, was open to attack as easily as to manipulation. In Bosnia all the locals, the Moslems, the Serbs, and the Croatians, assumed justice absolute, their cause sacred, their intentions misunderstood, and the United Nations irrelevant. As long as nothing was imposed, everything in time could be opposed. Thus the Americans and the others were stationed in the midst of those who could do harm if need be, would do harm for advantage, even short and transient advantage, but did not necessarily see a clash of interests so sure was each of justice. In Haiti the gunmen were finite, none with power had ideas much less ideals, none could summon up much more than a paid riot except Aristide, and his faithful for a time had no guns. Each mission was different and few aroused Pentagon enthusiasm.

The missions of the 1990s ordered by Bush and Clinton were diverse, across the entire spectrum of intensity, complex and simple, and mostly in formal terms unconventional. Some needed no troops. Economic and diplomatic pressure on those in government tolerant of the drug trade in Antigua and Aruba had an effect if less so in Mexico and Columbia. Washington deployed law enforcement assets, the DEA and the FBI, not soldiers to contain the cocaine cartels. And when force was deemed essential invariably the establishment sought to respond to the unconventional by shaping the mission as regular, seeking Noriega with the 82nd Airborne. This was, of course, appropriate during *Desert Storm* and enormously effective. As for Panama or earlier Grenada the seemingly large investment paid the required dividends—the operations turned mission profit. What continued to trouble the Pentagon were missions that were for the American doctrine unorthodox, participating in the Multinational Force in Lebanon or the United Nations interventions in Somalia, Haiti, and Bosnia. The services had no tradition of peacekeeping or nation building although ample experience. Even with the troops still on the ground in Bosnia, the Defense Department was examining the impact of the mission on combat capacity just as in Somalia the deployment of non-lethal weapons proved more problematical than imagined.

The military accepted that in the new world order such missions were inevitable and might even allow a display of competency; but acceptance as a policy matter, even as a first priority, did not greatly impose on American assumptions about war and military power, combat deployment, and the direct approach. The prospect of *Desert Storm* Two in the autumn of 1997 simply underlined the services priorities: even a weakened Saddam Hussein was dangerous and could be met most effectively by orthodox force. And America still had ample conventional force because of the establishment stewardship had not been diverted unduly by the irregular threats of the post-modern world.

Most of the unconventional missions were focused on the visible and explicable: even the Somali clan lords were deployed in Mogidishu as irregulars not as covert gunmen, were not engaged in an armed struggle. The dreams and ideals that at times drive armed struggle were present but shaped to other forms in the paramilitary militias of Bosnia in the process of becoming orthodox armies or the Grenada New Jewel armed party members armed with light weapons. Others were special, fit no easy category: the gangster assets of Noriega, the Tontons Macoute in Haiti, the armed factions of Africa.

When those pursuing an armed struggle emerged from the underground to attack the American military in Africa or Saudi Arabia, the enemy was as always, elusive, ruthless, covert, and uncompromising, each time inexplicable, each time a surprise—and not to the military but to all America horrified at the bomb in the World Trade Center as they would be by the more indigenous car bomb in Oklahoma City. And again and not without reason, the military saw such incidents as criminal as much as military, a matter of perimeter security as in Beirut or Saudi Arabia rather than in political intelligence and special operations. Psyops remained the province of propaganda and publicity not the core of a doctrine of deception and manipulation. Psyops was merely an add-on rather than a vital tool of unconventional tasking. Everyday unconventional missions were still outside psyops doctrine—and the exalted gunmen and bombers without banners had never been explicable to most Americans. How then could they be manipulated?

The result of a decade of unconventional missions was a nostalgia for the conventional, a focus on real war, and an acceptance that adjustments had to be made. Such adjustments, however, did not adjust American military preferences or attitudes: unconventional missions were as much as possible shaped to standard procedures and assumptions. Unconventional missions made the military uneasy, wary. They were necessary but unwelcome and were within the competence of the involved—as long as there were no surprises. And the crucial aspects of the unconventional is that there are always surprises—in Bosnia this might be no more than low-level recurrence of the unexpected or one morning it might be horror, feared, discussed but always a surprise, a surprise in Saudi Arabia and a surprise in Somalia. There can only be so much orthodox preparation for the unconventional. And for Americans rich, proud, competent, fully equipped, doctrinaire, and orthodox much is unconventional. Strange places, strange armies and allies, strange missions, all unconventional even after exposure, sometimes especially after exposure. And for Americans, for the national security establishment, for the intelligence community and for defense, as the century closed, the exposure was no longer unexpected.

Notes

1. *Washington Post*, December 6, 1992.
2. *New York Times,* December 5, 1992.
3. *Washington Post*, September 25, 1994.
4. *New York Times*, September 21, 1994.
5. *New York Time*s, September 21, 1994
6. *New York Times*, April 10, 1997.

Epilogue

As the end of the century approached, the military was hardly innocent of the prospects to come: more of the same, more uncertain missions that imposed unfamiliar roles. And as always the establishment sought to be prepared, to analyze the threats and vulnerabilities of the system, to underhand the alien, to shape doctrine for future unconventional conflict—while, as always, to maintain the real military capacity to wage real war. If there were to be a *Desert Storm,* Round Two, America must be prepared and so too if the Chinese proved aggressive or Russia made a come-back. The defense establishment had to be wary of marginal events no mater how spectacular: weapons of mass destruction in the hands of terrorists or peacekeeping in alliance with the United Nations.

In the post-modern world in the next century as in this one, Americans will remain American, not stereotype but various, complex, contradictory, and demonstrable themselves. They will continue to assume the world is one, the American way and the American agenda universal, believe that no task is beyond organization and perseverance, beyond reason, technology, and cooperation. And so with proper planning and great enterprise any challenge can be met: a man can be put on the moon, millions of immigrants can be absorbed and transformed, the world made safe for democracy, and even the budget balanced. Americans do, indeed, believe in will, do have ideals but no need of a secret army or recourse to terror. Since there is no need, for most Americans there has never been a need. History offers opportunity not further punishment. None can imagine a bandit career, the appeal of the gunman. The frontier is for films and paranoids, car bombs for the demented and alien.

At the end of the century much of the world is ripe for those who wish to change history, revenge grievance, find security in a new structure, or protect the old ways. Most of all those who are apt to destabilize order are not easily swayed for they seek not tangibles, are not the Marxist man, but seek the realization of a dream, the rewards of history. These aspirations are not easily accommodated. In fact in a global society with diluted power, many states can not perform adequately, many power structures fall outside the conventions: great transnational corporations, drug cartels, shifting populations seeking advantage, and assets that move at the speed of light—a blip worth billions and an image of murder on the kitchen television set. The next century, as the last, offers the prospect of war, the old wars of new high-tech lasers and weapon's platforms, but also the new wars that deploy assassins, plague, and the unconventional. These are wars waged against present reality, waged to martial music generated by dreams. These Dragonwars are armed struggles arising from the will deployed and so unconventional.

In most wars, for many unconventional missions, perception does not matter, coercive force will do. What makes a mission truly unorthodox is when perceptions matter, when the nature of the involved matters, when zealots with a vision determine events. Such visions arise in various conditions, when there is a fragile state as in Liberia or Columbia, when the state cannot cope with history as in German with the Euroterrorists, out of old aspirations as in Ireland, or a return to the fundamentals as in Algeria. What matters in all is what matters to the zealots who can evade brutal and efficient enemies or the accommodation of effective democratic societies. Sweden appears secure but so, once, did the Caucasus. Africa is a petrie dish of horror and India uneasy. China has billions to control and no sure future road. For those seeking an easy life even with luck, cunning and craft, accommodation and coercion, the post-modern world seems likely to be an arena crossed by zealots.

Not only do Americans fail to sympathize with such zealots, fail to empathize and so fail to understand, but also have failed to shape a response to such reality. Those responsible in the national security community are aware that there is such a reality and there are threats but the conventional mind is apt to balk at such convictions. So Americans will continue to impose their perceptions upon the world and proceed apace expecting the gunman to hew to a decent agenda or remain in Hollywood. There is public anguish over the terrorists dispersing the plague or the intentions of Islamic fundamentalists. There are seminars, classes, courses, and millions and millions spent on security and lectures. The result arising from American experience and predilections is the common wisdom and the conventions of the unconventional. The system can be refined to cope and somehow never does. This is a result of the American way of war, the American way of perception, the American vision.

Thus if there is to be a threat—and America recognizes that the unconventional challenges, unorthodox threats will be prominent in the next century, then many feel that an appropriate reaction can be organized properly. The theorists, those in responsibility, the national security establishment all want and assume possible the existence of a calibrated, seamless response to either unconventional danger or opportunity. Analysis will indicate the nature of the incident and so the required response will be in place ready to deploy according to doctrine. The governmental bureaucracy on all levels, the crisis managers and directors, will present a smooth interface adjustable by authority and circumstance to the external condition.

In the real world, of course, nothing is seamless, little foreseen contingency, planning erratic, and responsibility shared or evaded. Moreover what makes the unconventional difficult is the impact is often unanticipated in scope, content and location, most often is beyond assigning easy responsibility, and difficult to analyze. The threat is unconventional in that the conventions of the unorthodox accepted by the responsible are orthodox in manner, congenial to the involved, and rarely relevant to the dangers and opportunities beyond the beltway. Much is based on interpolating from perceived experience.

The threat is always a surprise, always appears when and where not expected, always engenders uncertainty and confusion over the responsibility, over perceived intensity and the scope of response. The unconventional pose a question that cannot be answered with a prepared answer rather the responsible must cope with limited intelligence, limited time, and limited assets to deploy. The nature of the unconventional is such that the conventional mind is often barren, the tangible assets usually inappropriate. The response is cobbled together from the available assets and assumptions—and we get Beirut or Grenada or the bomb in Saudi Arabia. A prepared mind flexible, familiar with the actual asymmetries of the real and perceived, coupled with experience in both the existing system and the history of the unconventional, offers the most effective preparation. Too much dogma, too much set doctrine, too many orthodox assumptions extrapolated into fashioning a model for all eventualities not only assures a flawed response but also reassures that America can each and every time react in a cunningly calculated, carefully calibrated commitment of tangible resources in a conflict that is a matter of perception that takes place in an environment shaped by ideas and conviction beyond easy reach.

The military if not the nation needs doctrine but the dogma of tactics and techniques does not offer a useful response to coping with an idea. Americans assumed ideas and convictions, the faith or the ideal, are subject to management and to coercive power. And since this is often the case—the conventions of the unconventional work, the underground is crushed or encouraged by orthodox power—there is often little interest in examining the reality that does not conform. Consequently, the dynamics of the armed struggle need not be understood if power can be exerted nor need the existing capacity and assumptions be questioned. And so largely they are not, even when power cannot be concentrated or mission easily assigned or a mission ordered to fit an arena shaped by perceptions. The mission in Dragonwar is focused on the war not the dragon, a doctrine for the military rather than a response to perceptions. The very nature of the threat makes a coherent response difficult for war by other means can at times not be thwarted simply by coercive force. So Americans face what is least liked, a problem without solution—and that is the problem for America. Doctrine won't work but only imagination and flexibility, creativity and the power to focus various assets in a special way. In Vietnam in pursuit of war the assets were deployed in ways congenial to America against targets sought by doctrine and thus the asymmetry was intense. Americans hardly noticed then or now. Americans persist in imposing their perceptions on reality.

And since the Pentagon is a most American institutions, any effort to introduce novel initiatives by diluting the national assumptions or by urging reforms based on transformed perspectives is apt to fail. Americas will stay Americans and so the Pentagon the Pentagon. Some Americans have an understanding of the unconventional. Those who know gunman reality often are not sufficiently charismatic or particularly well positioned to impose, even briefly, novel priorities on the policy makers. One short lecture on the nature of Somali aspirations or the vision of those who find all answers in Allah might well do if delivered to the proper people at the proper time by the proper expert that offered tangible advantage as well as realistic insight. This rarely has happened. So any effective response to unconventional challenge must be adjusted to national reality. More certainly is less in any response to terrorist provocation; but in two hundred years in America there has been no advantage, no purpose, no one to urge less when there is so much more available. No expert, no matter how informed or charismatic or properly placed, can effectively urge less as more. None can for long make the gunman real. Americans, the people, the politicians, and the Pentagon are going to respond to perceived reality as always: ahistorical, optimistic, trusting in organization and enterprise, devoted to technology and management, and without empathy for others. There is no grasp of the ideals of the denied and of the desperate and the limitless passions of greed and iniquity that lead to violence. The everyday Americans, the colonel in his Bosnian barracks or the Ranger flown over Mogidishu in a Cobra helicopter, all find the fanatics and the faithful alien, even the tribesman and the clan lord inexplicable, their motives mysterious, their ambitions bizarre. Only force or concession seem to be viable options.

Americans assume these Dragonwars need not necessarily be understood only won by deploying not only more but fieldcraft. Power can, often can, eliminate the dreamers and so the dream. When power and craft is ineffectual, for both must be limited by the area, limited by American standards and agenda, by imposed reality, there will as always be disappointment and occasionally disaster. Some will seek dragon ways misunderstanding the dynamics of conviction. The assumption that an unconventional can be countered by craft, secrecy and in so doing offer the freedom to discard convention arises from the American search for a tangible means to effect events. The focus is on the means, the rules of the war of the flea, the craft of the urban guerrilla, the techniques not the implications of perception. If power is ineffectual, the Americans still believe innocently in craft—and especially when such craft offers freedom from restraint, a return to frontier days. Neither power nor craft is always enough although almost always sufficient. The strong and righteous may not inevitably win, may be engage in the wrong war at the wrong time and certainly against a foe wrongly understood. What then? What if neither power or craft can be applied? In this American has been fortunate since there are so few indigenous dragons. The American Dragonwars are abroad. The nation may generate the aberrant, militiamen fleeing black helicopters or Charles Mason fleeing reason, may attract as stage the

Holy Jihad or the Palestinian nationalists but does not have to wage war only protect the heartland. Thus abroad there are the—a presence traditional imperial options may be needed.

If a presence does not pay, then evacuation is an option. If Beirut is a nightmare, American may simply withdraw. In fact, the strong or the sensible did withdraw, Reagan in Lebanon and Clinton in Haiti and Somalia. The prospect of either entanglement as in Vietnam or scandal as in Nicaragua will be a constant. And what if persistence is required? There are, of course, no answers but rather the desire not to be so entangled. And more power deployed is seen as assuring no entanglements but rather conventional victory: Panama or Grenada. Too little deployed for uncertain purpose produces disaster and withdrawal as in Vietnam or Lebanon or Somalia.

The Pentagon persists in assuming that more would have prevailed, more assets, more targets, more support, more of everything. Americans so assume. Small rescue teams become small task forces, small task forces require naval support, air support, a command-and-control center in the Pentagon, more of this and more of that until a real task force appears. Those zealots who seek to harm the great Satan in New York or Los Angeles are a tiny core of the faithful, ill-educated, naive, without experience who construct diabolic devices. Few, limited, innocents abroad to kill and yet they and their ilk have generated an enormous bureaucratic response to terror. Any survey would require flow charts and graphs, tables of responsibility, an index of legislation, a list of the previous reports and programs and responses. Over forty American bureaucracies are charged with a response to terrorism, limits of authority are defined, crisis management seminars offered, field exercises funded: and two or three illiterates emigrants with oddments bought over the counter can still frustrate the system—two men in Oklahoma City demonstrated the power of the few even without sugar sacks of anthrax or a nuclear device stolen from Vladivostok.

More will, in fact, often prove less in unconventional matters, not every time but often. The problem will be that there is no solution. More cannot be brought to bear. In the case of terror, there can be amelioration but not eradication. In the case of narcoterror the market is not amenable to interdiction. In Somalia all the gunmen could not be killed because all Somali men had guns. Dreams cannot be killed by weapons only dreamers and no mission will authorize killing all the dreamers, every gunmen, destroying those to be rescued. Destroying the village to save it cannot be a general policy. And simply sending more may crush the village without engaging much less endangering the actual enemy. It is difficult to bring weapons to bear on a dream—especially when the existence of the dream is imagined peripheral to the central purpose of waging war. In fact, in Dragonwars the asymmetry is not in assets but in perception, the conventional wage war and the zealots seek the truth through history and employ will over weapons. America is apt to focus on the AK-47s not the ideal, to seek practical accommodation, land reform or civil liberties, not corruption of the dream. The most appropriate policy is often not congenial to America, to the military and not likely to be achieved either by proposed institutional reforms or a shift in priorities or dictate. To respond to terror the vulnerable can be reassured without being overly penalized and the guilty may be intimidated but still the problem is there is no solution. And Americans will not easily accept a world without solutions a world vulnerable to terror. Americans insist on solution even if one must be imposed, invented, transported, erected, and guarded.

If five pounds of anthrax is a threat to American then a defense *must* be found. If there can be no solution, not even one imposed, Americans are apt to withdraw rather than persist, deny reality for their own dream of decency. Some may sleep badly with nightmares of the plague but most Americans will rest assured that a counter will be found. First, those responsible seek to deploy more, more troops, more power, more persuasion. A lack of a solution is an affront. If anthrax-terror is a threat, then more committees and conferences and missions will be the first and last resort. This is the American way.

Since America is not static but various, complex, shifting, and evolving, there are cross-currents, a nexus that includes the different: those who have a deeply felt sense of history, those pessimistic, those who empathize with the strange and the alien, as well as the majority who offer the stereotype. The later, the everyday, the typical, the average, determine the assumptions that fashion policy, fill the slots on the committees and draft Pentagon responses. The others that do not fit the stereotype have not necessarily been ignored, even by the military establishment. Reliance on those at the margins to pursue irregular war against the Apache or as special forces in larger war has brought returns and no threat to orthodoxy, the bureaucracy, career patterns, or doctrine. Since doctrine and orthodoxy have often proved poor guidance in irregular matters, since the establishment arrayed for war cannot pursue irregular war, then the alien have been dispatched and ignored, sought the Apache or the guerrillas of Nicaragua. So too might the contemporary Pentagon rely on shaping such assets, not merely because unconventional assets are advantageous but also because the major thrust of the Pentagon could be continued leaving the unorthodox as hostages to public demand. American hostages need rescue, starving children need to be fed or evacuated or protected, mad dog killers need to be arrested. Embassies need to be protected and plague fought even, especially, amid civil disorder. Special tasking is not congenial but necessary and the administration wants peacekeepers on call or the drug traffic interdicted.

There are those so attracted to unconventional tasking, who do not need nor want to be organized, administered, indoctrinated, restricted, and worse encouraged by offers of more. Bureaucracy and doctrine, national inclination and military practice impose regularity that hampers a response to the unconventional. If there is to be an effective response that falls outside such a pattern, then doctrine and practice need not be changed but only restraint introduced and rewards offered to the unorthodox. Admiration, authority, career advantage are wanted not new layers of control or more elegant systems. The unorthodox Americans should be tolerated, might best be admired, because they can cope not only with what the establishment cannot but also what the establishment seeks to avoid. They need not run free, must answer to the system, cannot play games out of the White House basement but be integrated into the institutions of response. Then there would be no need for games played by those who image secrecy is license. The price paid by the orthodox to the unconventional is restraint, stay small and disciplined. The key is acceptance of the special as special and worthy by the establishment. It is an acceptance seldom given so that those assigned irregular tasking often become corrupted by independence and isolation or denied approval.

What is necessary for the Pentagon and to Pentagon advantage and hardly un-American is that eccentricity and originality be accepted by the orthodox. Such assets are not to be transformed into the conventional, organized, administered, equipped, and trained to fit the existing orthodox agenda nor are they to be cast adrift in dirty wars. In the past those escaping orthodoxy have inevitably been assumed affront— outside convention and so to be ignored, expelled or denied or, if at all possible, transformed. In the American system, in most bureaucracies, the unorthodox is not tolerated but punished. Special operations, irregulars and elite units should rather be cherished as example of capacity, as lightening rods against public indignation, as means to response to the irregular.

In essence this means an acceptance by the culture of the military that there is a role for the irregular within the system because the system as a whole is not qualified to respond to unconventional challenges. To tolerate the minor investment not in funds and equipment and place but in awarding prestige and authority should be accepted as part of the military package: just as is control by civilians often driven by domestic priorities, as is public support often fickle and unrelated to the reality of sacrifice or the nature of war, as is the nature of war unforgiving to peacetime priorities. There need be almost no tangible institutional reform, no need for a conversion to the unorthodox or a sudden realization of the actual nature of the unconventional. Most of all there is no need to warp the major agenda of the establishment.

What would be required is to accept the margins, the unconventional warriors, as crucial to defense interests, a vital variant force, a respected career option and an effective response to the certain imposition of the unconventional on the American agenda. Such assets are scattered through the national security establishment, ready to deploy. There is no need to establish a centralizing system, to turn the small into the ineffectually grand, to fill slots, have an assistant secretary, manage and control. Small has charm. Deception can become doctrine without offices. Those who can operate independently, operate small at the margins need only be accepted. To reward the margins does not require institutional change.

The special may never rise to great command if they stay special—they will lack the experience great command is assumed to require. Those who can cope without more can most likely cope without such a career. What they are denied is not a traditional career but often the respect of the traditional. That could change. It would not be very costly but change, any change, even slight change is always painful—this change, however, is possible and does not alter the basics. And the basic needs of national defense, the basic nature of careers within the system, the priorities of purpose and habit need not be changed at all.

The establishment must accept the limits of orthodoxy: the Pentagon cannot as now construed respond effectively to the unconventional. If the "unconventional" is amenable to institutional practice then the challenge is no longer unconventional, falls within the conventions of anti-insurgency, commando intrusions, or hostage rescue operations. The generals and the admirals never get the war they want, the mission that requires what is available. Desert Storm was all but a miracle, the perfect Pentagon war courtesy of Saddam Hussein. All the other challenges of the century have been unexpected and often unwanted: Pearl Harbor or Pancho Villa, Vietnam and Somalia. What the military establishment wants is to prepare for war not to pursue it nor to be presented with missions that risk assets or require novelty. If the problem or the opportunity is unexpected without precedent or doctrinal definition—variants of subversion or psychological terror, non-violent strategies or zealots with strange weapons—then the system will not cope. Then the margins should be authorized to pursue an effective response however protracted or unorthodox the course.

More should not be offered or expected. To deploy the unconventional within doctrine and the priorities of the system is vital. More than most nations, America can deploy all sorts, various in perception and predilection, in culture and preference, descendants of the frontier, the slave ships of Africa and the ghettos of Europe, assets with strange languages, a taste of history, an understanding of the diverse, the poor, the best and the brightest, the aggrieved, the opportunists of the wilds and the slums. Many, of course, indoctrinated and assimilated, become as all the others, carry only wisps of the irregular. Others, however, are aberrant: the irregularities of choice and nature persist and so are available to be deployed. Americans are, indeed, moved by a dream but one mature, static and orthodox, but a dream that allows dissent and diversity. It is this diversity that the defense establishment must not only deploy but reward.

Conventional reward will not do for this imposes size, tiers of administration, the responsibilities of overhead at the price of capacity.[1] To be effective those at the margins must remain focused not on the usual, the systems and patterns of convention, on command and control and career, but on response to the special. Since less is more in unconventional matters, the margins should, of course, have on call the elegant and cunning but not be subject to offers that best be denied. The special must be kept so and as such neither denied prestige nor deluged with convention. And so in the end what need be done is adjust perceptual reward: those at the center of the circle must endow the unconventional with authority, freedom of action, and promise of honor and advancement. Then the American dream can be pursued underground as well as above, shaped to the special and various, can cope with the zealot, enlist the fanatic and not find war only in mass and convention.

The first danger is to ignore the special and the second to overload the mission so that orthodoxy is imposed. In Vietnam slow escalation allowed the orthodox to impose their perceptions on reality, ignore the special and the specialists, persist without career penalty until careers and the war and American society had been damaged. If Dragonwars are to be pursued effectively, the military, the orthodox and responsible, must neither deploy other dragons, nor craft without ideals, nor rely on additional power, on more when more may never be enough. Instead for Dragonwars, irregular wars, unconventional missions, the most effective response is to rely on American virtues and legitimacy, on the skills and talents of the special, and in the end not on power divorced from reality, not on perceptions that shape a congenial arena, but on a blend of tangible power, high ideals, and the grasp of the possible by those who prefer slaying the dragon with a single, cunning lance rather than a mass charge.

Far better to deploy a mix of assets against the dragons: intelligence, craft and stealth, special units as well as elegant weapons, technological systems, formal unit, conventional power. To mix and match is the responsibility of the commanders and to discard the particular for the grand, to rely on the congenial instead of the unconventional, to always seek more rather than less will sooner or later lead into a Dragonwar without easy exit. Then power cannot be deployed. Then orthodox assets are by underground alchemy turned into liabilities.

Many unconventional missions are simply what they appear: minor tasking amenable to specially trained units like the Special Forces or the 26th Marine Expeditionary Unit rescue force waiting off the Congo, fully trained, fully equipped, ready to go and most of all ready to withdraw mission completed. And if matters do not go according to doctrine, then those involved must rely on their wits as much as doctrine, on the contingent, on reading the arena, on skills not so easily taught but better deployed by a few rather than large formations. And, of course, some threats are easily met with American power, pure, inelegant, and massive. Grenada and Panama "worked"—even if less would have worked as well. There is every reason to assume that in the future as in the past, the establishment will insist on employing great power instead of special force. At least the special forces should be factored into any future war—Grenada put on offer to the unconventional—rather than ignored in favor the 82nd Airborne, specialists instead of the cobbled together rescue of the hostages in Teheran, and sensible arrangements with the exile cabals seeking to overthrow the monster in Baghdad. In 1998 Saddam Hussein was more vulnerable to deception and black propaganda than to airstrikes—and certainly to the plots of exile cabals funded out of Washington.

What is needed is to understand that unconventional missions are not as dangerous as the emerging and unexpected conditions of the arena. These are areas of the mind, strange, uncongenial, not amenable to power or tangible assets but not beyond the capacity of American ingenuity. Saddam Hussein is not at all like Americans or even as American imagine but vulnerable in his own special way. American assets and advantages, with proper preparation and planning, with good intelligence can pursue dragons to effect, might do so with Saddam Hussein or have done so in Panama. And the unconventional is almost always small in commitment, the risk in revelation and failure rather than in tangibles. The unconventional requires not mimicry or tradecraft but flexibility and guile: Rangers can be sent on a Ranger task with assurance but not into the back streets of Mogidishu. There have been and will be all kinds of missions and many unconventional tasks even in the most conventional of expeditions. Preparation is possible but the world of the covert, the fanatic, the alien and the duplicitous require flexibility, guile, and cunning. There are no guide books to guile and cunning, to strange arenas—that is why they are strange and require guile and cunning.

If wars were to be won by reading texts, then no general would ever lose. In the end, what works in Dragonwars works—and this is a basic American assumption easy to accept if less easy to incorporate into military priorities where what should work often assumes a reality that only combat may dispel. What will work best in Dragonwars, then, is not hired dragons or great armies but a meld of forces and assets, ideas and special operations and, of course, military power deployed to effect. The exception and special can be shaped beforehand, the prior planning and honed doctrine can be shaped beforehand, but victory or even accommodation depends not on a text but an effective response to reality, on the contingent and the unforeseen, on the slippage of mission, and on the errors of others.

And in the very end wars, Dragonwars, are won by power and art, by appropriate tactics and strategic imperatives, on the application of common sense and tangible assets. For this America is well prepared and in this the military should not deny the contribution possible from those who mimic not the dragon but understand the irregular, any more than underestimate the awesome impact of conventional weapons and forces. Victory in war, however defined, comes to power and to craft so why forgo craft and guile and special operations because such options are uncongenial: war is not meant to be congenial. War is brutal, harsh, ugly and costly. There are no good wars, not even *Desert Storm*.

Dragonwars are small and vicious, corrupt the involved and kill the innocent, are just beyond the reach of the orthodox but not beyond winning: grievance can be addressed, zealots isolated, the dream corrupted, gunmen countered, but not by applying the lessons of the text nor even by being prepared for all eventualities. How could Britain have been prepared for IRA hunger strikers acting to advantage within a culture alien to London, presenting impossible choices because they were willing to make unexpected sacrifices? How could the West have prepared for Salman Rushie to be forewarned, forearmed, ready for a fundamentalist assault beyond reach of missiles or the SAS? Washington is filled with those who offer maps to the terrorist country but know little of the actual terrain, never factor in the contingent and unforeseen, the illogical and the brutal. If the user manuals to unconventional war were valid, more than academic and analytical exercises, Castro would have died in Orient, Grivas in a dugout, Began and Mao would lie in unmarked grades—and perhaps the British would have had Washington on the gibbet. What is needed, rather, are the few, the eccentric, those with original minds and insight.

America must be prepared for another *Desert Storm*, for NATO wars, for real wars—and for whatever war is on offer and for small missions and murder from a ditch. To seek the congenial, the war imagined, to rely on systems and assumed experience is not unexpected, not a tendency of Americans alone, but this alone in a confused and uncertain world is inadequate preparation for the future. The establishment knows this, worries about anthrax and war short of war but not to best effect. For those not unmindful of the future, flexibility and curiosity, empathy with the alien offer useful tactical and strategic attributes. These are assets that can be deployed in wars that only rarely produce visible battles although often can be pursued with simple power. When power cannot come to battle, when the mission is beyond conventional roles, then the system, the establishment should be able to deploy existing unconventional attributes at low cost and to effect without discarding legitimacy. In such matters the dragons are to be sought to effect not by recourse to a text or plan, not by big battalions, but by those familiar with the indirect approach, with the power of dreams, with a grasp of the alien and a flexibility beyond form. Without vast institutional reform or new directions, without much change at all, America can deploy originality and reason, optimism, power, and competence—American virtues—by those charged with countering dragons, by those who should have a role in the system and so in the future a mission.

Note

1. For the curious the response of the American government to terror can be found outlined in the United States General Accounting Office report to Congress, *Combating Terrorism, Federal Agencies Efforts to Implement National Policy and Strategy*, GAO/NSIAD-97-254, or the course outline and reading *Operations Other than War (OOTW)*, United States Army Command and General Staff College, Fort Leavenworth, Kansas, January 2, 1995, for an indication of the nature of the national security establishment when deployed against the unconventional.

Sources

One of the more obvious by-products of contemporary events is paper, official and unofficial paper, documentation, enormous amounts of paper, tracing publicly and privately all aspects, potential ramifications, and evolving consequences of even the most transient and unimportant sequence. Since the book covers a long series of wars and complex responses, the paper trail is beyond detailing. Obviously the scholarly apparatus has been severely reduced, few notes and no check list of works consulted or works to consult. Even a list of single source works for the most important armed struggles would produce a considerable list: there have been at least a hundred important low-intensity conflicts most generating at least one authoritative study.

There is a surfeit. There have been nearly ten thousand books published on the Irish Troubles. There have been as many with terrorism in the title. For a subject like Vietnam even the major general histories are many and the source material, including that steadily emerging from Hanoi, massive. There are dictionaries and guides and Linda Reinberg's *In the Field: The Language of the Vietnam War* (New York: Facts on File, 1991), for those not exposed to the jargon of the war, do not have the facts on file. One can read a work, often a highly detailed work, on the Australians or the siege of Khe Shan, on the impact of the bombing of the North, on the year after Tet, on nearly any imaginable facet of the long war: river gunboats, lost prisoners, communist propaganda, Laotian refugees, surgical practice, or the underground world of the tunnels of Cu Chi. There is less for other campaigns and excursions but still much and even those long ago, forgotten wars in Aden or Borneo have their library shelf.

If the paper trail is enormous, there is still no consensus. There is still room for investigation, interpretation, room for additions and corrections. In this particular case, the armed struggle and America, every new catalog, each new publishing season generates more relevant works, books on irregular wars, on militant organizations, on national security issues, biographies of the involved, documents released and interviews given. The 1997 spring catalog of the University of Pittsburgh Press, a not especially famous scholarly house, for example, offers titles such as:

> Carment, David and Patrick James, eds., *Wars in the Midst of Peace, The International Politics of Ethnic Conflict*
> Michaels, Judith E., *The Presidents Call, Executive Leadership from FDR to George Bush*
> Elwarfally, Mahmound G., *Imagery and Ideology in United States Policy Toward Libya, 1969–1982*
> Lefebvre, Jeffrey A., *Arms for the Horn, United States Security Policy in Ethiopia and Somalia, 1953–1991*
> Renshon, Stanley A., ed., *The Political Psychology of the Gulf War, Leaders, Politics and the Process of Conflict*
> Ripley, Randall and James M. Lindsay, *United States Foreign Policy After the Cold War*
> Steinberg, Blema S., *Shame and Humiliation, Presidential Decision—Making on Vietnam*
> Cottam, Martha I., *Images and Intervention, U. S. Policies in Latin America*
> Heine, Jorge, ed., *A Revolution Aborted, The Lessons of Grenada*
> Everingham, Mark, *Revolution and the Multiclass Coalition in Nicaragua*
> Middlebrook, Kevin J. and Carlos Ricos, eds., *The United States and Latin America in the 1980s, Contending Perspectives on a Decade of Crisis*.

All or none may be crucial reading, but each is useful and any worth noting. A short work specifically focused on a comparative general issue that appeared as well in the spring of 1997 is Charles King's *Ending Civil War* (International Institute for Strategic Studies, Adelphi Paper 308), (New York: Oxford University Press), with a notes section filled with late, relevant works on low-intensity conflicts. The first sources King cites are:

> Ariel E. Levite, Bruce W. Jentleson and Larry Berman, eds., *Foreign Military Intervention: the Dynamics of Protracted Conflict* (New York: Columbia University Press, 1992)
> Arnold Kanter and Linton F. Brooks, eds., *United States Intervention Policy for the Post-Cold War World: New Challenges and Responses* (London: W. W. Norton, 1994)
> Richard N. Haass, *Intervention: The Use of American Military Force in the Post-Cold War World* (Washington: Carnegie Endowment, 1994)
> Roy Licklider, "The Consequences of Negotiated Settlements in Civil Wars, 1945–1993," *American Political Science Review*, vol. 89, no. 3 (1995), pp. 681–690, *Financial Times* and the *Washington Post*.

In other words or in too many words to count, the new world order is already filled with scholars, analysts, and journalists focused on the dynamics of low-intensity conflict and the appropriate responses, on the Irish peace process or the impact of non-lethal weapons on peacekeeping. There is too much to read, certainly too much to list as sources. Thus this essay on sources is really about what is not listed and what forms the basis of the book.

This is especially so because the most crucial sources for my Dragonwars have been those involved, a generation of discussions, interviews, queries, and conversations that are not always easy to list effectively. There have been all sorts of people queried, important people like Prime Ministers Menachem Begin or Yitzhak Shamir of Israel or General George Grivas, Archbishop Makarios of Cyprus, and Constantinos Karamanlis of Greece, and lesser known rebels—all but one or two Chiefs of Staff of the IRA over the last forty years and African nationalists like the late Eduardo Mondlane and Ruth First. There have been as well a great many nearly anonymous figures whose names mean nothing except to the involved.

Some are still engaged in wars in the southern Sudan or Lebanon, some seek to free Scotland or have found all the answers in Islam. I have talked with thousands of volunteers in crusades from Eritrea to Brittany. Again, some of the names have appeared in my other books and some cannot appear in any books. There are as well the numerous conventional officials, opponents to revolution, old soldiers and active operatives, administrators and policemen, any of whom might prefer not to be in a list heavy with guerrillas and terrorists. In America there have been those visible in policy matters and those who were less visibly engaged at home and abroad in such matters, diplomats and soldiers, national security officials, agents and observers and academics.

Most of my projects have produced interviews with cabinet ministers and colonels as well as carbombers and felons, trips to mean rooms and to the House of Lords but always discussions on the nature of the underground. The responsible are apt to have trained minds and formal hours, policemen, presidents and bureaucrats, journalists, scholars, and there have been as well transients, everyday people, and visiting firemen—the vox pop of the foreign correspondents. Mostly, however, I have sought the gunmen, articulate or no. After these primary contacts come all the other and more conventional sources: the paper, the primary sources and the secondary, the official and unofficial, the first cuts and the later histories, thousands of volumes read over a generation, whole issues of journals and journals wholly given to low-intensity conflict, terrorism or small wars, and newspapers in various languages.

As with all contemporary events, the first tide on any event contains the primary sources, the recollection of witnesses and the documentation of the involved, notes, diaries, memos, hand-written evidence, intimate government paper, telephone logs, initialed letters. Much of the paper, the most interesting paper is classified or secret or beyond reach. In time a few of the more salient examples will be edited and printed, but before then there may be access of a sort. The involved talk, paper is made available, hints and clues are distributed. Those in power in any case generate vast amounts of paper and those in the underground all that they can. The revolutionary rarely leaves heavy paper traces although even underground records are kept. I have held the entire IRA written record for two years in one hand before it was returned to a dump—for safekeeping but more likely to be forgotten. What is tangible is the grist of the underground propaganda mills churning endlessly, grievance and aspiration and little of substance.

In the underground, much of the daily routine, the operations and the maintenance, is done without words or without formal record—unlike even the most secretive governments. An armed struggle seldom has a historical memory. For an underground what survives from the underground is fragmentary, traces as much as sources. Only material published or widely distributed at the time ends in research libraries or bibliographies. Time, discretion, and bad fortune destroys much.

The Palestinians, possessed of a counter-state and a Lebanese haven that permitted the creation of real libraries, still lost their archives when the Israelis invaded Lebanon. In any case there is unlikely to be any data on Black September operations or money stolen, informers shot, or incompetence displayed. The more wanted paper is less likely to exist or, if existent, found. The IRA loses more to the authorities than the movement hides. There is not going to be a great cache of material on the Holy Jihad in Egypt any more than there was for the Moslem Brothers. What often is amazing is that so much does survive, including some of the survivors to write their own recollections, adjusting, as do all survivors, history to recollection.

Even when there are underground records more can often be discovered from trials, government papers, the data of the authorized.

The endless Italian trials have produced a wealth of data for the social scientists if not always swift justice. Some governments, of course, never reveal anything of note. Most security establishments never reveal anything—except the evidence necessary for open trials, where much can be found whether in South Africa or Italy. Mostly records of subversion, betrayal, insurrection, and conspiracy remained closed. This is true with many conflicts of greater intensity. The prospect of seeing, assuming such records still exist, the paper generated in most Middle Eastern crises is as unlikely as reading the Vatican's financial records. Much material that even relates to political or militarily sensitive matters in the hands of governments, if not classified or destroyed, is simply held out of sight—what is not revealed can do no harm is basic to any bureaucracy.

In fact the denial of access to such documentation even when a half-century has passed indicates such precautions are not without purpose. Subversion and insurrection have long roots. The names and addresses from a generation ago may still be relevant in matters of subversion. Dozens of people in Northern Ireland were interned in 1971 because their names were found in my *Secret Army, A History of the IRA Since 1916*. History has no closure. No one considering informing for the state wants to be on any list kept by the state. No one wants grandfather to turn up having been that informer. In any case or rather in most cases, if there is such documentation, it ordinarily remains classified, closed, for years and, as a matter of fact, in many cases is lost forever. Intelligence matters are routinely shredded or hard discs destroyed. And many "sources" are not just delicate but also false, shaped to deceive, to contaminate over time. Even the most democratic and elegant governments have been know to adjust documentation for future historians, alter, edit, discard, add, erase all evidence. The state cannot, however, rest assured that any hidden document will always, forever remain so hidden.

The evidence and the event have long lives—reality becomes history very slowly. This may be true with both high policy matters—murder authorized by the premier—as well as the details—the list of payments to secret agents in provincial cities. Someone may keep copies. Someone may simply tell, show notes, give evidence. And sometimes the state is destroyed, implodes, or is occupied by aliens who read the mail, give guided tours, publish the secrets. So at times the interested wait forever and at other times the unexpected opens the most closed archives: the KGB after 1989 shopping intelligence records to the highest bidder or the secrets of Nazi Germany or Fascist Italy appearing in scholarly tomes, translated into English and French with footnotes and guides to additional data. Despite the concerns about excess classification, the United States government publishes much and often quickly and the involved are equally quick to print once office is resigned. This is most useful for the American response but is rarely as helpful in understanding the nature of the irregular.

Always the state's view of the Dragonworld is determined by perception, assumption, and need whether or not those views are noted, printed, distributed, or in time available to the authorized or interested. The state and the responsible see the underground from above ground. Those who creep about on the ground engaged in the craft of the unconventional often assume, as do their superiors, that the underground is simply a mirror image of the conventional. Thus the evidence of the state often reveals more about the state than the underground. So what happens during the relevant conflict is often available in time but the explanation is devised by those denied any but forced entry into the Dragonworld.

For much that goes on underground, the most useful source remains the involved. The underground paper seized by the state or hidden by the gunmen is fragmentary at best. The public paper is propaganda rather than insight in the dynamics of the movement. Thus the recollections and reflections out of the underground are often, usually, crucial to an understanding of the asymmetrical lethal dialogue. And evidence that lives in memoirs always presents the investigator with complex problems. In time these tales are best approached by the guidelines provided by the methodological management of oral history as taught in graduate school. With oral history, the risks and gains are known. At other times when the struggle is hardly over or continues all about, the investigator is more anthropologist, a political ethnographer observing the customs of the involved, as much as someone conducting scholarly interviews.

In any case, in the midst of an armed struggle, few of the involved have time for disinterested reflection. Surrounding this primary core of the involved, even and particularly at the time of the events, a vast paper storm printed and distributed for varying purposes appears, has appeared in general and in particular: instant government or organizational material, policy statements, leaks and journalistic probes, first-person accounts, pop history and hurried explanation. This is the primary commentary and explanation of the moment. All of this, early or late, tends to be raw data, immediate, skewed to purpose, fragmentary and unreflective—part of the event in most instances. Most will have a short shelf life or be filed for future reference. The concerned must read it if with a jaundice eye. This is the conventional raw material for historical and political studies, the traditional primary data.

Then, at one more remove, come the first chroniclers, those who array the events in chronological order, weave in a story, indicate motives and results. Some gunmen even give public interviews, contemplate a book. More often than not, the more visible evidence of this process appears reworked in the swiftly published articles and then books of the journalists, the paid and supposedly disinterested witnesses of the contemporary panorama. Thus, in the case of Bosnia, an early sweep through recent history by a journalists is Robert D. Kaplan's *Balkan Ghosts* (New York: St. Martins, 1993) is an effort to give historical coherence to a crisis far from finished. Edgar O'Ballance, who has made a career of recent small war, offered *Civil War in Bosnia 1992–1994* (New York: St. Martins, 1995) even before there was a satisfactory terminal date. And when the shooting seemed to have stopped, Laura Silber and Allan Little of the *Financial Times* in 1997 updated their 1995 *Yugoslavia, Death of a Nation* (New York: Penguin, 1997). This means that neither the comparative scholar nor the general reader need, unless desirable, face the rooms of paper, the files of newspapers, the transcriptions of the media or the statements of the times: there is a solid summary, based on people and on paper.

At this stage the more adventurous of the scholars and analysts begin to produce highly focused academic articles or more general surveys of events. These first essays appearing in journals of opinion, specialist tracts, and small monograph series in time may actually become an aspect of the event, indicate the perceptions and suppositions of observers. The general works may last if attention moves on and no one cares anymore about the war in Yemen or the Eritrean armed struggle. Then one reads O'Ballance or Silber and Little because no one has replaced them with more recent, more scholarly, or more reflective work.

In matters of low-intensity conflict, all struggles are not equal, not equal in general concern and not equal in access. It is easier to deploy the orthodox sources than rely on the fragments from underground and so easier to write on small wars, expeditions, rescue operations than on the dynamics of the covert and illicit. The fact that there was a book on the decision-making process of the PLO in Beirut (Rashid Khalidi, *Under Siege, P.L.O. Decisionmaking During the 1982 War* [New York: Columbia University Press, 1986]) indicated how far the PLO had moved from Arafat and the other Abus sitting about a table, visible but deeply underground, shaping Fatah. In Italy for a decade during the years of lead with murder present in the streets, no one could write a proper survey because no one knew anything useful about *Brigate rosse* or the neo-fascists. Tens of thousands of pages of speculative journalism and propaganda and tracts appeared but little of substance. There was nothing like Silber and Little on the implosion of Yugoslavia because the sources did not exist even if the interest did. On the other hand, nothing seriously involving the Pentagon goes unnoted nor unchronicled—the Greek involvement still attracts analysts as does Grenada or the pursuit of Pancho Villa. The first generation is always special, often part of the event as much as the gunman or police constable. The occasional quick summary or immediate insight have lasting value: mostly not, however. (*cf.* "The Chroniclers of Violence in Northern Ireland: The First Wave Interpreted," *The Review of Politics*, vol. XXV, no. 3 [July 1973], pp. 398–411; followed by similar surveys in the same review in October 1974 and 1976, when exhaustion and the realization that there was to be no visible end point aborted further publication). Even the fiction, the thrillers and the films, the made-for-television or the popular songs, the Troubles as Trash, are source as much as comment.

With a few exceptions the chroniclers move on to other crises, other books along with the television cameras, the enthusiasm of editors and producers, and the tastes of the public. The stage is now open for the full invasion of contemporary scholars, academics familiar with the arena or the crisis or an investigatory means of potential use. The academic are often joined by analysts for governments, for commissions and foundations, for distant institutions concerned with applications and lessons and not tenure and publication like the professors. The Europa Hotel in Belfast was for years a nexus of social scientists, government investigators, visiting scholars, lawyers, advocates of conflict management, social workers from Scandinavia and political scientists from America, the media from anyplace and free-lance writers of all sorts seeking out the usual suspects, collecting data and exchanging insight, trying out their paradigms and surveys. Much the same was true for the Palestinians or the Italians during the years of lead—an arena of early analysis that turned out to have no due date. Quick expeditions and special operation swoops tend to produce quicker pop studies and less focused analytical investigation: Grenada has come and gone generating less than might be expected because there have been other more dramatic expeditions. Sometimes there were big books but often there were specialized works and quick studies for the paperback trade.

As the more formal evidence of the involved continues to dribble out in published memoirs and special pleading, the new material will be incorporated into general history and specific analysis. In time the investigators, except for the few who find a career in the particular event, also drift away to other matters. In the case of armed struggles, the classics like Vietnam, Palestine, or Ireland persist as subject. And so does published concern with American and Western responses to low-intensity conflict, trouble by various names. There is opportunity for new generations of experts and specialists and observers. And thousands and thousands of pages is the result: each decade generates not only more but also specific authoritative monuments—*the* history or *the* interpretation. Even when there is a stopping point, as in Vietnam or Algeria or South Africa, there are new sources and new prospects to enliven old concerns and capacities. Thus the process of analysis for the popular and accessible conflict, for conflicts in general, for responses and prospects is continuous.

The published result grows enormous. In the case of America and Dragonwars, even a list of United States government publications relevant to the theme would run to book size. There is a veritable government library on Vietnam chronicling each stage culminating with an on-going multivolume military history and the *Economist* undertook R. B. Smith's multi-volume *An International History of the Vietnam War*. There are, especially in the very recent years when a Vietnam reexamination has been underway, good, solid books that are no longer embedded in the issue, including one-volume histories like:

Phillip B. Davidson, *Vietnam at War, The History 1946–1975* (Novato, CA: Presidio Press, 1988)
Stanley Karnow, *Vietnam, A History* (New York: Viking, 1983)
Stanley Maclear, *The Ten Thousand Day War, Vietnam 1945–1975* (New York: St. Martin's, 1981)

and in the case of Neil Sheehan's *A Bright Shining Lie, John Paul Mann and America in Vietnam* (New York: Random House, 1988), a history of the American period wrapped around a single individual. And there will be more to come and most of these will have to be adjusted or replaced or amplified as more material continues to arrive from Hanoi and the Viet Cong.

There is almost always more in matters of contemporary affairs even if in covert matters, in the case of Dragonwars. There is never enough, however, from the center of the revolutionary galaxy. On armed struggles there have been three major clusters of analytical work. The first mostly by Western and American scholars is the array of anti-insurgency studies that were only temporarily discredited by the impact of Vietnam. After 1975, although the vocabulary changed and the optimism was tempered, the focus and assumptions remain much the same. Second, there were the explanations and texts from the underground that overlapped the Cold War anti-insurgency texts, the guerrilla guides beginning with Mao's various redbooks and tapering away into the religious tracts of the new generation. These two foci merged in the crucible of Vietnam—the great cold war for Americans. Even before Vietnam was over, the third current, the rise and proliferation of terrorist studies, almost all written from the perspective of the threatened, became a publishing phenomena. Most explained the threat, the vulnerability, and suggested appropriate responses. And nearly all of this library until the implosion of the Soviet Union was shaped by Cold War assumptions and context. Some of these assumptions have yet to be discarded.

Not until the unexpected arrival of the postmodern world has there been a broad examination of the dynamics of low-intensity conflict and the appropriate Western—especially American—response that is not shaped by an agreed global context. In fact many of these newly discovered low-intensity conflicts and unconventional threats are deployed as analytical evidence to proposed world contexts: a clash of civilization, the rise of zealotry, factionalism and failed states, new nationalism and internecine conflict. There is strong professional interest in missions short of war and small wars driven by fanatics. Because of the expansion of academia over the Cold War years, the rise of sponsored analysis, the multiplication of venues, the last decade has generated a huge library.

There are journals dedicated to small wars, to conflict and terror, to international crime, military journals and law-enforcement journals, not to mention the old faithfuls focused on world politics or foreign policy or international affairs. Thus to the ten thousand books on terrorists—and the concomitant articles—there are those thousands and thousands more in most languages representing most ideological postures and professional experience probing every aspect of contemporary conflict. Much is, by nature, academic, much is, for sponsors, shaped to assumed conclusion but much is also fascinating, convincing, innovative, deploys new methodologies or new concepts, and even at times new sources.

In some part I have been able to avoid the problems of excessive printed sources for very contemporary movements by dealing with the principles. This means very limited hard, quantitative data. Hard data is hard to bring out of the underground: even the average age of Beirut militia gunmen is difficult to determine exactly, if simple to estimate by observation. Those who want to count, to examine inter-office communications, read the mail or make charts tend to find contemporary rebels an uncongenial subject. Few who write on terror can operate across cultures, time, and customs. Few know in real life those very different from their own nor can imagine the poorly educated, the fanatic, those with limited education driven by the revealed truth. Most academics are apt to assume their agenda, their values and assumptions, are not greatly different from those of the terrorists or the gunmen—the reasonable terrorists or the rational gunmen pursuing specific goals with carefully chosen means. They model but do not imagine, offer guidebooks to unknown and unseen venues. Academic product is suppose to be academic.

Much analytical work remains very academic, not so much valid as useful to the responsible who need authorization or congenial insight. Those who do know the gunmen have difficulty in translating to effect without rigorous data or the display of authorized methodologies. Either one can pursue the underground with acceptable means and produce work divorced from reality or rely on what is possible given the nature of access and produce results if not rigorous social science. The access problem is patent to the serious scholars who find that most work on the terrorists and gunmen to be marginal and methodologically dubious and so move on to more amenable subjects. The dragons are not so much badly served which is the case, but ignored by the competent, approached by the curious to no great effect, and imagined by others. And there is not much to be done but to wait until sufficient time passes to allow conventional methodologies to be deployed—which, given the attrition rate of the underground, will mean few witnesses. This, however, has never hampered medievalists relying on worn coins and scraps of vellum.

The rebels are left to area specialists or those with singular access. My own effort On *Revolt, Strategies of National Liberation* (Cambridge, MA: Harvard University Press), written a generation ago in 1976, to explain a single generation of various rebels against the British empire from the Irgun and LEHI in Palestine in 1944 to the IRA in Northern Ireland in 1969 simply reveals all the problems in print. The book gave shape to very few useful generalizations. I knew something about a series of movements but was apt to see the differences rather than discovered similarities. This was in part the residue of training as a historian but in greater part came from the acceptance that the worlds of Begin and the Mau Mau, of Grivas and the Egyptian Moslem Brothers were very different, the Arabs are not Irish even if the British were apt to remain very British over a generation.

There is always this difficulty with detailed comparative work that goes beyond parallel chapters by parallel scholars on set assignments: amending constitutions or voting regulations. The result has been that those most concerned with the rebel ecosphere, the inhabitants and their opponents, have tended to dominate the discussion from entrenched positions. One can read Mao or *The Marine Corps Gazette* and both are interesting, one ethereal and one technical, but neither gives much of the flavor of the underground or the nature of that covert gunman's world. When that world can be made explicit and more general, there is still the problem of describing the asymmetrical dialogue of most low-intensity conflict. Neither the underground nor the state understands the other or deploys assets against reality. Thus what has been done and what best might be done by America can best, or certainly most easily, be examined in isolation. And in many cases when mere power is deployed, when the medium and the arena do not matter—or when matters go well, the dynamics of the underground either do not exist or do not matter. And so in much analysis there is no great need to understand the gunman or the bandit. When power cannot be deployed effectively, then the nature of the armed struggle matters.

The nature of the American response to revolutionary challenge is far more closely documented, Pentagon guerrillas move forward on memos through a jungle of conferences, leaking as they go. There is a veritable library on American defense policy, on American technowar and American vital interests and American anti-insurgency strategies or lack of them. And there are as well those thousands of relevant works on terrorism and low-intensity conflict and ethnic violence. In each case there are schools of thought, ideological positions, the assumptions of those who fear power and those eager to use it, those who are idealists and those more pragmatic. At nearly any point over the last fifty years, one can find a cluster of works that display the concerns of the concerned. In the mid-1980s amid the glory years of the Reagan administration, there were those who feared a decay in capacity and those who insisted upon it. This short list from the 1980s when times were actually good for the Pentagon touches most of the bases for the genre with a focus on the defense establishment as a whole. A similar list—at times with similar authors—can easily be compiled for any period: those who urge more or less, want more for less, fear drawdown or an aggressive capacity—whatever, lots of work often elegantly plowing the same field:

> Choates, James and Michael Kilian, *Heavy Losses, The Dangerous Decline of American Defense* (New York: Viking, 1985)
> Fallow, James, *National Defense* (New York: Random House, 1981)
> Halloran, Richard, *To Arm a Nation, Rebuilding America's Endangered Defenses* (New York: Macmillian, 1986)
> Kaldor, Mary, *The Baroque Arsenal* (New York: Hill and Wang, 1981)
> Luttwak, Edward N., *The Pentagon and the Art of War* (New York: Simon and Schuster, 1985)
> Rasor, Dina, *The Pentagon Underground* (New York: Times Books, 1985).

These follow similarly concerned in the previous decade and those so focused in this decade—again the Reagan-Bush years at the bottom line were boom times for the defense establishment's big systems and spending. Even during the era of scarcity, there has been debate over missiles to defend against possible threats, advanced airplanes to defend against potential rivals and transport craft that are larger, more elegant, possessed of all the desirable bells-and-gongs that do not rival smaller, cheaper models. And very few aspects of the American involvement in Dragonwars have over the years been ignored. During the last decade even more work has been focused on the low-intensity end of the conflict spectrum, on expeditions and intervention, on American capacities and expectations—if as always the gunmen have to be colored in hastily. It is simpler to focus on the special responses—the troops on the ground—and the grand picture of the new world order. Thus there are monographs and collections on special operations—and government paper (*Special Operations Forces, Force Structure and Readiness Issues,* GAO/NSIAD-94-105, Report to the Chairman, Committee on Armed Services, House of Representative [Washington, DC: United States General Accounting Office, 1994])—and scattered over a dozen years the following representative and works:

> Bank, Aaron, *From OSS to Green Berets, The Birth of Special Forces* (Novato, CA: Presidio Press, 1986)
> Barnett, Frank R., B. Hugh Tovar, and Richard H. Shultz, eds., *Special Operations in US Strategy* (Washington, DC: National Defense University Press and National Strategy Information Center, Inc., 1984)
> Klare, Michael T. and Peter Kornbluh, eds., *Low-Intensity Warfare, Counterinsurgency, Proinsurgency, and Antiterrorism in the Eighties* (New York: Pantheon, 1987)
> Marquis, Susan L., *Unconventional Warfare: Rebuilding U. S. Special Operations Forces* (Washington, DC: Brookings, 1997)
> Paddock, Alfred H., *US Army Special Warfare, Its Origins, Psychological and Unconventional Warfare, 1941–1952* (Washington, DC: National Defense University Press, 1982).

There are as well as all sorts of analysis that suggest using such forces may lead to undesirable entanglements, that all presidents have deployed special operations seemingly not to advantage—or one can find a popular literature on elite units and special operations—as well as texts for the civilian, the survivalists, the curious, or the militia volunteer. Such paperbacks can be found in far parts, the pockets of rural insurgents who have read no other book or tidily on shelves in hand-built cabins in Idaho.

There have always been fewer studies of malpractice and operational blunder. Why Che failed or for that matter why most deep penetration raids fail? What are the limits of the SAS in Northern Ireland or the unpleasant operational lessons of Vietnam? It is easier to find splendid probes into what may go wrong on a larger scale like James William Gibson, *The Perfect War, Technowar in Vietnam* (Boston and New York: The Atlantic Monthly Press, 1986) or with special events: the library of books on the Iran-Contra affair, on the attempt to rescue the hostages in Teheran, and on the incident in Beirut in 1983.

In a muddled world at the end of the century American hegemony seems assured for the immediate future. It is a hegemony that assures great responsibility but also local obstacles to the deployment of power, military, diplomatic, or moral. There may be no challenge but there is not enough power to impose general order. In fact the new world order is mostly visible as stable disorder. A world composed of action and sharp edges: the a clash of cultures, ethnic violence, small intractable wars, weapons of mass destruction, new zealots with bombs, and all sorts of low-intensity dangers. The world of the gunmen has never lacked for observers and commentators and with the disappearance of the great strategic foci there are more to focus on the emerging dragons.

At the end there is sufficient further reading to startle even the most eager. There are good works, violating Gresham's Law, that have not been driven out by the bad. There are important works, lasting works on nearly every facet of the Dragonworld and the Dragonwars,. There is no work quite like *Dragonwars*. There is, however, an enormous literature on such wars and on the American responses. The contemporary violence and turmoil, an unconventional constant, continues to inspired more work, more good work, than can be listed neatly or appropriately for general readers or for specialists at the end of the text on Dragonwars. And the old dragons of the armed struggle have been rivaled as incubators of turmoil by those lacking other means who pursue special, narrow visions or the old ways by recourse to low-intensity conflict. Terror, insurgency, ethnic cleansing, rebellion is with us for the foreseeable future. The gunman's vocation and the irregular appears on the agenda of the state regularly. There are new conventions of the unconventional and an enormous, variegated and expanding literature: often high-intensity analysis, history, sociology, political journalism and bureaucratic paper, action agendas and notebooks of the day. There is something for everyone and at times it seems something by everyone.

In the heel of the hunt, as my MIT student, now Dr. Ernest Evans a specialist in such matters, once long ago said, "Go see *The Battle of Algiers*." And it is true: most of the nature of the dragon and the risks of response are there in black and white, still valid, still applicable in the postmodern world.

Index

Acheson, Dean, 268
Afghanistan, 7, 8, 86, 100, 108, 154
   Carter administration, 331
   Reagan administration, 343–44, 348–49
Africa, 4, 7, 69, 101, 108, 182. *See also specific countries*
   and armed struggle
     arena for, 133–34
     organization of, 143
   Clinton administration, 409–10, 411
   cults, 76
Aidid, Mohammed Farah, 390, 395, 398
Albania, 267, 270
Algeria, 7, 8, 182
   campaigns, 161, 164
   ideology, 79, 86, 89
   popular support, 96
Allen, Richard, 335
Allies, 165–66
Angola, 88, 133–34, 164, 167
   Reagan administration, 343–44
Arenas, conflict and, 133–34
Aristide, Jean-Bertrand, 398–99, 400, 402–3
Armed struggle, dynamics of. *See also* Ideology: Revolutionary ecosystems, armed struggle and
   allies, 165–66
   arena, 133–34
   asymmetry, 167–72, 197–99
   campaigns, 160–65
     as armed struggle, 166–67
     central structure of, 101–3
   command/control, 149–52
   commitment, 129–30
   communications, 155–57
   enemies, 172–73
   intelligence, 157–60
   leadership, 135–38
   maintenance, 152–54
   organization, 141–49
   recruits, 134–35
   tradecraft, 154
   volunteers, 138–41
Armed struggle, responses to. *See also* Perception; United States, conventional war and; United States, unconventional war and
   armed struggle limitations, 203–6
   and legitimacy, 205–6
   armed struggle meaning, 206–13
   and asymmetry, 167–72, 197–99
   attraction, 199–203
   and United States, 201–3
   coercion, 220–26, 418, 420–23
   conventional, 226–28
   by governments, 178–83
     and ideology, 180–81
     and legitimacy, 179–80
     results of, 182–83
   and terrorism, 5, 182, 330, 336, 339–40, 362–63
   and transitory violence, 181–82
   interpretation, 190–95
   and legitimacy, 71, 87–88, 142
     and armed struggle limitations, 205–6
     and governments, 179–80
   by media, 195–97
   psychological context, 213–20
   and results
     failures, 5, 7, 12
     governmental response, 182–83
     Lebanon, 29–30
     successes, 5, 7, 70
   variety of, 177–78, 231–34
Army of the Republic of Vietnam (ARVN)

Vietnam (1945–61), 283, 291
Vietnam (1961–68), 292, 301
Vietnam (1968–72), 312, 313, 318, 319, 320–22
Aspin, Les, 398
Asymmetry, 167–72, 197–99
Attraction, unconventional war, 199–203
and United States, 201–3

Baader, Andreas, 139
Balkan States, 267–68, 269, 270–71. *See also specific countries*
Clinton administration, 403–9, 411
Barrios de Chamorro, Violeta, 373
Bay of Pigs, 281–82, 285, 309n.1
Begin, Menachem, 136, 137–38, 142, 167
Beirut. *See* Lebanon
Bir, Cevic, 394
Black September, 18, 23
Bolivia, 121
Bosnia, 8, 81, 86, 403–9, 411
Boutrous-Ghali, Boutrous, 393
Brzezinski, Zbigniew, 331, 335
Bulgaria, 270
Burma, 100, 162
Bush, George, 42, 362
administration of (1988–92), 364–79
and CIA, 364–65
and El Salvador, 373
and Iran, 370, 374, 375–78
and Iraq, 370, 375–78
and Nicaragua, 373
and Panama, 364–68
and South Africa, 374
Soviet Union dissolution, 368–69, 373

Cambodia, 161, 162, 285, 323, 324
Campaigns, 160–65
as armed struggle, 166–67
central structure of, 101–3
Carter, Jimmy, administration of (1977–81), 330–34
and Afghanistan, 331
and Iran, 331–34
Casey, William, 42, 201, 336, 342, 344
Castro, Fidel, 89–90, 136, 137, 146, 161, 162, 164, 182
armed struggle perception, 184–85
Causes, armed struggle, 61–62, 83–86.

See also Ideology
cults, 75–79, 84
in Lebanon, 26–28, 32–33
nationalism, 4–5
religion, 4–5, 7
truth, 4–5, 7, 8–9, 11–14
Cedras, Raoul, 398, 399, 400–401
Central Intelligence Agency (CIA)
Bush administration, 364–65
Clinton administration, 402–3
Reagan administration, 341–42
Cheney, Richard B., 391–92
China, 5, 7, 73, 135, 167, 266–67
and Philippines, 275, 276
and Vietnam, 282
Christian Phalangists, 20, 25, 38–39
Civil Operation and Rural Development Support (CORDS), 311–12
Clark, William, 43, 335
Clarridge, Duane, 342
Clinton, Bill, administration of (1992–), 379–414
and Africa, 409–10, 411
and Balkan States, 403–9, 411
and CIA, 402–3
and Colombia, 381–82
and drug trade, 381–82
and Haiti, 398–403, 411
and Mexico, 382
military budget, 379–81
and Peru, 381–82
and Somalia, 389–98, 411
and Soviet Union, 379
unconventional war response, 382–89, 396–98, 411–14
Coercion, 220–26, 418, 420–23
Colby, William, 311
Cold War. *See* Soviet Union
Colombia, 69, 100, 104, 162
Clinton administration, 381–82
Revolutionary Armed Forces of Colombia (FARC), 157
Command/control, 149–52
Commitment, 129–30
Communications, 155–57
Communism
and Greece, 267–68, 269–71
Kommouniskiki Komma Ellados (KKE), 269, 271
and Philippines, 275, 276, 277, 278
and Vietnam, 282

Connolly, James, 136
Conventional war, 62–63, 69, 71, 75. *See also* United States, conventional war and
  as response, 226–28
Cuba, 7, 12, 89, 100, 101, 103, 104, 145–46, 161, 162, 164, 182
Cu Chi tunnels (Vietnam), 299–309
  Limited Warfare Laboratory (LWL), 305–6
  Tunnel Exploration Kit (TEK), 305–6
  Tunnel Explorer Locator and Communications System (TELACS), 305–6
Cults, 75–79, 84
Curcio, Mario, 136

Democracy, 70
Drug trade, 381–82
Druze sect (Lebanon), 21, 28–29, 39, 43, 52, 54, 72, 73
  National Movement, 29
  Progressive Socialist Party, 29
Dulles, John Foster, 286
Duncan, Dale E., 202–3
Durate, Jose Napoleon, 340–41
Duvalier, Jean Claude, 399

East Timor, 162
Egypt, 86, 89, 92, 96, 154
  Moslem Brothers, 79, 190–91
Eisenhower, Dwight D., 274, 282–83, 284, 286
el-Qaddafi, Muammar, 27, 166
El Salvador
  Bush administration, 373
  Reagan administration, 336, 340–41, 344, 345–46, 348–49
Enders, Thomas O., 342
Enemies, 172–73
Eritrea, 164
Ethnic cleansing, 69

Fishel, Welsey, 283
Ford, Gerald, 323
Forrestal, James V., 266
France, 232
  and Vietnam, 282–83
Front for the Advancement and Progress (FRAPH), 402

Gemayel, Amin, 20, 36–37, 43, 44
Gemayel, Bashir, 36–37
Gemayel, Pierre, 38
Geraghty, Timothy J., 22, 55
Germany, 90, 97, 105, 214–15
  armed struggle response, 233–34
  and Greece, 267
  Rote Armee Fraktion (RAF), 77–78, 86
Goldwater, Barry, 359, 360
Gorbachev, Mikhail, 368–69
Gorman, Paul F., 342
Great Britain, 3
  armed struggle perception, 183–84
  armed struggle response, 231, 232–33
  and Greece, 267
Greece
  1946–48, 267–74, 275, 277, 278, 288, 295
    American response to, 272–74
    and Balkan war, 267–68, 269, 270–71
    and Communism, 267–68, 269–71
    and Germany, 267
    and Great Britain, 267
    and Soviet Union, 267–68, 270
  Enosis, 96, 144
  EOKA, 136, 144–45, 166, 167
Grenada, 354–55
Grivas, George, 136, 138, 144–45, 149–50, 166–67
Guatemala, 100
Guevara, Che, 136, 137–38, 146, 182
Gulf of Tonkin Resolution, 296
Gulf War, 253–57, 262, 370, 375–78
Guzm(n, Abimael, 137

Habash, George, 137–38
Habib, Philip, 43
Haig, Alexander, 18, 42, 201, 335, 336, 339, 340
Haiti, 398–403, 411
Headquarters, 91–93
Henderson, Loy, 268
Hoar, Joseph, 396
Holcomb, Staser, 51
Holloway, James L., III, 362
Holy Jihad, 79
Honduras, 346
Hope, Bob, 308–9
Hussein, Saddam, 331, 370, 375–78

Ideology
  as armed struggle cause, 62, 64–65, 66–68, 70–71, 79–91, 122–25
  as armed struggle dynamic, 127–29, 130–33, 174–76
  governmental response to, 180–81
India, 3, 7
Indochina, 168
Intelligence, 157–60
Interpretation, 190–95
Iran, 12, 108
  Bush administration, 370, 374, 375–78
  Carter administration, 331–34
  and Lebanon, 25, 27, 31, 50–51
  Mojahedin, 138, 148–49
  Reagan administration, 334–35
    Iran-Contra affair, 342–45
Iraq
  Bush administration, 370, 375–78
  Gulf War, 253–57, 262, 370, 375–78
Ireland, 7, 12, 70, 81, 92, 93–94, 104
  armed struggle response, 232–33
  Easter Rising (1916), 96, 97
  Irish Free State, 107
  Irish Republican Army (IRA), 7, 71, 75, 85–86, 88, 100, 103, 105, 106, 113
    and armed struggle arena, 133
    and asymmetry, 168, 169, 170, 171
    and campaigns, 162, 163, 165
    and command, 151
    and communications, 155–56, 157
    and recruits, 134
    and terrorism, 182
    and volunteers, 140
  Troubles, 69
Israel
  armed struggle response, 233–34
  and Lebanon, 18–19, 21, 25, 29, 31, 33, 36, 50–51, 55, 71, 72
  Operation Peace for Galilee (1982), 18
Italy, 8, 70, 81
  armed struggle response, 233–34
  Brigate rosse, 68, 69–70, 80, 90, 96, 102–3, 113
    and command, 151
    and organization, 148
    psychological context, 215–16
    and recruits, 134
    and terrorism, 182
    and tradecraft, 154

Japan
  Aum Shinrikyo, 76–77
  Sekigun-ha, 77
Johnson, Louis A., 266
Johnson, Lyndon, 294, 295, 296, 311, 314, 315, 318
Joint United States Military Advisory and Planning Group (JUSMAPG), 268–69
Joint United States Military Assistance Group (JUSMAG), 277
Jordan, 18
Jumblatt, Kamal, 28–29

Kennedy, John F., 286–87, 294
Kenya, 120–21, 143
Khartoum, 89
Khomeini, Ayatollah Ruhollah, 27, 33, 138
Kissinger, Henry, 288, 314, 318, 320, 323, 335
Kitson, Frank, 279
Kommouniskiki Komma Ellados (KKE), 269, 271
Korean War, 249–50, 280–81, 289
Kurdistan, 68

Lansdale, Edward G., 277, 279, 283, 284, 289
Laos, 161, 285, 324
Latin America, 7, 12, 69, 70, 101, 103, 121, 146, 162. *See also specific countries*
Laurel, Jose, 276
Leadership, 135–38
Lebanon
  armed struggle
    causes of, 26–28, 32–33
    results of, 29–30
  Black September, 18, 23
  and Christian Phalangists, 20, 25, 38–39
  and Druze sect, 21, 28–29, 39, 43, 52, 54, 72, 73
  National Movement, 29
  Progressive Socialist Party, 29
  and Iran, 25, 27, 31, 50–51
  and Israel, 18–19, 21, 25, 29, 31, 33, 36, 50–51, 55, 71, 72
  Operation Peace for Galilee (1982), 18

Index   451

and Jordan, 18
and Multinational Force (MNF), 19, 20, 21, 23, 33–34, 36–37, 39, 44
and Palestine, 18, 20, 24, 25, 26, 29, 71
   Palestine Liberation Organization (PLO), 36–37, 134
and Shi'ite Moslems, 18, 27, 33, 39, 54
and Syria, 22, 25, 29, 33, 37, 39, 43, 51, 55
United States response
  acceptance of, 20, 22–23, 33
  ambiguity of, 18–19, 43–44
  armed struggle understanding, 15–16, 30–33, 35, 37–38, 49, 56–57, 63–67
  Beirut (1958), 17–18
  Beirut (1960–70), 24
  Beirut (1970–80), 24–25
  Beirut (1982), 19–20, 21, 31, 36–37
  Beirut (1983), 19–20, 21–24, 33–34, 38–39, 41–52
  failure of, 52–58
  and Soviet Union, 31
  and unconventional wars, 39–41, 42, 45, 46, 48–50, 52, 57, 63–67
  unpreparedness of, 34–35, 39, 40–42, 45–48, 50, 51–53
  War Powers Act, 43, 331
Legitimacy, 71, 87–88, 142, 205–6
  governmental response to, 179–80
Lenin, V. I., 91, 92–93, 147
Liberia, 73
Libya, 356
Limited Warfare Laboratory (LWL), 305–6
Linebarger, Paul, 277
Lodge, Henry Cabot, 296
Longhofer, James F., 201–3

Macedonia, 68
  Black Hand, 70
Magsaysay, Ramon, 277–78, 280, 288–89
Mahdi, Mohammed Ali, 390
Maintenance, armed struggle, 152–54
Malaya, 101, 105, 121, 147–48, 167, 170, 176n.2, 275, 276
Mandela, Nelson, 374
Marshall, George, 268

Marxism, 190–91, 192
McCarthy, Joseph, 286
McFarlane, Robert, 43, 335, 344
McNamara, Robert, 253, 254–55, 292, 294, 314, 321
Media, 195–97
Meese, Ed, 42
Mexico, 73, 86
  Clinton administration, 382
Meyers, E. C., 358, 361
Military Assistance and Advisory Group (MAAG), 283, 293
Military budget
  Clinton administration, 379–81
  Reagan administration, 335, 336, 358–60
Milosevic, Slobodan, 407
Montgomery, Thomas, 395
Multinational Force (MNF), Lebanon and, 19, 20, 21, 23, 33–34, 36–37, 39, 44
Mundy, Carl E., 392

National Security Council, 22, 42, 43, 201–3
Nation-state
  and democracy, 70
  development of, 68–69
  power of, 69–70, 74, 80–82
New world order, 5, 7
  and United States, 3–4, 261, 368–69, 373, 378–79
Ngo Dinh Con, 294
Ngo Dinh Diem, 283–84, 285, 290–91, 294
Ngo Dinh Nhu, 294
Ngo Dinh Thuc, 294
Nicaragua, 12, 70, 88
  Bush administration, 373
  Reagan administration, 340, 341, 343–44, 346, 348–49
Nixon, Richard, 286, 314, 315–16, 317–18, 320, 323
Noriega, Manuel Antonio, 364–68
North, Oliver, 201–2, 342, 344, 361
North Atlantic Treaty Organization (NATO) (1949), 4, 280, 281, 404–8, 410, 411
Nunn, Sam, 359, 360

Oklahoma City, 132

452 Dragonwars

Okomoto, Kozo, 139
Operation Peace for Galilee (1982), 18
Organization, armed struggle, 141–49
Organization of American States, 281
Orthodoxy, 74–75, 80–81

Palestine, 4, 7, 10–11, 165
   and Lebanon, 18, 20, 24, 25, 26, 29, 71
   Popular Front for the Liberation of Palestine, 137–38, 139
Palestine Liberation Organization (PLO), 36–37, 134
Palmer, Bruce, Jr., 361
Panama, 364–68
Peoples Will, 78
Perception
   and armed struggle response, 183–90, 206–13
   and interpretation, 190–95
   of revolutionary ecosystems, 107–22
      basic assumptions, 109–10
      and intensity, 110–14
      and time, 115–18
      and velocity, 118–22
   by United States, 418–21
   of Vietnam, 316–26
Peru
   Clinton administration, 381–82
   Sendero Luminoso, 137
Philippines (1946–54), 274–82, 288–89, 295
   American response to, 278–82
   and China, 275, 276
   and Communism, 275, 276, 277, 278
   Huks, 70, 121, 275–78, 279, 280
   and Soviet Union, 274–75, 280, 281
   Trade Act (1946), 276
Poindexter, John, 335, 342, 344, 361
Popular Front for the Liberation of Palestine, 137–38, 139
Powell, Colin, 358, 391, 392, 393
Psychological context, armed struggle, 213–20

Quirino, Manuel, 276–77

Reagan, Ronald, 18, 31, 37–38, 51, 53–54, 55
   administration of (1981–88), 334–64
   and Afghanistan, 343–44, 348–49
   and Angola, 343–44
   appointments, 335–36
   and CIA, 341–42
   and El Salvador, 336, 340–41, 344, 345–46, 348–49
   and Grenada, 354–55
   and Honduras, 346
   and Iran, 334–35
   and Iran-Contra affair, 342–45
   and Libya, 356
   military budget, 335, 336, 358–60
   and Nicaragua, 340, 341, 343–44, 346, 348–49
   and Soviet Union, 336, 337, 339–40, 343
   unconventional war response, 337–40, 346–54, 355–58
Recruits, 134–35
Regan, Donald, 42
Revolutionary Armed Forces of Colombia (FARC), 157
Revolutionary ecosystems, armed struggle and. *See also* Armed struggle, dynamics of; Armed struggle, responses to: Lebanon
   causes of, 61–62, 83–86
   cults, 75–79, 84
   ideology, 62, 64–65, 66–68, 70–71, 79–91, 122–25, 127–29, 130–33, 174–76, 180–81
   nationalism, 4–5
   religion, 4–5, 7
   truth, 4–5, 7, 8–9, 11–14
   characteristics of, 5–11, 71–74
   speciality, 79–82
   and conventional war, 62–63, 69, 71, 75
   and legitimacy, 71, 87–88, 142, 179–80, 205–6
   and nation-state
      and democracy, 70
      development of, 68–69
      power of, 69–70, 74, 80–82
   and new world order, 3–4, 5, 7, 261, 368–69, 373, 378–79
   and orthodoxy, 74–75, 80–81
   results of
      failures, 5, 7, 12
      successes, 5, 7, 70
   tactics for, 5–6, 83
      guerrilla warfare, 5

insurrection, 5
terrorism, 5, 182, 330, 336, 339–40, 362–63
and unconventional war, 63–67
Revolutionary ecosystems, design/ determinants
design
  armed struggle, 98–99
  beginning, 100–101
  central campaign structure, 101–3
  central core, 94–95
  ending, 104–7
  life cycle, 99–100, 107–9
  popular support, 96–98
  protraction, 86–87, 103–4
  secrecy, 123–24
headquarters, 91–93
and legitimacy, 71, 87–88, 142
perception, 107–22
  basic assumptions, 109–10
  intensity, 110–14
  time, 115–18
  velocity, 118–22
tactics, 83
and unconventional war, 84
Rhodesia, 143
Rote Armee Fraktion (RAF), 77–78, 86
Roxas, Manuel, 276
Russia. *See* Soviet Union

Saavedra, Daniel Ortega, 341
Sadr, Mussa, 27
Sandinistas, 88
Savimbi, Jonas, 343
Sharon, Arik, 31, 36
Shi'ite Moslems, 18, 27, 33, 39, 54
Shultz, George, 18, 42, 53, 55
Singapore, 68
Slovakia, 68
Somalia, 4, 7, 8, 73
  Clinton administration, 389–98, 411
Somoza, Anastasio, 341
South Africa, 7, 107
  Bush administration, 374
South-East Asia Treaty Organization (SEATO), 281
Soviet Union
  Clinton administration, 379
  Cold War as conventional, 237, 238, 239, 241, 242, 248–49, 265
  dissolution of, 3, 5, 7

Bush administration, 368–69, 373
and new world order, 3–4, 5, 7, 261, 368–69, 373, 378–79
and Greece, 267–68, 270
and Lebanon, 31
and Philippines, 274–75, 280, 281
Reagan administration, 336, 337, 339–40, 343
Russia, 90, 94, 147
and Vietnam, 282, 284
Special Operations Division (U.S.), 201–3
Stern Group, 78, 167
Sudan, 73, 89
Symbionese Liberation Front, 78
Syria, Lebanon and, 22, 25, 29, 33, 37, 39, 43, 51, 55

Tactics, 5–6, 83
  guerrilla warfare, 5
  insurrection, 5
  terrorism, 5, 182, 330, 336, 339–40, 362–63
Taruc, Luis, 276, 278
Taylor, Maxwell, 294
Tebbits, Norman, 140
Temporary Equipment Recovery Mission (TERM), 283
Terrorism, 5, 182
  American response to, 330, 336, 339–40, 362–63
Thayer, Paul, 360
Tradecraft, 154
Trapnall, Thomas J. H., Jr., 287–88, 289
Truman, Harry, 266, 268, 274
Tunnel Exploration Kit (TEK), 305–6
Tunnel Explorer Locator and Communications System (TELACS), 305–6
Turkey, 89
Turner, Stansfield, 331

Uganda, 76
Ulster, 86, 89, 170
UNITA, 88
United Nations, 4
United States, conventional war and attitude toward
  avoidance, 236–37, 240, 241, 242, 247–48
  foreign perspective, 240–41

and image, 244
and imperialism, 243
and national character, 245–46
and values, 244–45, 247
Cold War, 237, 238, 239, 241, 242, 248–49, 265
Gulf War, 253–57, 262, 370, 375–78
Korean War, 249–50, 280–81, 289
new world order, 3–4, 5, 7, 261, 368–69, 373, 378–79
preparedness, 239, 242, 249–52
  cost of, 255–56
  intelligence, 257–58
  management, 236–37, 252, 254–55
  Pentagon transformation, 253–55
  technology, 236–37, 252–53, 256–57
results of, 237
sociocultural history, 234–38
United States, unconventional war and attraction to, 201–3
  Bush administration (1988–92), 364–79
    and CIA, 364–65
    and El Salvador, 373
    and Iran, 370, 374, 375–78
    and Iraq, 370, 375–78
    and Nicaragua, 373
    and Panama, 364–68
    and South Africa, 374
    Soviet Union dissolution, 368–69, 373
  Carter administration (1977–81), 330–34
    and Afghanistan, 331
    and Iran, 331–34
  and China, 266–67
  Clinton administration (1992–), 379–414
    and Africa, 409–10, 411
    and Balkan States, 403–9, 411
    and CIA, 402–3
    and Colombia, 381–82
    and drug trade, 381–82
    and Haiti, 398–403, 411
    and Mexico, 382
    military budget, 379–81
    non-military response, 382–89
    and Peru, 381–82
    and Somalia, 389–98, 411
    and Soviet Union, 379

Greece (1946–48), 267–74, 275, 277, 278, 288, 295
  and Balkan war, 267–68, 269, 270–71
  and Communism, 267–68, 269–71
  and Germany, 267
  and Great Britain, 267
  response to, 272–74
  and Soviet Union, 267–68, 270
Lebanon, 39–41, 42, 45, 46, 48–50, 52, 57, 63–67
  acceptance of, 20, 22–23, 33
  ambiguity of, 18–19, 43–44
  Beirut (1958), 17–18
  Beirut (1960–70), 24
  Beirut (1970–80), 24–25
  Beirut (1982), 19–20, 21, 31, 36–37
  Beirut (1983), 19–20, 21–24, 33–34, 38–39, 41–52
  failure of, 52–58
  and Soviet Union, 31
  understanding of, 15–16, 30–33, 35, 37–38, 49, 56–57, 63–67
  unpreparedness for, 34–35, 39, 40–42, 45–48, 50, 51–53
Philippines (1946–54), 274–82, 288–89, 295
  and China, 275, 276
  and Communism, 275, 276, 277, 278
  Huks, 275–78, 279, 280
  response to, 278–82
  and Soviet Union, 274–75, 280, 281
  Trade Act (1946), 276
Reagan administration (1981–88), 334–64
  and Afghanistan, 343–44, 348–49
  and Angola, 343–44
  appointments of, 335–36
  and CIA, 341–42
  and El Salvador, 336, 340–41, 344, 345–46, 348–49
  and Grenada, 354–55
  and Honduras, 346
  and Iran, 334–35
  and Iran-Contra affair, 342–45
  and Libya, 356
  military budget, 335, 336, 358–60
  and Nicaragua, 340, 341, 343–44, 346, 348–49
  and Soviet Union, 336, 337, 339–

40, 343
  unconventional war response, 337–40, 346–54, 355–58
  response to, 14–16, 239–40, 251, 258–67
    by Clinton administration, 382–89, 396–98, 411–14
    coercion, 418, 420–23
    future agenda, 417–18, 419, 428–30
    Greece (1946–48), 272–74
    and perception, 418–21
    Philippines (1946–54), 278–82
    by Reagan administration, 337–40, 346–54, 355–58
    special forces, 423–28
    terrorism, 330, 336, 339–40, 362–63
    Vietnam (1945–61), 285–91
    Vietnam (1961–68), 291–99, 306–8
    withdrawal, 422–23
  Vietnam (1945–61), 282–91
    Army of the Republic of Vietnam (ARVN), 283, 291
    and Cambodia, 285
    and China, 282
    and Communism, 282
    Dien Bien Phu (1954), 282–83
    and France, 282–83
    and Laos, 285
    response to, 285–91
    and Russia, 282, 284
    Saigon Military Mission, 283
  Vietnam (1961–68), 291–309
    Army of the Republic of Vietnam (ARVN), 292, 301
    Cu Chi tunnels, 299–309
    response to, 291–99, 306–8
  Vietnam (1968–72)
    Army of the Republic of Vietnam (ARVN), 312, 313, 318, 319, 320–22
    lessons of, 322–30
    perception of, 316–26
    Tet offensive, 311–16
United States Economic Mission (USECOM), 283
Uruguay
  Montoneros, 70

Vance, Cyrus, 331, 332

Venezuelan Communist Party, 101
Vessey, John W., Jr., 357
Vietnam, 4, 7, 8, 12, 15, 32, 164, 182
  1945–61, 282–91
    American response to, 285–91
    Army of the Republic of Vietnam (ARVN), 283, 291
    and Cambodia, 285
    and China, 282
    and Communism, 282
    Dien Bien Phu (1954), 282–83
    and France, 282–83
    and Laos, 285
    and Russia, 282, 284
    Saigon Military Mission, 283
  1961–68, 291–309
    American response to, 291–99, 306–8
    Army of the Republic of Vietnam (ARVN), 292, 301
    Cu Chi tunnels, 299–309
  1968–72
    Army of the Republic of Vietnam (ARVN), 312, 313, 318, 319, 320–22
    lessons of, 322–30
    perception of, 316–26
    Tet offensive, 311–16
Volunteers, 138–41

War Powers Act, 43, 331
Wazzan, Shafik, 36
Weinberger, Caspar, 18, 22, 42, 53, 55, 336, 358, 360
Westmoreland, William, 295–96, 301–2, 307, 308, 314, 321
World Trade Center, 151, 158, 159, 163, 165, 414

Yeltsin, Boris, 369
Yemen, 7
Yugoslavia, 69, 86, 267, 270, 403–9

Zaire, 68, 73, 76, 133–34, 162, 164, 182, 409–10
Zakhariadis, Nikos, 269
Zimbabwe, 104, 143